Long Noncoding RNAs

Riki Kurokawa
Editor

Long Noncoding RNAs

Structures and Functions

Editor
Riki Kurokawa
Division of Gene Structure and Function
Research Center for Genomic Medicine
Saitama Medical University
Saitama, Japan

ISBN 978-4-431-55575-9 ISBN 978-4-431-55576-6 (eBook)
DOI 10.1007/978-4-431-55576-6

Library of Congress Control Number: 2015944101

Springer Tokyo Heidelberg New York Dordrecht London

Printed on acid-free paper

Springer Japan KK is part of Springer Science+Business Media (www.springer.com)

Preface

In this book the aim is to provide an outline of the molecular mechanism of long noncoding RNAs (lncRNAs). Only a small percentage of the human genome stores information on proteins expressed in living cells, whereas more than 95 % of the genome bears "noncoding" DNA sequences. The major transcripts from the non-coding DNA are lncRNAs whose length is greater than 200 bases. These noncoding DNA sequences are heterogeneous—for instance, retrotransposons such as SINE and LINE, pseudogenes, and introns. Therefore, lncRNAs have diverse sequences and should have multivalent secondary structures that still have not been revealed. More than 35,000 lncRNAs are estimated to have been transcribed from the genome, but just 200 of them have been annotated. However, these lncRNAs have been reported to play various roles—for instance, in structures of nuclei, transcriptional regulations, and epigenetic regulations. So far, no one has explained the common principle of the actions of lncRNAs behind such divergent roles of lncRNAs. As presented in this book, the quest is for the solution of the elusive question of whether there is a common principle to explain those actions. The question is approached from five points of view, as follow.

In Part I, "Bioinformatics and Other Methodologies for lncRNAs", two chapters describe these methodologies. Schein and Carninci present recent progress in deep sequencing analysis, and Sugiyama et al. use chemically synthetic approaches to study lncRNAs.

In Part II, "Atomic and Molecular Structures of lncRNAs", Katahira presents data about RNA aptamers as an example of the functions of noncoding RNAs. Oyoshi shows the specific structure of RNA around the telomere G-quadruplex related to its binding proteins.

Kurokawa emphasizes the importance of transcriptional initiation for the origin of lncRNAs in Part III, "Molecular Functions of lncRNAs". Kumon and Ohta describe prototypic principles of lncRNAs in the yeast system. Mannen, Chujo, and Hirose present structural functions of lncRNAs.

In Part IV, "Biological Actions of lncRNAs", Hasegawa and Nakagawa elucidate the biology of imprinting and related lncRNAs. Ogawa, a pioneer of the XIST studies, presents with Yamada the leading-edge story of X-chromosome inactivation and lncRNAs.

In Part V, "Potential Outcomes for Clinical Medicine", Kotake and Kitagawa reveal the pRB and p53 pathways and lncRNAs with intriguing issues related to tumor formation. Takayama and Inoue, who initially identified androgen-dependent expression of a lncRNA, report their work here. Allison and Glass examine macrophage biology and recently revealed enhancer RNA functions, and Jin and Rosenfeld explicate the lncRNA functions regarding the biology of steroid hormone receptors.

These points of view provide thoroughly new insights into the functions of lncRNAs in living cells. Previously, separate publications regarding various analyses of lncRNAs have appeared, but this book presents these topics in a single volume, thus providing a great benefit for readers. Recently, the ENCODE project has demonstrated that 80 % of the human genome has a unique function, in which case most of the transcripts from the genome are assumed to be lncRNA. Therefore, the need and opportunity to understand the function of lncRNAs has been growing and the timing of the publication of this book is perfect.

As editor, I am grateful to all the authors who have joined in contributing to the book. These contributions have warmly encouraged me as I encountered the harsh reality that many scientists are unable to find time for writing a book chapter or review article in the highly competitive environment of academic work. I would also like to thank Dr. Christopher K. Glass for his valuable comments. Last, but not least, I would express my thanks to Ms. Ritsuko Tanji for her excellent support and also to the editorial staff members of Springer Japan for their great service in completing the editing. All these endeavors have created this valuable volume.

Saitama, Japan Riki Kurokawa

Contents

Part I
Bioinformatics and Other Methodologies for lncRNAs

Chapter 1
Complexity of Mammalian Transcriptome Analyzed by RNA Deep Sequencing

Aleks Schein and Piero Carninci

Abstract Genetic information in most living organisms on Earth is stored in the form of a chemical structure, known as deoxyribonucleic acid (DNA). Researchers discovered that pieces of long DNA molecules, called genes, are recognized by the nuclear multi-subunit complex of ribonucleic acid (RNA) polymerase, which then produces molecules of RNA, complementarily mirroring the original DNA. Some of these RNA molecules carry information that can be used to produce polypeptide chains with pre-defined amino acid sequences. These molecules have been named messenger RNAs (mRNAs). Others, such as ribosomal RNAs, transfer RNAs, and small nuclear RNAs, have been found to drive and regulate production of proteins. They are sometimes referred to as housekeeping or structural RNAs.

However, sequences of mRNAs together with structural RNAs account for less than 10 % of animal and plant genomes. The rest of the genome was considered silent and non-functional, until on-going research revealed that about 80 % of DNA might be transcribed, producing numerous long noncoding RNA molecules with important functions. This chapter gives an overview of mammalian transcriptome research in recent decades. It discusses the main technology platforms, comparing their strong sides and disadvantages. Some of the most important findings are summarized, with an overview of the future prospectives in long noncoding RNA research.

The chapter shows that the current understanding of what is a gene should be revised, in order to clearly define the complex relationship between product-coding regions, regulatory sequences, and the organism's phenotype.

Keywords Noncoding RNA • lncRNA • Mammalian transcriptome • Cap-analysis gene expression • Transposable elements • Enhancer sequences • Gene annotation

---

A. Schein
Division of Genomic Technologies, RIKEN Center for Life Science Technologies, 1-7-22 Suehiro-cho, Tsurumi-ku, Yokohama, Kanagawa 230-0045, Japan

Department of Clinical Medicine, Center for RNA Medicine, Aalborg University, A.C. Meyers Vaenge 15, Bldg. Fk10b. 217, DK-2450 Copenhagen, Denmark

P. Carninci (✉)
Division of Genomic Technologies, RIKEN Center for Life Science Technologies, 1-7-22 Suehiro-cho, Tsurumi-ku, Yokohama, Kanagawa 230-0045, Japan
e-mail: carninci@riken.jp

R. Kurokawa (ed.), *Long Noncoding RNAs*, DOI 10.1007/978-4-431-55576-6_1

1.1 Introduction

The genome of mammalian cell encodes all necessary information for survival, proliferation, development and function in the context of the multi-cellular organism. Although all cells in the individual have an identical genome, more than 400 different cell types are recognized in mammals (Matera et al. 2007; Vickaryous and Hall 2006). Furthermore, each separate cell contains its unique regulated set of active and inactive genes. These facts show that even if genes are obviously important by themselves, DNA regions that control gene expression are crucial for the cells' and the organism's fate. It was long considered that about 23,000 protein-coding genes represent most of the transcribed part of the genome, even though they occupy less than 2 % of the total genomic DNA (International Human Genome Sequencing Consortium 2004). A number of functional non-protein-coding RNAs discovered and characterized during the twentieth century include well-known examples of ribosomal RNA, snRNA, snoRNA, tRNAs and miRNAs. The biogenesis, localization and function of these noncoding RNAs have been extensively studied. With the emergence of next-generation sequencing technologies, even more transcribed noncoding genomic regions have been discovered, though the number of annotated protein-coding genes actually decreased (International Human Genome Sequencing Consortium 2004; Lander et al. 2001).

As the total number of protein-coding genes is similar in most animals, including simple worms and sponges, the extent of genes encoding for long noncoding RNA (lncRNA) grows with increasing developmental and cognitive complexity. Moreover, the vast majority of these sequences are dynamically transcribed to produce different classes of short and long non-protein-coding RNAs. In addition, expression of these RNAs shows precise cell- and tissue-specific patterns and subcellular localizations. The current transcriptome model claims that as much as 80 % of the human genome is being transcribed (International Human Genome Sequencing Consortium 2012; Carninci et al. 2005). Some recent studies, however, disagree with this claim, arguing that transcriptionally active regions, defined by transcription start site (TSS) and transcription termination site (TTS) activity and biochemical activity, such as binding to transcription factors, might not be really functional, since the number of evolutionarily conserved (and thus functional) regions in the genome is much smaller (Graur et al. 2013; Kellis et al. 2014). At the same time, conservation itself is not an obligatory feature of a functional element. Studies such as the one by Dermitzakis et al. (2002, 2004) have found many conserved sequences with unknown functions in the human genome. Other sequences, such as repeat elements, have been shown to be significantly less conserved than protein-coding regions but are still functional and important (Cloonan et al. 2008; Lv et al. 2013; Sakurai et al. 2014). In fact, Alu repeat elements, known to be only primate-specific, have a conserved genomic context and editing patterns (Bazak et al. 2014), suggesting that evolutionary conservation may be expressed on levels other than simple nucleotide sequence similarity. These arguments show that even though numerous putative long noncoding RNAs are being discovered (International Human Genome Sequencing Consortium 2012; Forrest et al. 2014), the importance and functionality of these findings need to be further experimentally validated.

1.2 Definition and Types of Transcripts Encoded by the Genome

Transcripts are genomic regions, surrounded by TSS and TTS. Between these features lie exons, marked by splice donor and acceptor sites. Exons of protein-coding genes may construct an open reading frame (ORF), which can be translated into an amino acid sequence, giving rise to a polypeptide chain. Such transcripts are known as messenger RNAs (mRNAs). mRNAs in eukaryotes are encoded as independent transcription units and are transcribed by RNA polymerase II. If no ORF is found, the transcript is designated as noncoding RNA (ncRNA). These are additionally divided into short noncoding RNAs (microRNAs, piwiRNAs, snRNAs, tRNAs, etc.), which are shorter than 200 nucleotides in their mature form, and long noncoding RNAs (lncRNAs), which are longer than 200 nucleotides (an arbitrary size selection, based on cloning and sequencing protocol technical details). In mammalian cells, RNA is produced by three RNA polymerase complexes. Ribosomal 18S and 28S RNAs are the most well-known and clearly most abundant lncRNAs, accounting for more than 90 % of all cellular RNA, together with short 5S and 5.8S. rRNA is encoded by ~43 kbp units, which include 18S, 5.8S, and 28S precursors, as well as intergenic spacer regions. These units are arranged in tandem repeat clusters, transcribed by RNA polymerase I (30–40 units per cluster) on chromosomes 13, 14, 15, 21, and 22 (Worton et al. 1988).

Most snRNAs and miRNAs, together with mRNAs, are encoded as independent genomic units, transcribed by RNA polymerase II (Matera et al. 2007). The majority of snoRNAs are not transcribed independently; rather, they originate from processing of spliced introns of mRNAs and noncoding transcripts. RNA polymerase III makes 5S rRNA, encoded by separate units, consisting of several hundred tandemly arranged genes; tRNAs and other short functional RNAs, such as 7SK, 7SL, and U6 snRNA. All mentioned classes of noncoding "housekeeping" RNA are, to some extent, characterized in terms of their cellular function. tRNAs are parts of the protein translation machinery; snRNAs build the spliceosome core; snoRNA guides chemical modification of other molecules. Functions have also been also established for other types of small RNAs.

Still, for a growing number of long noncoding transcripts, no established function has been found, though their expression has been predicted or even confirmed to have tissue or cell-specific patterns. These RNAs are found in both the nucleus and the cytoplasm; they have a broad range of sizes, morphological characteristics, and expression levels. Different classes of long noncoding RNAs and their known or proposed functions are further discussed in the following chapters.

Overall, 14,880 transcripts, originating from 9,277 loci in the human genome, are defined as lncRNAs by GENCODE v7 annotation (Derrien et al. 2012; Harrow et al. 2012) and further classified into four sub-groups, based on their localization and orientation, as outlined in Fig. 1.1. Yet, some transcripts of the GENCODE data set could not be assigned to either group, because of their complexity (alternative splicing, multiple TSSs, etc.).

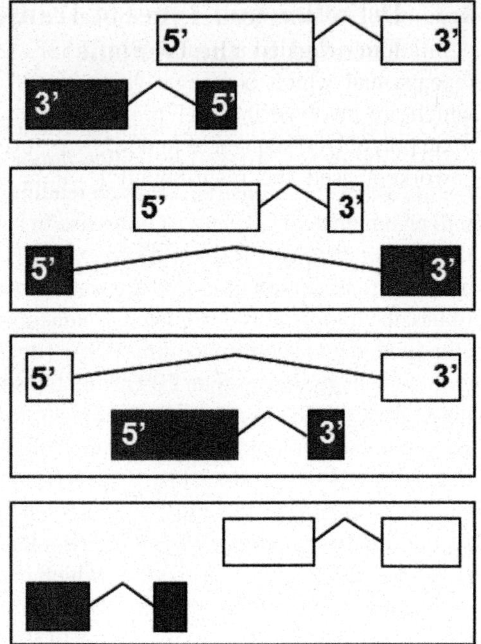

Fig. 1.1 Gencode v7 Classification of lncRNA in respect to intersection with protein-coding genes. Protein-coding transcripts are shown in *white* and lncRNAs in *black*. *Box shapes* represent exons; *lines* represent introns. The orientation of each transcript is shown by indication of the 5 and 3′ ends. The orientation signs have been removed from the intergenic panel, as they are not relevant for this type of transcript (Adapted from Derrien et al. (2012))

1.3 Transcriptome Discovery Technology Overview

1.3.1 Pre-NGS Era: From Genetic Studies and cDNA Isolation to Microarrays

The discovery and functional studies had been centered on mRNAs for a long time, because their ORFs could be easily detected and analyzed. Nevertheless, genetic and biochemical studies also suggested that cells contain numerous additional transcripts, which do not encode for proteins. Many abundant ncRNAs have been isolated, cloned, and studied (Brockdorff et al. 1992; Gupta et al. 2010; Paul and Duerksen 1975; Rinn and Chang 2012; Salditt-Georgieff and Darnell 1982; Salditt-Georgieff et al. 1981; Weinberg and Penman 1968). In fact, full-length cDNA cloning and analysis possibly remains the most accurate means of transcriptome discovery. However, this approach is highly laborious in terms of laboratory operation and data analysis, producing data sets of hundreds of thousands of individual clones, requiring manual curation. Microarray technology made it possible to investigate gene expression on the whole-genome level with more convenience. Yet, arrays could only detect known transcripts complementary to their specific probe set, which made impossible detection of previously unknown molecules. This limitation was removed with the introduction of tiling arrays (Cheng et al. 2005; Kapranov et al. 2002; Rinn et al. 2003). Still, array technology suffered from high background levels, cross-hybridization, and exclusion of repetitive elements, which represent about 50 % of the genome in humans (Treangen and Salzberg 2012). In addition, whole-genome-covering tiling array studies were highly expensive and required large volumes of starting material (total RNA).

1.3.2 High-Throughput Sequencing Platforms to Study Mammalian Transcriptomes

The advance of sequencing technologies in the twenty-first century has led to the revolutionarily increased number of known lncRNAs (Carninci et al. 2005; Rinn and Chang 2012). The FANTOM3 project, using high-throughput sequencing to support cDNA cloning and annotation, identified ~35,000 noncoding transcripts in the mouse genome. Additional noncoding RNAs have been discovered and annotated by further work of the FANTOM Consortium (Djebali et al. 2012; Forrest et al. 2014). A number of methods to prepare cDNA libraries compatible with downstream high-throughput sequencing are described below. These methods use different input materials and produce distinct outputs. Thus, an appropriate technique is being selected as required by each individual project.

1.3.2.1 RNAseq

RNAseq is aimed at reconstruction and quantitative analysis of full-length transcripts. Usually, library preparation is preceded by removal of extremely abundant ribosomal RNAs to prevent accumulation of their corresponding reads in the sequencing output. A general overview of the method is shown in Fig. 1.2. Input RNA is usually fragmented, resulting in 200–400 bp fragments. These fragments are converted into cDNA by either random or dT-primed RT-PCR. Then adaptor sequences are introduced (Fig. 1.2). Numerous methods of library construction have been developed, differing in exact techniques of fragmentation, tagging sequence incorporation, and intermediate purification steps. The original protocols have been modified to provide strand specificity to resulting DNA reads. For details on RNAseq technologies, the reviews by Zhang et al. (Wang et al. 2009) and Levin et al. (2010) are highly recommended. Overall, RNAseq techniques aim at uniform amplification of all transcripts (random priming) or polyadenylated transcripts only (oligo-dT priming). RNAseq methods usually produce poor coverage of extreme 5′ and 3′ ends of transcripts, thus complicating the definition of exact genomic boundaries. Also, it might be difficult to distinguish between numerous isoforms and splice variants in complex transcriptomes, as well as to accurately compare expression levels, especially of low-abundant transcripts. In addition, some sequences show higher affinity for priming and amplification, creating unpredictable biases in quantitation of gene expression levels (Lahens et al. 2014).

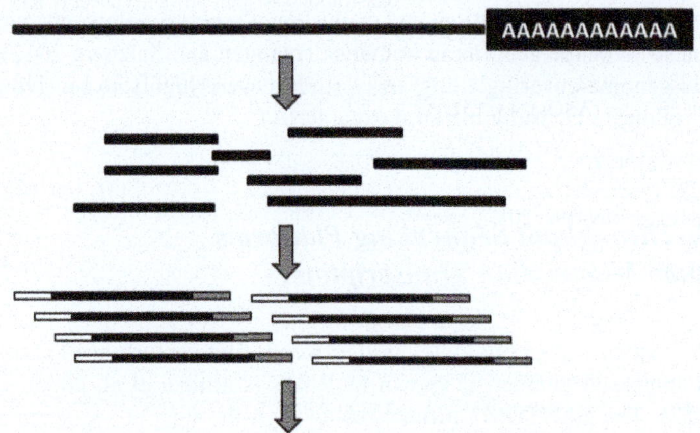

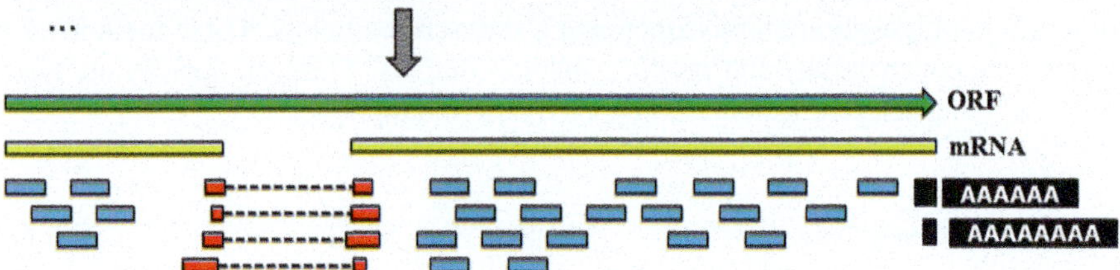

Fig. 1.2 Outline of a typical RNAseq experiment. The RNA molecule (*top*) is fragmented onto ~200 nucleotide fragments. The resulting pieces are reverse transcribed into the cDNA library, containing adaptors (*white* and *gray* ends). Sequencing reads, produced by NGS technology, are mapped to the reference genome. Three types of reads are shown: exonic (*blue*), junction mapping on both sides across the splice junction (*red*, connected with *dotted lines*), and poly(A)-end reads (*black*) (Based on Wang et al. (2009))

1.3.2.2 Cap-Analysis Gene Expression (CAGE)

The CAGE Next Generation Sequencing (NGS) library construction technology (Fig. 1.3), developed for genome-wide expression analysis, focuses specifically on capped 5′ ends of RNA molecules (Kodzius et al. 2006; Takahashi et al. 2012). This method identifies TSSs on a single-nucleotide level and allows quantitative measurement of transcript expression values. The TSS-centered approach allows researchers to overcome technical issues associated with other NGS library construction, such as low coverage of transcript 5′ ends and difficulties in distinguishing multiple isoforms and splice variants. In addition, mapping of sequencing reads back to the genome provides a transcription landscape that is unbiased to known annotation models and concentrates the analysis on a small portion of the genome (27 nucleotides down to TSSs). This allows effective pooling of barcoded samples into complex libraries, significantly reducing the subsequent sequencing cost. The outline of the method is shown in Fig. 1.3.

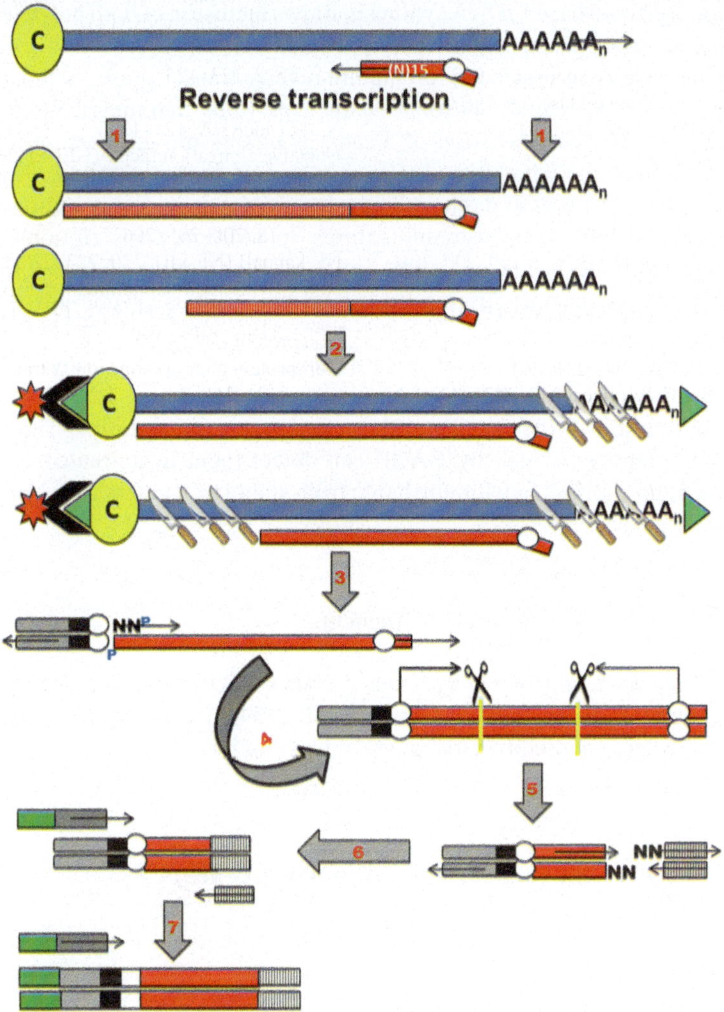

Fig. 1.3 Workflow of CAGE. cDNA is reverse transcribed with random primers, containing an EcoP15I recognition site (*white rounded shape*). A newly synthesized first DNA strand is indicated by a pattern-filled *box shape*. Full-length cDNA, reaching the capped 5′-end of the RNA (stage 1, *top*) or premature terminated cDNA (stage 1, *bottom*) can be produced. At stage 2, RNA 5′ caps and 3′-ends are biothinylated (*triangular shapes*), and the non-hybridized RNA strand is cleaved by RNase I (knives). Complete cDNAs are then captured by streptavidin (*black V-shapes*), connected to magnetic beads (*star shapes*). The cDNA (*red*) is released from RNA and ligated to double-stranded linkers containing barcode sequences and EcoP15I sites (stage 3, *black boxes* and *white rounded shapes*, respectively). Linkers provide priming sites for second-strand DNA synthesis with cDNA as a template (stage 4). The resulting double-stranded DNA is cleaved by EcoP15I (scissors). Actual cleavage sites, 27 bp downstream of the recognition sequence, are shown with yellow lines. Next, a 3′-linker with Illumina primer sequence (white pattern-filled shape, stage 5) is ligated to the 3′ end. The resulting 96 bp CAGE tags are amplified by PCR with forward and reverse primers, compatible with the Illumina flow cell surface. The 5–3′ direction is indicated by a *line arrow*, when relevant. Workflow stages mentioned in the legend text are represented with *block arrows* with numbers

Random primer-based cDNA synthesis allows inclusion of both polyadenylated and non-polyadenylated transcripts. 5′ CAP-assisted purification eliminates incomplete cDNAs produced by premature reverse transcriptase termination and removes over-represented ribosomal sequences (thus making rRNA depletion unnecessary). Stepwise addition of 5′ and 3′ linkers provides transcript directionality. It has to be noted that CAGE only reveals genomic points corresponding to capped 5′ RNA ends defining TSSs. It does not provide any information on the transcript length or downstream features. In addition, transcription initiation and 5′ cap addition is not necessary followed by creation of functional RNA molecules. Many recently discovered transcripts, such as cryptic unstable transcripts (CUTs) and promoter upstream transcripts (PROMPTs) (Preker et al. 2011; Thompson and Parker 2007) were found to be degraded co-transcriptionally or shortly after transcription. Normally, knockdown of RNA degradation factors is required to detect these RNAs in sequencing libraries. The extreme sequencing depth provided by CAGE can detect them in untreated cells, thus making unstable RNAs indistinguishable from stable transcripts with low expression levels.

1.3.2.3 Paired-End Tag (PET) Sequencing

NGS DNA sequencing of both 5′ and 3′ ends of cDNAs was developed at the Genome Institute of Singapore (Fullwood et al. 2009). Though PET produced short tags, complicating subsequent mapping, and had high bias due to plasmid amplification, it was useful to understand the borders of polyadenylated transcripts.

1.3.2.4 Poly(A) Tail-Guided Sequencing (Poly(A)-Seq)

Poly(A)-seq is a method allowing sequencing of genomic regions immediately upstream of cleavage and polyadenylation sites (Derti et al. 2012). Genome-wide analysis of such regions defines a “3′-terminome” of the cell (Chang et al. 2014). cDNA is produced by RT-PCR, primed by nested oligo-dT primer, binding to the extreme 5′ end of the poly(A) tail and the last nucleotide of the transcript body. This allows direct sequencing of RNA 3′ ends. Another technique developed by Chang et al. (2014) is based on biotinylated adaptor ligation to the poly(A) tail 3′ end and RNase T1 fragmentation. This allows 3′-end sequencing and measurement of a poly(A) tail length, as well as identification of heteropolymeric tails (Chang et al. 2014). Apparently, nested oligo-dT primer can also bind to adenosine-rich RNA sequences, giving rise to false TTSs. Also, this method only provides a set of TTSs without reference to the transcript to which they belong. Nevertheless, having a number of such termination sites within an annotated transcript would strongly suggest multiple 3′ ends for the corresponding RNA.

1.3.2.5 Further Technology Development

The methods described above are being modified constantly to improve their sensitivity and specificity, and face various research needs. CAGE technology, originally based on CAP-assisted purification, requires a large amount of starting material. This has limited its usage for many applications, such as in clinical studies, where only often single nanograms of RNA can be obtained from the sample. Thus, a modified method, called nano-CAGE has been developed (Salimullah et al. 2011), using template switching instead of CAP trapping. Nano-CAGE allows library preparation from as little as 50 ng of total RNA. Another CAGE modification, named CAGE-scan allows profiling of TSSs but also identification of full-length transcripts (Plessy et al. 2010).

In complex tissues, such as the brain, liver, or bone marrow, it may be difficult to detect gene expression changes, because of the "obscuring average" factor. Thus, it is extremely important to enable isolation and analysis of a small sub-population of affected cells or even single cells. Accordingly, single-cell sequencing protocols are constantly being developed and improved (Saliba et al. 2014).

Additional new technologies, such as global run-on sequencing (GRO-seq) enable researchers to study transcription at specific time points. GRO-seq is based on labeling, isolation, and analysis of nascent transcripts by nuclear run-on, detecting transcription events initiated at or shortly before the original cell had been collected. This technique provides an insight into transcription events, associated with a specific developmental stage or an outside factor, such as a drug or environmental stress (Core et al. 2008).

1.4 Downstream Applications and Result Overview

The library preparation platforms described above facilitate conversion of cellular RNA into cDNA with simultaneous addition of sequencing tags. The choice of chemistry and platform depends largely on the technical availability and research needs. After the sequencing is done, obtained reads are filtered by quality and mapped back to the reference genome. Longer transcripts could be later assembled from mapped reads, either de novo or using reference annotation. This is done by a number of software packages, such as Cufflinks (Roberts et al. 2011; Trapnell et al. 2012). Other analysis algorithms, such as Trinity (Grabherr et al. 2011), first build transcripts from sequencing reads and then map these assembled transcripts to the genome. Further work identifies novel RNA molecules and different isoforms of previously known transcripts. CAGE data is used to make detailed maps of promoters, TSSs, enhancer regions and other regulatory sequences. In complex projects, data obtained from different platforms is being integrated to fully reconstruct complex transcripts. As was mentioned above, RNA sequencing leaves 5 and 3′ ends poorly covered. Integrating CAGE data makes it possible to precisely map TSSs, including those with low abundance, and to distinguish between multiple TSSs positioned close to each other. Software packages for integration of sequencing data from different platforms have been developed and successfully used for large-scale transcriptome analysis (Boley et al. 2014; Brown et al. 2014). The following sections provide an outline of the significant transcriptome studies that have used multiple platforms and data sources.

1.4.1 FANTOM Consortium

During the last decade, a number of multi-scale studies have been conducted, aiming at building detailed landscapes of eukaryotic transcriptomes. Comprehensive transcriptional maps of the mouse (Carninci et al. 2005), drosophila (Brown et al. 2014), zebrafish (Kelkar et al. 2014), worm (Yook et al. 2012) and human (Forrest et al. 2014) genomes were constructed, covering most of the classic model organisms. In mammals, this work was initiated by the FANTOM Consortium, resulting in detailed reconstruction of the mouse transcriptome by the FANTOM3 collaborative project (Carninci et al. 2005). This work utilized a combined approach, based mainly on full-length cDNA technology. More than 150,000 cDNA clones and expression tags were isolated and sequenced (Carninci et al. 2000; Ng et al. 2005). CAGE was extensively used to support the data by identifying and verifying TSSs. A summary of the FANTOM3 resource data set is shown in Table 1.1. Sequenced cDNAs were mapped to the genome and analyzed with respect to their identity and overlap with each other and with known genomic features. Elements sharing the same TSSs, TTSs, and/or splicing patterns were grouped into bigger groups, termed transcriptional units (TUs) and transcriptional frameworks (TKs), gradually collapsing overlapping elements. The result of this clustering is summarized in Table 1.2. Further on, TKs were additionally clustered to produce genomic regions transcribed without gaps on either strand, named as transcription forests (TFs). These regions corresponded to 62.5 % of the mouse genome. Significant conservation was observed between mouse and human mRNAs and ncRNAs and their promoters (Fig. 1.4a, b).

Table 1.1 FANTOM3 source data set summary

	Total	Number of libraries	Safely mapped
RIKEN full-length cDNAs	102,801	237	100,313
Public (non-RIKEN) mRNAs	56,009		52,119
CAGE tags (mouse)	11,567,973	145	7,151,511
CAGE tags (human)	5,992,395	24	3,106,472
GIS ditags	385,797	4	118,594
GSC ditags	2,079,652	4	968,201
RIKEN 5′ESTs	722,642	266	607,462
RIKEN 3′ESTs	1,578,610	265	907,007
5′/3′EST pairs of RIKEN cDNA	448,956	264	277,702

Reproduced from Carninci et al. (2005)
The total number of features, number of containing libraries and mapped elements are shown

Table 1.2 FANTOM3 transcript grouping and classification

	Total	Average per TU cluster	Average per TK cluster
Total number of transcripts	158,807	7.59	7.30
RIKEN full-length	102,801		
Public (non-RIKEN) mRNAs	56,006		
GFs	25,027	1.20	1.15
Framework clusters	31,992	1.53	1.47
TUs	44,147	2.11	2.03
With proteins	20,929	1.00	0.96
Without proteins	23,218	1.11	1.07
TK	45,142	2.16	2.07
With proteins	21,757	1.04	1.00
Without proteins	23,385	1.12	1.07
Splicing patterns	78,393	3.75	3.60

Reproduced from Carninci et al. (2005)
The extent of splice variation was calculated by excluding T-cell receptor and immunoglobulin genes from the transcripts. The remaining 144,351 transcripts were grouped into 43,539 TUs, of which 18,627 (42.8 %) consist of single-exon transcripts, 8,110 (18.6 %) contain a single multi-exon transcript, and the remaining 16,802 TUs (38.6 %) contain at least two spliced transcripts. Among these TUs, 5,862 (34.9 %) show no evidence of splice variation, whereas 10,940 (65.1 %) contain multiple splice forms

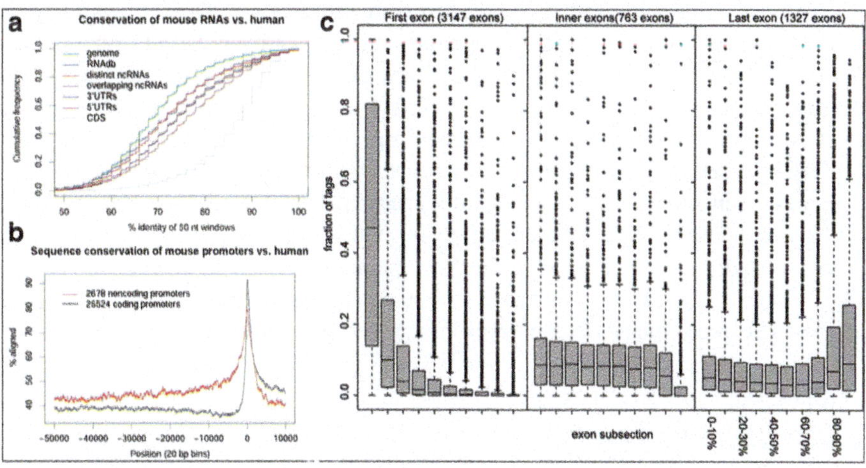

Fig. 1.4 Analysis of the FANTOM3 data set. (**a**) Human–mouse conservation of coding and non-coding RNAs compared with a random genome sequence (marked as "genome"). (**b**) Conservation of promoter, based on human–mouse sequence alignment. The positions of −50 Kb to +10 Kb from TSS (point 0) are shown. (**c**) *Box plot*, showing distribution of CAGE tags across indicated exons. For each analyzed exon, the fraction of tags mapped to ten equally large subsections of the exon was calculated. (*Left*) CAGE tags mapping to the first exon are prevalently located in the first part of the exon. (*Middle*) CAGE tags mapping to internal exons are uniformly distributed. (*Right*) The last exons show a distinct overrepresentation of CAGE tags mapping close to the 3′ end (Adapted from Carninci et al. (2005))

As shown by the CAGE tags distribution, TTSs were originated primarily at the annotated 5′ ends of the transcripts. However, internal and 3′-terminal CAGE clusters were also detected. Surprisingly, though distribution of TSSs across the first exons was apparently shifted toward the 5′ ends, in terminal exons the number of clusters increased toward the 3′ end (Fig. 1.4c). Out of 102,801 FANTOM3 clones, 34,030 contained no ORFs and thus were recorded as ncRNAs. A significant number of those had their TSSs at 3′-UTRs of protein-coding genes, accounting for the CAGE tag distribution shift mentioned above (Fig. 1.4c). Cross-validation by different subsets of the resource data set (Table 1.1), filtering by expression level and collapsing overlapping sequences, produced a list of 2,886 high-confidence ncRNAs, highly conserved in the mouse, human, and chicken genomes. One thousand eighty-nine members of this set were found to be noncoding variants of protein-coding transcripts.

Since publication of the main FANTOM3 results in 2005 (Carninci et al. 2005), the FANTOM Consortium has been continuing its collaborative approach to study mammalian transcriptomes. The FANTOM5 Project (http://fantom.gsc.riken.jp/5/) broadly analyzed transcription in human cells and tissues, as well as additional data from mice. The results of stage 1 of FANTOM5, released in 2014 (Andersson et al. 2014; Forrest et al. 2014), provided comprehensive sets of human and mouse promoters and enhancer regions. The full set contains data from primary cells, tissues, and cell lines (Fig. 1.5a). A summary of CAGE profiling is shown at Fig. 1.5b. The project discovered and characterized the activity of promoters, corresponding to 95 % of known protein-coding genes in the human genome. The remaining genes might be expressed in rare cell types or only for restricted periods of time not covered by the sample set, or might have very low expression levels. Stage 1 also characterized housekeeping and cell-specific promoters and quantified conservation of promoters and transcripts between humans and mice. As mentioned in Sect. 1.3.2, CAGE data cannot discriminate between unstable and lowly expressed RNAs, and provides no information about a transcript's length. Therefore, stage 2 of FANTOM5 will also concentrate on characterization and analysis of RNAseq libraries produced from human tissues and primary cells. Specific attention will be given to noncoding transcripts. Another focus of stage 2 will be on the dynamics of promoter usage in time-course experiments on cell development and proliferation.

1.4.2 ENCODE Project

The project was launched in 2003, aiming to characterize all functional elements of the human genome. The pilot stage focused on only 1 % of the total genome sequence (ENCODE Project Consortium 2007), after which the work was extended to the whole-genome study (Harrow et al. 2012). GENCODE Version 7, the reference genome annotation of the ENCODE Project, released in 2012, is considered to be the most complete and comprehensive listing of transcripts produced by human cells. Transcript annotation in ENCODE is based on both physical evidence and biochemical markers, such as DNA exposure to DNase treatment, transcription

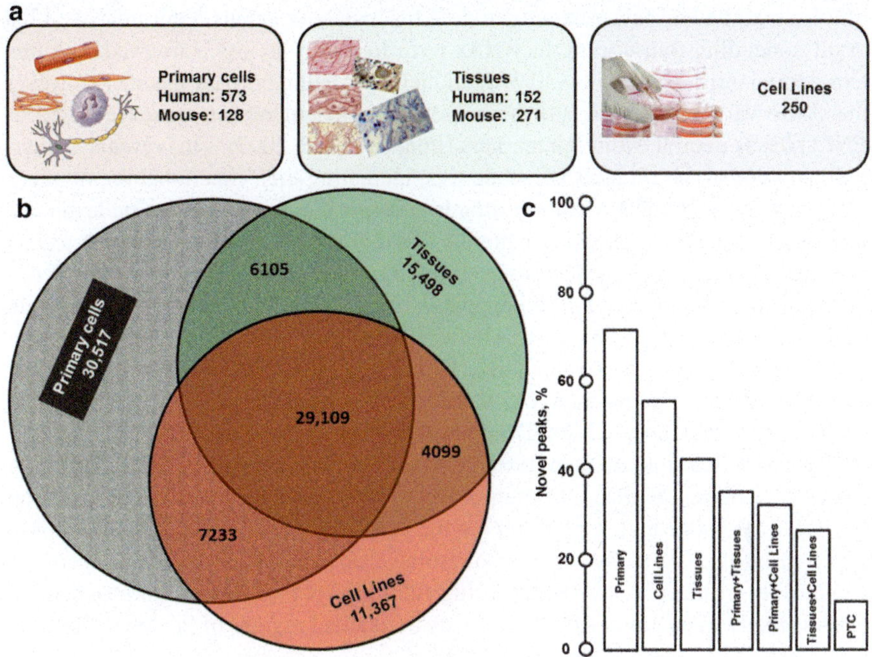

Fig. 1.5 (**a**) Sample list profiled in FANTOM5. (**b**) Venn diagram showing CAGE peaks expressed at ≥10 TPM (tags per million) in primary cells (*gray*), tissues (*green*), and cell lines (*red*). (**c**) Fraction of unannotated peaks observed in subsets of (**b**). PTC correspond to peaks found in all three sample types (Data obtained from Forrest et al. (2014))

Intergenic (9,518)			Genic (5362)		
Converge nt	Diverg ent	Same strand	Exonic	Intronic	Overlapping
1937	3416	4165	Sense: NA Antisense: 2411	Sense: 563 Antisense: 2221	Sense: 52 Antisense: 115

Fig. 1.6 Gencode lncRNAs, with numbers per sub-category, defined by Fig. 1.1. Intergenic transcripts were additionally divided on the "same strand", convergent (tail-to-tail) or divergent (head-to-head) (Data obtained from Derrien et al. (2012))

factor binding and chromatin modifications, marking promoters, elongated regions, and transcription terminators. GENCODE 7 contains 20,687 protein-coding and 9,277 lncRNA loci (not containing miRNAs and transcripts shorter than 200 nucleotides). lncRNAs were additionally classified as summarized in Fig. 1.1. The total number and subtypes of GENCODE lncRNAs is shown in Fig. 1.6. Transcribed regions were found to produce numerous RNA products, with average ~6 transcripts per locus for protein-coding genes and 1.6 per noncoding locus. The smaller number of noncoding transcripts per locus may be explained by lncRNA being shorter than mRNAs and having fewer exons (mode 2 for lncRNAs versus 5 per mRNA; 42 % of all noncoding transcripts). lncRNAs were found to be less conserved than protein-coding transcripts, with ~30 % of all lncRNAs being primate specific. However, the conservation level was still higher than in ancient repeat sequences used by ENCODE as neutral evolution markers (Ponjavic et al. 2007). This reveals the evolutionary constraint on lncRNA sequences, indicating their functional importance.

Compared with mRNAs, noncoding transcripts had lower expression levels and more tissue-specific expression patterns. These features suggested putative regulatory functions. Indeed, most of the recently discovered lincRNAs were located in the nucleus and associated with epigenetic regulation of gene expression (Mondal et al. 2010), similarly to previously characterized molecules, such as HOTAIR (Rinn et al. 2007; Tsai et al. 2010) and XIST (Brockdorff et al. 1992; Sado and Brockdorff 2013). In addition, some lncRNAs were found to be precursors for smaller molecules—in particular, snoRNAs (Derrien et al. 2012; Harrow et al. 2012).

Overall, transcriptome complexity is likely to increase while GENCODE annotation continues. Of note, the number of protein-coding genes decreased from 22,500 to 20,700 from the year 2009 to the year 2011 (version 3c versus version 7) (Derrien et al. 2012). The number of isolated and annotated lncRNAs increased from 6,000 to more than 10,000 during the same period. Thus, more noncoding RNA is likely to be found when previously untouched regions of the human genome are analyzed and more cell types are studied.

1.5 Future Directions in ncRNA-Omics

As described in this chapter, most of the effort in twentieth century molecular biology was focused on products of protein-coding genes, representing about 2 % of the total genome size in humans. During recent decades, however, this view has dramatically changed. With about 80 % of the genome being transcribed and thousands of noncoding RNAs identified, not many sequences could be safely considered totally silent any longer. The absence of common TSS, TTS, and regulatory sequences can be explained by non-canonical transcription regulation. The lack of expression tags (ESTs or sequencing reads) may suggest weak expression levels restricted to only selected cells and/or developmental or functional stages. In addition, a high turnover rate can lead to silent appearance of selected genomic loci.

The current view of the human (or other eukaryote) genome can be well reflected by the "iceberg tip" concept. This vision puts known noncoding RNAs into the tip of an iceberg, while the vast majority of them are still "below the surface", waiting to be discovered and/or characterized. First, only a small proportion of known lncRNAs have clearly established functions. These will be discussed in more detail in further chapters. Next, novel lncRNAs are still being discovered in old and newly produced data sets. One example is a class of antisense RNA produced from the strand opposite the one being the template for an mRNA (or noncoding transcript). About 20 % of mammalian transcripts have been estimated to participate in sense/antisense (S/AS) interaction (Chen et al. 2004; Yelin et al. 2003). This number may underestimate the actual scale of antisense transcription, as shown by additional studies. In fact, more than 50 % of all mammalian RNAs may overlap an opposite strand transcript in a divergent, convergent, or full-length configuration. Generally, there is some correlation (either positive or negative) in the expression of S/AS pair members (Katayama et al. 2005).

Furthermore, it is now becoming evident that genomic sequences that were excluded from previous studies on purpose may, in fact, be extremely important and have numerous biological functions. Examples of such sequences are transposable elements, which can compose up to 90 % of the genome (McClintock 1953). First reported in 1953 in *Zea mays* (corn) by Barbara McClintock, these repetitive sequences were long considered "genomic parasites" with no function in the cell besides self-propagation and genome predation (McClintock 1953). In fact, it took scientists 30 years to realize the importance of those sequences, acknowledged by awarding to McClintock a Nobel Prize in 1983. Transposable elements exist in thousands of copies across eukaryotic genomes. Because of their repetitive origin, it is hardly possible to map each particular EST or sequencing read to its original genomic locus. Therefore, these elements have been excluded from analysis in most RNA or DNA sequencing data sets by "removing" them from the reference genomes, using cleverly designed algorithms, such as Repeat Masker (Smit AFA, Hubley R, Green P. *RepeatMasker Open-3.0*; 1996–2010; http://www.repeatmasker.org). It is now obvious that transposable elements are part of the cellular expression program, associated with important functions in growth, development, and determination of cell faith (Criscione et al. 2014; De Cecco et al. 2013; Fort et al. 2014). Furthermore, repetitive sequences originate from common ancestors, such as tRNAs (MIR) (Smit and Riggs 1995), or 7SL RNA (Alu) (Quentin 1992). Thus, repeat-containing RNA may find a base-pairing target sequence not only in its original locus but also in numerous points across the genome, creating virtually endless and unpredictable regulatory possibilities. An example of such a complex network was shown by Holdt et al. (2013) for ANRIL lncRNA, which, with no interaction partners in close proximity to its genomic locus, was found to produce an Alu-mediated effect on expression of numerous atherosclerosis-related genes in diverse loci (Holdt et al. 2013).

An additional source of novel lncRNA may be provided by pseudogenes. These sequences are either duplicated copies of coding genomic regions or processed reverse-transcribed elements integrated into new genomic locations. Over 11,000 pseudogenes were identified by GENCODE (Harrow et al. 2012; Siggens and Ekwall 2014). Far from being silent, "broken" copies of coding sequences, these elements were recently shown to be involved in regulation of important cellular processes, including gene expression, cell maintenance, and cancer (Frith et al. 2006; Johnsson et al. 2013; Kalyana-Sundaram et al. 2012; Pei et al. 2012).

Finally, enhancer RNAs (eRNAs) have received growing attention in recent years (Andersson et al. 2014; Fort et al. 2014; Kim et al. 2010; Ronnerblad et al. 2014). These noncoding RNAs range from 50 to 2,000 bp and are produced from DNA sequences known as transcription enhancers. As RNA polymerase II can be recruited to numerous genomic sites, eRNAs may simply be produced by random "leaky" transcription. However, the notions that some enhancers have tissue-specific transcription patterns and that eRNA transcription correlates with enhancer-specific activity support the idea that individual eRNAs carry distinct and relevant biological functions. Also, knockdowns of enhancer RNAs may show phenotype changes (Feng et al. 2006; Wang et al. 2008). Future research will need to clarify these issues and establish the final status for eRNA.

1.6 Conclusion

The studies mentioned in this chapter have shown that the noncoding part of the human genome is far from being 'junk'—rather, it represents a complex, sophisticated RNA-based regulatory system controlling cell growth, development, and function, which has a tremendous impact on all stages of the organism's life cycle and global biological processes. Discoveries of numerous and diverse noncoding RNAs bring scientists to the challenge of re-thinking the meaning of a "gene". Originally, genes were defined as DNA fragments that code for a polypeptide or for an RNA chain that has a function in the organism. An additional functional definition claimed that any genomic sequence capable of changing the organism's phenotype, if modified, should be counted as a gene (or part of a gene). Thus, promoters, enhancers, and other regulatory elements had to be recognized as "genes". Today, considering that many regulatory sequences are actually being transcribed (and thus produce RNA "products"), the difference between a coding sequence and a regulatory element becomes even more blurred. ncRNAs transcribed from the "intergenic" DNA sequence may have an effect on expression of proteins and other RNAs. This effect is not limited to transcriptional control but can be executed on other levels, such as translation, localization, trafficking, or stability (Ambros 2004; Bartel 2004; Carrieri et al. 2012; Chen and Carmichael 2009; Leucci et al. 2013; Lim et al. 2005). Therefore, a modern working definition of a gene would rather be "a locatable region of a genomic sequence, corresponding to a unit of inheritance, which is associated with regulatory regions, transcribed regions, and/or other functional sequence elements" (Pearson 2006; Pennisi 2007). Being extremely broad, this description may lead to a dramatic decrease in the length of "intergenic regions" in complex eukaryotic genomes. Further research will provide more details on the number of genes in a given cell and on the regulatory networks that can be controlled and modulated for the benefit of humans, animals, and the Earth's biosphere.

References

Ambros V (2004) The functions of animal microRNAs. Nature 431:350–355

Andersson R, Gebhard C, Miguel-Escalada I, Hoof I, Bornholdt J, Boyd M, Chen Y, Zhao X, Schmidl C, Suzuki T et al (2014) An atlas of active enhancers across human cell types and tissues. Nature 507:455–461

Bartel DP (2004) MicroRNAs: genomics, biogenesis, mechanism, and function. Cell 116:281–297

Bazak L, Haviv A, Barak M, Jacob-Hirsch J, Deng P, Zhang R, Isaacs FJ, Rechavi G, Li JB, Eisenberg E et al (2014) A-to-I RNA editing occurs at over a hundred million genomic sites, located in a majority of human genes. Genome Res 24:365–376

Boley N, Stoiber MH, Booth BW, Wan KH, Hoskins RA, Bickel PJ, Celniker SE, Brown JB (2014) Genome-guided transcript assembly by integrative analysis of RNA sequence data. Nat Biotechnol 32:341–346

Brockdorff N, Ashworth A, Kay GF, McCabe VM, Norris DP, Cooper PJ, Swift S, Rastan S (1992) The product of the mouse Xist gene is a 15 kb inactive X-specific transcript containing no conserved ORF and located in the nucleus. Cell 71:515–526

Brown JB, Boley N, Eisman R, May GE, Stoiber MH, Duff MO, Booth BW, Wen J, Park S, Suzuki AM et al (2014) Diversity and dynamics of the Drosophila transcriptome. Nature 512:393–399

Carninci P, Shibata Y, Hayatsu N, Sugahara Y, Shibata K, Itoh M, Konno H, Okazaki Y, Muramatsu M, Hayashizaki Y (2000) Normalization and subtraction of cap-trapper-selected cDNAs to prepare full-length cDNA libraries for rapid discovery of new genes. Genome Res 10:1617–1630

Carninci P, Kasukawa T, Katayama S, Gough J, Frith MC, Maeda N, Oyama R, Ravasi T, Lenhard B, Wells C et al (2005) The transcriptional landscape of the mammalian genome. Science 309:1559–1563

Carrieri C, Cimatti L, Biagioli M, Beugnet A, Zucchelli S, Fedele S, Pesce E, Ferrer I, Collavin L, Santoro C et al (2012) Long non-coding antisense RNA controls Uchl1 translation through an embedded SINEB2 repeat. Nature 491:454–457

Chang H, Lim J, Ha M, Kim VN (2014) TAIL-seq: genome-wide determination of poly(A) tail length and 3′ end modifications. Mol Cell 53:1044–1052

Chen LL, Carmichael GG (2009) Altered nuclear retention of mRNAs containing inverted repeats in human embryonic stem cells: functional role of a nuclear noncoding RNA. Mol Cell 35:467–478

Chen J, Sun M, Kent WJ, Huang X, Xie H, Wang W, Zhou G, Shi RZ, Rowley JD (2004) Over 20% of human transcripts might form sense–antisense pairs. Nucleic Acids Res 32:4812–4820

Cheng J, Kapranov P, Drenkow J, Dike S, Brubaker S, Patel S, Long J, Stern D, Tammana H, Helt G et al (2005) Transcriptional maps of 10 human chromosomes at 5-nucleotide resolution. Science 308:1149–1154

Cloonan N, Forrest AR, Kolle G, Gardiner BB, Faulkner GJ, Brown MK, Taylor DF, Steptoe AL, Wani S, Bethel G et al (2008) Stem cell transcriptome profiling via massive-scale mRNA sequencing. Nat Methods 5:613–619

Core L, Waterfall J, Lis JT (2008) Nascent RNA sequencing reveals widespread pausing and divergent initiation at human promoters. Science 322(5909):1845–1848

Criscione SW, Zhang Y, Thompson W, Sedivy JM, Neretti N (2014) Transcriptional landscape of repetitive elements in normal and cancer human cells. BMC Genomics 15:583

De Cecco M, Criscione SW, Peterson AL, Neretti N, Sedivy JM, Kreiling JA (2013) Transposable elements become active and mobile in the genomes of aging mammalian somatic tissues. Aging 5:867–883

Dermitzakis ET, Reymond A, Lyle R, Scamuffa N, Ucla C, Deutsch S, Stevenson BJ, Flegel V, Bucher P, Jongeneel CV et al (2002) Numerous potentially functional but non-genic conserved sequences on human chromosome 21. Nature 420:578–582

Dermitzakis ET, Kirkness E, Schwarz S, Birney E, Reymond A, Antonarakis SE (2004) Comparison of human chromosome 21 conserved nongenic sequences (CNGs) with the mouse and dog genomes shows that their selective constraint is independent of their genic environment. Genome Res 14:852–859

Derrien T, Johnson R, Bussotti G, Tanzer A, Djebali S, Tilgner H, Guernec G, Martin D, Merkel A, Knowles DG et al (2012) The GENCODE v7 catalog of human long noncoding RNAs: analysis of their gene structure, evolution, and expression. Genome Res 22:1775–1789

Derti A, Garrett-Engele P, Macisaac KD, Stevens RC, Sriram S, Chen R, Rohl CA, Johnson JM, Babak T (2012) A quantitative atlas of polyadenylation in five mammals. Genome Res 22:1173–1183

Djebali S, Davis CA, Merkel A, Dobin A, Lassmann T, Mortazavi A, Tanzer A, Lagarde J, Lin W, Schlesinger F et al (2012) Landscape of transcription in human cells. Nature 489:101–108

Feng J, Bi C, Clark BS, Mady R, Shah P, Kohtz JD (2006) The Evf-2 noncoding RNA is transcribed from the Dlx-5/6 ultraconserved region and functions as a Dlx-2 transcriptional coactivator. Genes Dev 20:1470–1484

Forrest AR, Kawaji H, Rehli M, Baillie JK, de Hoon MJ, Lassmann T, Itoh M, Summers KM, Suzuki H, Daub CO et al (2014) A promoter-level mammalian expression atlas. Nature 507:462–470

Fort A, Hashimoto K, Yamada D, Salimullah M, Keya CA, Saxena A, Bonetti A, Voineagu I, Bertin N, Kratz A et al (2014) Deep transcriptome profiling of mammalian stem cells supports a regulatory role for retrotransposons in pluripotency maintenance. Nat Genet 46:558–566

Frith MC, Wilming LG, Forrest A, Kawaji H, Tan SL, Wahlestedt C, Bajic VB, Kai C, Kawai J, Carninci P et al (2006) Pseudo-messenger RNA: phantoms of the transcriptome. PLoS Genet 2:e23

Fullwood MJ, Wei CL, Liu ET, Ruan Y (2009) Next-generation DNA sequencing of paired-end tags (PET) for transcriptome and genome analyses. Genome Res 19:521–532

Grabherr MG, Haas BJ, Yassour M, Levin JZ, Thompson DA, Amit I, Adiconis X, Fan L, Raychowdhury R, Zeng Q et al (2011) Full-length transcriptome assembly from RNA-Seq data without a reference genome. Nat Biotechnol 29:644–652

Graur D, Zheng Y, Price N, Azevedo RB, Zufall RA, Elhaik E (2013) On the immortality of television sets: "function" in the human genome according to the evolution-free gospel of ENCODE. Genome Biol Evol 5:578–590

Gupta RA, Shah N, Wang KC, Kim J, Horlings HM, Wong DJ, Tsai MC, Hung T, Argani P, Rinn JL et al (2010) Long non-coding RNA HOTAIR reprograms chromatin state to promote cancer metastasis. Nature 464:1071–1076

Harrow J, Frankish A, Gonzalez JM, Tapanari E, Diekhans M, Kokocinski F, Aken BL, Barrell D, Zadissa A, Searle S et al (2012) GENCODE: the reference human genome annotation for The ENCODE Project. Genome Res 22:1760–1774

Holdt LM, Hoffmann S, Sass K, Langenberger D, Scholz M, Krohn K, Finstermeier K, Stahringer A, Wilfert W, Beutner F et al (2013) Alu elements in ANRIL non-coding RNA at chromosome 9p21 modulate atherogenic cell functions through trans-regulation of gene networks. PLoS Genet 9:e1003588

International Human Genome Sequencing Consortium (2004). Finishing the euchromatic sequence of the human genome. Nature 431:931–945

International Human Genome Sequencing Consortium (2012) An integrated encyclopedia of DNA elements in the human genome. Nature 489:57–74

Johnsson P, Ackley A, Vidarsdottir L, Lui WO, Corcoran M, Grander D, Morris KV (2013) A pseudogene long-noncoding-RNA network regulates PTEN transcription and translation in human cells. Nat Struct Mol Biol 20:440–446

Kalyana-Sundaram S, Kumar-Sinha C, Shankar S, Robinson DR, Wu YM, Cao X, Asangani IA, Kothari V, Prensner JR, Lonigro RJ et al (2012) Expressed pseudogenes in the transcriptional landscape of human cancers. Cell 149:1622–1634

Kapranov P, Cawley SE, Drenkow J, Bekiranov S, Strausberg RL, Fodor SP, Gingeras TR (2002) Large-scale transcriptional activity in chromosomes 21 and 22. Science 296:916–919

Katayama S, Tomaru Y, Kasukawa T, Waki K, Nakanishi M, Nakamura M, Nishida H, Yap CC, Suzuki M, Kawai J et al (2005) Antisense transcription in the mammalian transcriptome. Science 309:1564–1566

Kelkar DS, Provost E, Chaerkady R, Muthusamy B, Manda SS, Subbannayya T, Selvan LD, Wang CH, Datta KK, Woo S et al (2014) Annotation of the zebrafish genome through an integrated transcriptomic and proteomic analysis. Mol Cell Proteomics: MCP 13:3184–3198

Kellis M, Wold B, Snyder MP, Bernstein BE, Kundaje A, Marinov GK, Ward LD, Birney E, Crawford GE, Dekker J et al (2014) Defining functional DNA elements in the human genome. Proc Natl Acad Sci U S A 111:6131–6138

Kim TK, Hemberg M, Gray JM, Costa AM, Bear DM, Wu J, Harmin DA, Laptewicz M, Barbara-Haley K, Kuersten S et al (2010) Widespread transcription at neuronal activity-regulated enhancers. Nature 465:182–187

Kodzius R, Kojima M, Nishiyori H, Nakamura M, Fukuda S, Tagami M, Sasaki D, Imamura K, Kai C, Harbers M et al (2006) CAGE: cap analysis of gene expression. Nat Methods 3:211–222

Lahens NF, Kavakli IH, Zhang R, Hayer K, Black MB, Dueck H, Pizarro A, Kim J, Irizarry R, Thomas RS et al (2014) IVT-seq reveals extreme bias in RNA sequencing. Genome Biol 15:R86

Lander ES, Linton LM, Birren B, Nusbaum C, Zody MC, Baldwin J, Devon K, Dewar K, Doyle M, FitzHugh W et al (2001) Initial sequencing and analysis of the human genome. Nature 409:860–921

Leucci E, Patella F, Waage J, Holmstrom K, Lindow M, Porse B, Kauppinen S, Lund AH (2013) microRNA-9 targets the long non-coding RNA MALAT1 for degradation in the nucleus. Sci Rep 3:2535

Levin JZ, Yassour M, Adiconis X, Nusbaum C, Thompson DA, Friedman N, Gnirke A, Regev A (2010) Comprehensive comparative analysis of strand-specific RNA sequencing methods. Nat Methods 7:709–715

Lim LP, Lau NC, Garrett-Engele P, Grimson A, Schelter JM, Castle J, Bartel DP, Linsley PS, Johnson JM (2005) Microarray analysis shows that some microRNAs downregulate large numbers of target mRNAs. Nature 433:769–773

Lv J, Liu H, Huang Z, Su J, He H, Xiu Y, Zhang Y, Wu Q (2013) Long non-coding RNA identification over mouse brain development by integrative modeling of chromatin and genomic features. Nucleic Acids Res 41:10044–10061

Matera AG, Terns RM, Terns MP (2007) Non-coding RNAs: lessons from the small nuclear and small nucleolar RNAs. Nat Rev Mol Cell Biol 8:209–220

McClintock B (1953) Induction of instability at selected loci in maize. Genetics 38:579–599

Mondal T, Rasmussen M, Pandey GK, Isaksson A, Kanduri C (2010) Characterization of the RNA content of chromatin. Genome Res 20:899–907

Ng P, Wei CL, Sung WK, Chiu KP, Lipovich L, Ang CC, Gupta S, Shahab A, Ridwan A, Wong CH et al (2005) Gene identification signature (GIS) analysis for transcriptome characterization and genome annotation. Nat Methods 2:105–111

Paul J, Duerksen JD (1975) Chromatin-associated RNA content of heterochromatin and euchromatin. Mol Cell Biochem 9:9–16

Pearson H (2006) Genetics: what is a gene? Nature 441:398–401

Pei B, Sisu C, Frankish A, Howald C, Habegger L, Mu XJ, Harte R, Balasubramanian S, Tanzer A, Diekhans M et al (2012) The GENCODE pseudogene resource. Genome Biol 13:R51

Pennisi E (2007) Genomics. DNA study forces rethink of what it means to be a gene. Science 316:1556–1557

Plessy C, Bertin N, Takahashi H, Simone R, Salimullah M, Lassmann T, Vitezic M, Severin J, Olivarius S, Lazarevic D et al (2010) Linking promoters to functional transcripts in small samples with nanoCAGE and CAGEscan. Nat Methods 7:528–534

Ponjavic J, Ponting CP, Lunter G (2007) Functionality or transcriptional noise? Evidence for selection within long noncoding RNAs. Genome Res 17:556–565

Preker P, Almvig K, Christensen MS, Valen E, Mapendano CK, Sandelin A, Jensen TH (2011) PROMoter uPstream Transcripts share characteristics with mRNAs and are produced upstream of all three major types of mammalian promoters. Nucleic Acids Res 39:7179–7193

Quentin Y (1992) Origin of the Alu family: a family of Alu-like monomers gave birth to the left and the right arms of the Alu elements. Nucleic Acids Res 20:3397–3401

Rinn JL, Chang HY (2012) Genome regulation by long noncoding RNAs. Annu Rev Biochem 81:145–166

Rinn JL, Euskirchen G, Bertone P, Martone R, Luscombe NM, Hartman S, Harrison PM, Nelson FK, Miller P, Gerstein M et al (2003) The transcriptional activity of human Chromosome 22. Genes Dev 17:529–540

Rinn JL, Kertesz M, Wang JK, Squazzo SL, Xu X, Brugmann SA, Goodnough LH, Helms JA, Farnham PJ, Segal E et al (2007) Functional demarcation of active and silent chromatin domains in human HOX loci by noncoding RNAs. Cell 129:1311–1323

Roberts A, Pimentel H, Trapnell C, Pachter L (2011) Identification of novel transcripts in annotated genomes using RNA-Seq. Bioinformatics 27:2325–2329

Ronnerblad M, Andersson R, Olofsson T, Douagi I, Karimi M, Lehmann S, Hoof I, de Hoon M, Itoh M, Nagao-Sato S et al (2014) Analysis of the DNA methylome and transcriptome in granulopoiesis reveals timed changes and dynamic enhancer methylation. Blood 123:e79–e89

Sado T, Brockdorff N (2013) Advances in understanding chromosome silencing by the long noncoding RNA Xist. Philos Trans R Soc Lond B Biol Sci 368:20110325

Sakurai M, Ueda H, Yano T, Okada S, Terajima H, Mitsuyama T, Toyoda A, Fujiyama A, Kawabata H, Suzuki T (2014) A biochemical landscape of A-to-I RNA editing in the human brain transcriptome. Genome Res 24:522–534

Salditt-Georgieff M, Darnell JE Jr (1982) Further evidence that the majority of primary nuclear RNA transcripts in mammalian cells do not contribute to mRNA. Mol Cell Biol 2:701–707

Salditt-Georgieff M, Harpold MM, Wilson MC, Darnell JE Jr (1981) Large heterogeneous nuclear ribonucleic acid has three times as many 5′ caps as polyadenylic acid segments, and most caps do not enter polyribosomes. Mol Cell Biol 1:179–187

Saliba AE, Westermann AJ, Gorski SA, Vogel J (2014) Single-cell RNA-seq: advances and future challenges. Nucleic Acids Res 42:8845–8860

Salimullah M, Sakai M, Plessy C, Carninci P (2011) NanoCAGE: a high-resolution technique to discover and interrogate cell transcriptomes. Cold Spring Harbor Protoc 2011:pdb.prot5559

Siggens L, Ekwall K (2014) Epigenetics, chromatin and genome organization: recent advances from the ENCODE project. J Intern Med 276:201–214

Smit AF, Riggs AD (1995) MIRs are classic, tRNA-derived SINEs that amplified before the mammalian radiation. Nucleic Acids Res 23:98–102

Takahashi H, Lassmann T, Murata M, Carninci P (2012) 5′ end-centered expression profiling using cap-analysis gene expression and next-generation sequencing. Nat Protoc 7:542–561

Thompson DM, Parker R (2007) Cytoplasmic decay of intergenic transcripts in Saccharomyces cerevisiae. Mol Cell Biol 27:92–101

Trapnell C, Roberts A, Goff L, Pertea G, Kim D, Kelley DR, Pimentel H, Salzberg SL, Rinn JL, Pachter L (2012) Differential gene and transcript expression analysis of RNA-seq experiments with TopHat and Cufflinks. Nat Protoc 7:562–578

Treangen TJ, Salzberg SL (2012) Repetitive DNA and next-generation sequencing: computational challenges and solutions. Nat Rev Genet 13:36–46

Tsai MC, Manor O, Wan Y, Mosammaparast N, Wang JK, Lan F, Shi Y, Segal E, Chang HY (2010) Long noncoding RNA as modular scaffold of histone modification complexes. Science 329:689–693

Vickaryous MK, Hall BK (2006) Human cell type diversity, evolution, development, and classification with special reference to cells derived from the neural crest. Biol Rev Camb Philos Soc 81:425–455

Wang X, Arai S, Song X, Reichart D, Du K, Pascual G, Tempst P, Rosenfeld MG, Glass CK, Kurokawa R (2008) Induced ncRNAs allosterically modify RNA-binding proteins in cis to inhibit transcription. Nature 454:126–130

Wang Z, Gerstein M, Snyder M (2009) RNA-Seq: a revolutionary tool for transcriptomics. Nat Rev Genet 10:57–63

Weinberg RA, Penman S (1968) Small molecular weight monodisperse nuclear RNA. J Mol Biol 38:289–304

Worton RG, Sutherland J, Sylvester JE, Willard HF, Bodrug S, Dube I, Duff C, Kean V, Ray PN, Schmickel RD (1988) Human ribosomal RNA genes: orientation of the tandem array and conservation of the 5′ end. Science 239:64–68

Yelin R, Dahary D, Sorek R, Levanon EY, Goldstein O, Shoshan A, Diber A, Biton S, Tamir Y, Khosravi R et al (2003) Widespread occurrence of antisense transcription in the human genome. Nat Biotechnol 21:379–386

Yook K, Harris TW, Bieri T, Cabunoc A, Chan J, Chen WJ, Davis P, de la Cruz N, Duong A, Fang R et al (2012) WormBase 2012: more genomes, more data, new website. Nucleic Acids Res 40:D735–D741

Chapter 2
Synthetic Strategies to Identify and Regulate Noncoding RNAs

Ganesh N. Pandian, Junetha Syed, and Hiroshi Sugiyama

Abstract RNA plays a central role in cell development and differentiation by regulating the flow of essential genetic information into the dynamic molecular machinery within the cell. Recently, large-scale sequencing techniques to analyze mammalian transcriptomes have substantiated the importance of the diverse population of the unannotated regions of the genome called noncoding RNAs (ncRNAs) in numerous biological processes and diseases. The ncRNAs demonstrate tissue-specific expression and sequence conservation across species. Therefore, ncRNAs are a highly desirable target for small-molecule modulators. However, therapeutic ncRNAs are considered "undruggable" because of the lack of understanding about the RNA secondary structural motifs, which are the preferred binding sites of small molecules. Strategies to design bioactive compounds based on the RNA secondary structure and sequence would enable researchers to develop novel therapeutic strategies and would aid the elucidation of the intricate translational machinery. In this chapter, we give a chemical perspective of ncRNAs and detail the synthetic strategies available to modulate novel RNA structures, which have been identified as therapeutic targets in disease and development.

Keywords Chemical probes • DNA-based small molecules • Gene expression profiling • Noncoding RNAs • RNA therapeutics • Synthetic strategies • Triple helix formation

2.1 Introduction

Ribonucleic acid (RNA) is a prime intermediary macromolecule that transcribes the essential biological information between the other two fundamental macromolecules—deoxyribonucleic acid (DNA) and proteins—found in all forms of life. Cellular RNAs play a critical role in a multitude of transgenerational biological processes, including transcription, translation, protein synthesis, and gene

G.N. Pandian • J. Syed • H. Sugiyama (✉)
Department of Chemistry, Graduate School of Science, Institute for Integrated Cell Material Sciences (WPI-iCeMS), Kyoto University, Sakyo, Kyoto 606 8502, Japan
e-mail: hs@kuchem.kyoto-u.ac.jp

R. Kurokawa (ed.), *Long Noncoding RNAs*, DOI 10.1007/978-4-431-55576-6_2

regulation (Fedor and Williamson 2005). RNA is chemically similar to DNA and is transcribed from only four nucleobases even though there are about 100 naturally occurring modified nucleotides (Czerwoniec et al. 2009). However, several more classes of small molecules can bind DNA, compared with the number of those molecules that can bind RNA. The major reason behind this striking difference is attributed to the minor groove in the DNA helical structure, which has been harnessed as the target for bioactive small molecules that bind to DNA through intercalation (Bischoff and Hoffmann 2002). By contrast, there are fewer examples of small molecules capable of targeting the RNA groove. This is because RNA is distinguished from DNA by the presence of a hydroxyl group at the 2′ position of the ribose sugar. This structural feature causes the RNA helix to adopt A-form geometry, which results in a deeper major groove that is narrower than that observed in B-form DNA, and a minor groove that is shallower (Hermann and Patel 2000). Interestingly, the three-dimensional structures adopted by DNA and RNA under in vivo conditions also differ. While double-stranded DNA (dsDNA) adopts a helical structure, single-stranded RNA adopts folds similar to those of a protein structure, which minimizes energy and conforms into diverse structures. It is important to note here that this diversity in RNA structure can be exploited in the design of small molecules capable of targeting specific RNAs of interest, because most RNAs harbor unique binding pockets for small molecules. Some well-documented classes of compounds known to modulate RNA function include antibiotics such as aminoglycosides, macrolides, tetracyclines, and oxazolidinones, and their derivatives, which have moderate affinities and selectivities for RNA (Guan and Disney 2012). Recently, strategies have been reported to harness the privileged scaffolds to design ligands for bulged RNA secondary structures (Meyer and Hergenrother 2009) and those capable of recognizing double helices (Zengeya et al. 2011).

With the rise of high-throughput sequencing technologies in the past decade, the gene-regulatory function of the noncoding transcripts representing the 'dark matter of the genome', classified as non-protein-coding RNAs (ncRNAs), has been recognized (Derrien et al. 2012). There has been an exponential increase in our knowledge about the functional role of diverse classes of ncRNAs in regulating cell development (Orkin and Hochedlinger 2011) and in controlling the normal and disease states of the cell (Ng et al. 2013; Heward and Lindsay 2014; Maass et al. 2014; Takahashi et al. 2014). ncRNAs occur in nearly every cell type and are expressed in a more tissue-specific manner than their protein-coding counterparts (Ulitsky et al. 2011; Washietl et al. 2014). Therefore, ncRNAs are a preferred therapeutic target for developing functional modulators such as small molecules (Lee et al. 2009; Guan and Disney 2012). However, fewer small-molecule modulators of function have been developed for targeting therapeutically important ncRNAs, compared with those targeting DNA and protein (Rask-Anderson et al. 2011; Disney et al. 2014). This difference is attributed to the poorer understanding of the structure, diverse biogenesis, and functional properties of ncRNAs. ncRNAs can be categorized as follows: (1) structural ncRNAs, such as the constitutively expressed ribosomal and transfer RNAs (tRNAs); and (2) regulatory ncRNAs, which can be classified further on the basis of the transcript size into small ncRNAs like microRNAs (miRNAs), medium ncRNAs, and long ncRNAs (lncRNAs). Recently, large-scale analytical techniques have suggested the biological significance of the different classes of RNAs. In particular, the functionally diverse lncRNAs (longer than 200 bases) have been gaining importance because of their generally accepted role as the cellular address codes in disease and development (Batista and Chang 2013). The recent ENCODE (Encyclopedia of DNA Elements) Project mapped the regions of transcription and chromatin modification to annotate an overwhelming number of 9,640 lncRNA loci in comparison with 20,687 protein-coding genes in 15 human cell lines to substantiate the variety of biological effects mediated by lncRNAs (Birney 2012). lncRNAs function at the genetic level by regulating themselves and all RNA-based processes. lncRNAs also function at the mechanistic level by serving as scaffold platforms to either recruit RNA–protein complexes to target genes or as traps by binding and sequestering key regulatory proteins away from their target DNA sequences (Fig. 2.1a). The functional roles of several ncRNAs in diseases and cellular development have been identified (Fig. 2.1b). Therefore, designing small molecules capable of targeting the structurally and functionally unique lncRNAs may rectify diseases characterized by the dysregulation of gene expression. In this chapter, we summarize the known ncRNAs associated with disease and development, and we briefly discuss the RNA dynamics and mechanisms that facilitate the identification and specific regulation of lncRNAs.

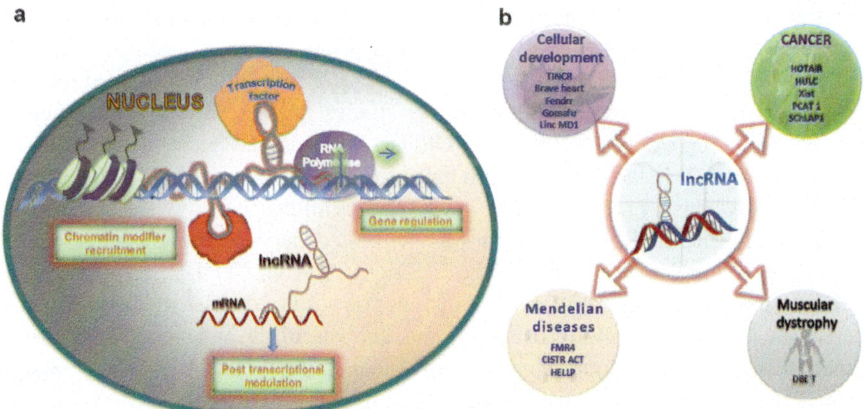

Fig. 2.1 (**a**) Long noncoding RNAs (lncRNAs) play a critical role in gene expression by functioning as transcriptional activators and post-transcriptional modulators, and by recruiting epigenetic modifiers (Graphics derived from Wang and Chang 2011). (**b**) Several lncRNAs have been implicated in cellular development and diseases (Graphics derived from Hauptman and Glavac 2013)

2.2 lncRNAs in Disease and Development

Interestingly, lncRNAs, which are more stable than mRNA transcripts, can be detected as free nucleic acids in urine and blood, and have been harnessed in clinically approved molecular diagnostic tests for prostate carcinoma (Shappell 2008). Summarizing all of them is beyond the scope of this review, and only a brief overview of certain prominent lncRNAs is given here. A lncRNA called terminal differentiation-induced lncRNA (TINCR), which is capable of regulating somatic tissue differentiation by binding directly to the STAU1 protein and stabilizing the differentiation-associated mRNAs, has been identified (Kretz et al. 2013). Klattenhoff et al. (2013) identified a heart-associated lncRNA termed 'Braveheart' (or Bvht), which mediates the epigenetic activation of the core cardiac gene network by functioning upstream of mesoderm posterior 1 (MesP1) and interacting with SUZ12, an essential subunit of polycomb-repressive complex 2 (PRC2). Similarly, a loss-of-function study revealed 'Fendrr' as a tissue-specific lncRNA, which is a critical factor in the regulation of the heart and body wall development. Fendrr modulates the chromatin signature by binding to PRC2 and trithorax group (TrxG)/MLL complexes and Fen (Grote et al. 2013). Recently, Barry et al. (2014) profiled the early transcriptomic responses to neuronal activation and identified the involvement of a lncRNA called 'Gomafu' in schizophrenia-associated alternative splicing. Gomafu was found to be downregulated in postmortem analysis of cortical gray matter from the superior temporal gyrus from patients with schizophrenia, which suggests that its dysregulation may contribute to neurological disorders (Barry et al. 2014). A muscle-specific lncRNA, linc-MD1, regulates muscle differentiation by functioning as a competing endogenous RNA that regulates the distribution of miRNAs and is associated with the pathogenesis of Duchenne muscle dystrophy (Cesana et al. 2011). Khalil et al. (2008) discovered a primate-specific ncRNA transcript called FMR4, which is silenced in fragile X syndrome, and demonstrated its anti-apoptotic function, using siRNA knockdown studies. Maass et al. (2012) discovered the dysregulation of an ncRNA called CISTR-ACT (named because of its *cis* and *trans* interactions) in two families with the autosomal-dominant Mendelian disorder of chondrodysplasia brachydactyly type E. The significance of the genome conformation and gene–lncRNA interface was suggested because the chromosome 12 translocation was shown to perturb the CISTR-ACT from PTHLH, a key chondrogenic factor (Maass et al. 2012). Loss-of-function studies of Xist RNA show that it can cause leukemia, myelofibrosis, sarcoma, and vasculitis, and suggest the importance of Xist RNA in suppressing hematological cancer (Yildrim et al. 2013). The role of the metastasis-associated lung adenocarcinoma transcript 1 lncRNA, called MALAT1, in regulation of the expression of oncogenic transcription factors, such as p53 and B-MYB, is known (Schimdt et al. 2011). Triple helix formation, which will be discussed in detail in a later section, is suggested as a key mechanism. Gupta et al. (2010) demonstrated the importance of HOTAIR, a lncRNA, in promoting cancer metastasis by differentially regulating genes of the PRC2 complex through chromatin remodeling. Quagliata et al. (2013) showed that increased expression of the lncRNA HOTTIP/HOXA13 is associated with the progression of hepatocellular carcinoma in vivo and may be useful as a marker for predicting the disease outcome. The HULC (highly upregulated in liver cancer) lncRNA is known to accelerate hepatoma cell proliferation by mediating the downregulation of tumor-suppressing p18 and altering the expression of the transcription factor CREB by modulating miRNAs such as miR-372 (Du et al. 2012). Prensner et al. (2011, 2013) discovered unannotated prostate cancer-associated PCAT-1, whose overexpression may be implicated in disease progression, and later reported a lncRNA, SChLAP1, that was capable of antagonizing the tumor-suppressing functions of the SWI/SNF chromatin-remodeling complex. An androgen-responsive region of C-terminal binding protein 1 (CTBP1-AS) was shown to promote prostate cancer progression by interacting with chromatin-modifying enzymes such as histone deacetylases and the transcriptional repressor PSF (Takayama et al. 2013). A polyadenylated lncRNA was shown to be associated with HELLP (hemolysis, elevated liver enzymes, low platelets) syndrome, a recessively inherited pregnancy complication (Van Dijk et al. 2012).

Facioscapulohumeral muscular dystrophy (FSHD) is an autosomal-dominant hereditary disorder characterized by facial and shoulder girdle weakness. Unlike other common muscular dystrophies characterized by mutant protein, FSHD is accompanied by a large deletion in the D4Z4 repeat region at chromosome 4q35 (Cabianca and Gabellini 2010). The other common feature of FSHD patients is the altered chromatin structure and the derepression of dystrophy-associated genes in the 4q35 chromosomal region, compared with those in healthy individuals (Gabellini et al. 2002). Recent studies have demonstrated the role of a lncRNA, DBE-T, in the derepression of the disease-associated genes and the 4q35 chromatin structure. In healthy individuals, the genes in the 4q35 region are silenced by a polycomb group (PcG) of complex proteins recruited by D4Z4 repeats. In FSHD patients, deletion of the D4Z4 repeat fails to recruit the PcG complex and initiates the transcription of the lncRNA DBE-T, which, in turn, recruits ASH1L to facilitate the derepression of 4q35 genes (Cabianca et al. 2012). This was the first report to demonstrate the role of a lncRNA in the epigenetic regulation of a human genetic disorder. It is important to understand the mechanistic aspect of the increasing number of ncRNAs identified as critical components in any biological process under investigation.

2.3 Identification and Analysis of Noncoding RNA

2.3.1 Synthetic Strategies to Profile RNA Secondary Structure

RNA levels are highly dynamic and change markedly in response to certain stimuli. The dynamics of a given RNA play a critical role in many cellular functions and are closely intertwined with its folding. RNA folding occurs over timescales ranging from picoseconds to seconds and follows a hierarchical pattern but in a disorderly fashion. The large-scale secondary structural transitions occur at a timescale of about >0.1 s, base-pair/tertiary dynamics occur at a microsecond-to-millisecond timescale, stacking dynamics occurs at timescales ranging from nanoseconds to microseconds, and other motions occur at timescales ranging from picoseconds to nanoseconds (Mustoe et al. 2014). The hairpin ribozyme, a small catalytic RNA, undergoes a sizable multidomain rearrangement to perform catalysis and substantiate the effects of RNA dynamics on RNA function (Kladwang et al. 2014). The secondary profile of RNA can be elucidated using enzyme probes, chemical probes, and phylogenetic analysis (Novikova et al. 2013). Among them, chemical probing to determine the RNA structure has many advantages. This technology does not impose restrictions on the size, quantity, and heterogeneity of the RNA molecules to be analyzed. Also, chemical probes could be used to tackle RNA sequences in vitro and in vivo (Spitale et al. 2013). Wong et al. (2007) showed that pause sites could facilitate the folding of ncRNAs created as transcription-induced nonnative structures. Base-specific reagents such as dimethyl sulfate (DMS) react with the single-stranded adenine and cytosine. Similarly, kethoxal modifies guanosine, and 1-cyclohexyl-(2-morpholinoethyl)carbodiimide metho-*p*-toluenesulfonate (CMCT) targets uracil (Fig. 2.2a) (Novikova et al. 2013). Weeks and coworkers developed selective 2′-hydroxyl acylation analyzed by primer extension (SHAPE) reagents, which react with the RNA backbone, allowing the probing of all four nucleotides in a single experiment (Wilkinson et al. 2006; McGinnis et al. 2012). Some of the existing SHAPE reagents include NMIA (*N*-methylisatoic anhydride), 1M7 (1-methyl-7-nitroisatoic anhydride), and BzCN (benzoyl cyanide), which vary in their hydrolysis half-lives (Fig. 2.2b) (Weeks 2010). RNA folds when being transcribed in the natural cellular environment, and the folding is determined by key parameters including crowding, RNA–protein interactions, temperature, pH, and external stimuli (Pan and Sosnick 2006; Kilburn et al. 2010; Novikova et al. 2012). Among the wide spectrum of chemical reagents, base-specific DMS is frequently used to study intact cells (Wells et al. 2000). Two electrophilic SHAPE reagents, FAI (2-methyl-3-furoic acid imidazolide) and NAI (2-methylnicotinic acid imidazolide), with extended half-lives and better solubility, gave consistent results in determining 5S rRNA in different cell lines and modified nuclear RNAs (SNORD3A) and U2 RNA (Spitale et al. 2013). Therefore, these reagents could be used in the future for investigating nuclear-retained transcripts.

2.3.2 *Genome-Wide Techniques to Profile the Structure and Expression of ncRNAs*

Conventional techniques using one molecule at a time are inadequate to meet the large demand to profile ncRNAs at the genome-wide gene level. Rapid evolution of lncRNAs also occurs because of the regulatory roles assumed by freely available transcripts that do not interact with the protein conservation (Pang et al. 2006). Also, it is not straightforward to apply RNA prediction programs that incorporate sequence homology for lncRNAs, as they lack diverse sequences and conservation of sequences. Consequently, RNA prediction programs such as RNAalifold, which

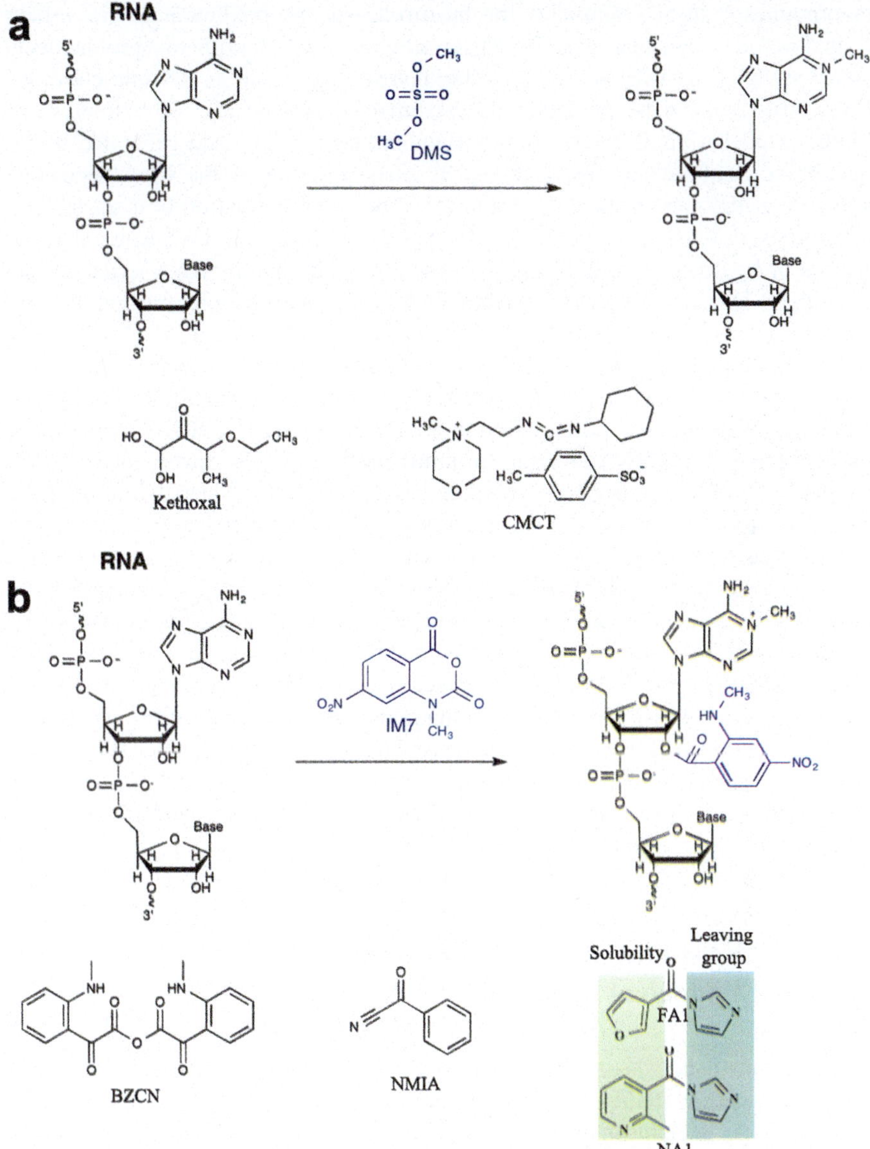

Fig. 2.2 Chemical probing of RNA structure. (**a**) Base-specific chemical probes dimethyl sulfate (DMS) (Methylation of adenine is shown) (Wells et al. 2000), Kethoxal and CMCT (1-cyclohexyl-(2-morpholinoethyl)carbodiimide metho-*p*-toluenesulfonate). (**b**) SHAPE reagents 1M7 (1-methyl-7-nitroisatoic anhydride) (2′ adduct formation is shown), NMIA (N-methylisatoic anhydride), BzCN (benzoyl cyanide), FAI (2-methyl-3-furoic acid imidazolide), and NAI (2-methylnicotinic acid imidazolide) (Modified from Novikova et al. (2013) and Spitale et al. (2013))

incorporates sequence homology for improved structure prediction accuracy, have been developed (Bernhart et al. 2008). By integrating an experimental probing technique, such as SHAPE, and a computational strategy, such as RNAstructure, the RNA structure can be predicted successfully (Deigan et al. 2009; Reuter and Mathews 2010). Similarly, the high-throughput sequencing data generated by the integrative SeqFold package enables the reconstruction of the RNA secondary structure at the genome scale and allows for accurate prediction of a set of short RNA transcripts (Ouyang et al. 2013). Approaches involving UV/chemical cross-linking or hydroxyl radical probing are used to determine the RNA tertiary structure, and 3-D structures can be attained by X-ray crystallography for homogenous RNA systems and NMR to study particular motifs (Ben-Shem et al. 2011). However, questions about the potential of lncRNAs to adopt higher-order tertiary organization and the stability and the number of tertiary configurations are yet to be clarified. A method called FragSeq utilizes RNase P1 (a single-stranded RNA nuclease) to probe the nuclear ncRNA structure in mouse cell lines (Underwood et al. 2010). Unlike PARS, FragSeq does not include an alkaline hydrolysis step that shortens the transcripts to allow analysis of short ncRNAs. Multiplexed RNA structure characterization using SHAPE-Seq, which integrates a chemical probing protocol with a deep-sequencing platform, has taken us a step closer to a vast structural analysis of entire transcriptomes in the natural cellular environment (Lucks et al. 2011). The dynamic gene expression profile of ncRNAs could be investigated by real-time PCR studies (Chen et al. 2005). Immunoprecipitation (IP) assays such as RIP-Seq (RNA-IP) and RNA-ChIP (RNA-chromatin IP) allow identification of protein–RNA interaction sites, polycomb-associated RNAs, and cotranscriptional RNA processing, respectively (Zhao et al. 2010; Bittencourt and Auboeuf 2012). However, in IP approaches, the results can be influenced by the specificity and affinity of the antibodies.

2.4 ncRNA-Induced Triplex Formation and Biological Implications

The mechanistic details of how an ncRNA can regulate these biological functions is intriguing. Recent studies have shed light on the role of triple helix formation of these ncRNAs, which can actively participate in regulating epigenetics and mRNA stability (Wilusz et al. 2012; Geisler and Coller 2013). The triple helix is a complex formed by sequence-specific base pairing between three oligonucleotide strands, such as a dsDNA and a single-stranded RNA or DNA. The sequence-specific base pairing occurs through the recognition of homopolypurine/homopolypyrimidine sequences in the major groove of the duplex DNA by the third single-stranded triplex-forming oligonucleotide (TFO) and forms Hoogsteen hydrogen bonding with the purines in the target duplex DNA (Fig. 2.3a). The binding of the TFO to the polypurine stretch of the duplex DNA can occur in either parallel or antiparallel orientation, depending on the type of triple helix motifs present in the

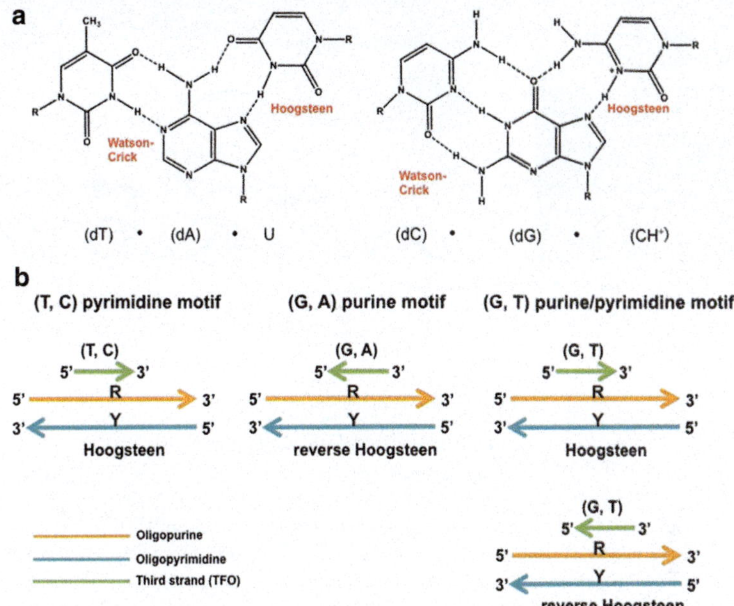

Fig. 2.3 (**a**) Chemical structure depicting the formation of T-A•U and C-G•C base triplexes by Hoogsteen hydrogen bonds to the Watson-Crick base-paired helix. (**b**) Three basic triple-helix motifs depicting the parallel and antiparallel orientations (Reproduced from Duca et al. 2008)

corresponding TFO (Morgan and Wells 1968; Buske et al. 2011). The TFO has three basic triple helix motifs: (T, C) pyrimidine motif, (G, A) purine motif, and (G, T) purine/pyrimidine motif. The pyrimidine motif binds in the parallel orientation, and the purine motif binds in the antiparallel orientation, whereas the purine/pyrimidine motifs can bind in either the parallel or the antiparallel orientation with respect to the purine stretches of the duplex DNA. Reverse Hoogsteen bonds favor the antiparallel binding of the TFO to the duplex DNA (Fig. 2.3b).

Bioinformatics demonstrates the existence of several TFO target sequences (TTS) across the whole human genome, especially in the important regulatory regions on the gene promoter, which suggests a role of triplexes in controlling gene expression (Goñi et al. 2004). The evidence implies the in vivo existence of triplexes whose exact role in biological function is yet to be confirmed. Some dyes, such as thiazole orange, are specific in binding to noncanonical DNA structures such as triplexes and exhibit a distinct staining pattern in the U2OS cell nucleus, which demonstrates the occurrence of triplexes in vivo (Lubitz et al. 2010). In a similar way, antibodies developed against the DNA triple helix structures stain positive for triplexes in the chromosomes of *Drosophila melanogaster*, *Homo sapiens*, and other species (Burkholder et al. 1991; Agazie et al. 1994). These antibodies also have higher affinity for RNA–DNA•DNA triple helix structures (Buske et al. 2011).

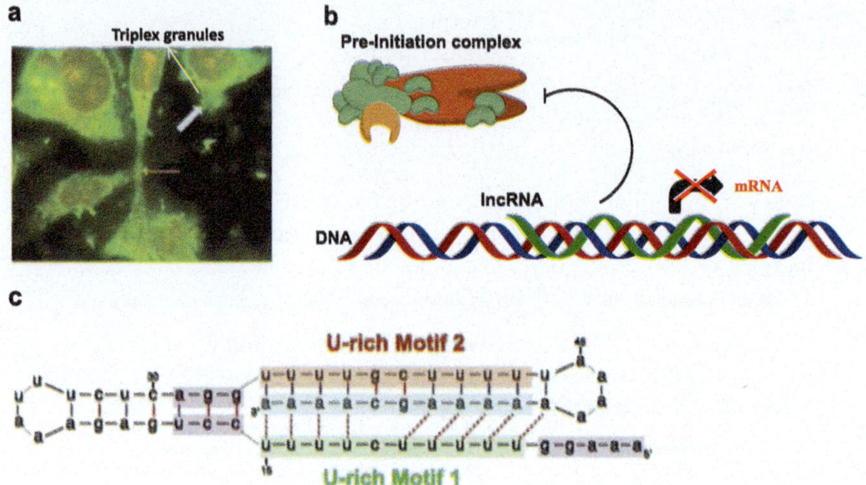

Fig. 2.4 (**a**) HeLa CD4 cells transfected with triplex-forming human miRNAs complementary to HIV-1 are stained with antitriplex monoclonal antibodies. Indirect immunohistochemistry reveals the presence of green granules in the cytoplasm, representing the potential triple helix structure formation (Reproduced from Kanak et al. 2010). (**b**) lncRNA forming a triple helix with the promoter region of the gene, thereby preventing binding of the transcriptional pre-initiation complex and inhibiting the transcription of the respective gene (Reproduced from Geisler and Coller 2013). (**c**) The U- and A- rich motifs present at the 3′ end of the MALAT1 ncRNA forms a stable triple helix, which helps in preventing RNA decay (Reproduced from Wilusz et al. 2012)

These findings are supported by the observation that human miRNA (*has-miR*) complementary to HIV-1 forms a stable triple helix with the HIV-1 proviral sequence of the preintegration complex (Kanak et al. 2010), thereby inhibiting its replication and providing resistance to HIV-1 in HeLa-CD4 cells (Fig. 2.4a). The abovementioned evidence suggests the existence of active RNA–DNA•DNA triplex regions in vivo. There is also growing evidence suggesting the direct interaction of lncRNAs with duplex DNA, resulting in triple helix formation through base-pair interactions, which thereby control the expression of genes by recruiting chromatin-remodeling proteins. One report demonstrating the function of ncRNA-induced triplex formation in vivo concerned the transcriptional regulation of the gene dihydrofolate reductase (DHFR), which contains two alternative promoters: a major and a minor promoter. The major promoter is responsible for 99 % transcription of the gene, whereas the minor promoter codes for an ncRNA that has been found to form a triple helix within the major promoter of the DHFR gene in a sequence-specific fashion (Martianov et al. 2007). In U2OS cells, this ncRNA-induced triplex has been shown to cause the dissociation of the pre-initiation complex from the major promoter by interacting with the TFIIB, thereby repressing the expression of the DHFR gene (Fig. 2.4b). Further recent evidence supporting the role of ncRNA-induced triplex formation in gene regulation concerns the epigenetic control of the ribosomal DNA (rDNA) promoter by a promoter-associated ncRNA (pRNA). The 20 nt stretch of this pRNA interacts and forms a stable RNA–DNA•DNA triple helix structure with the binding site of a transcription factor named TTF-I in the rDNA promoter region. The triplex structure formed in the rDNA promoter recruits the DNA methyltransferase DNMT3b, which induces DNA methylation and silencing of rRNA genes (Schmitz et al. 2010). This study demonstrated the mechanism of ncRNA-mediated de novo CpG island methylation by forming stable triplex structures with the duplex DNA.

The stability of the mRNA transcribed by RNA polymerase II is regulated by the addition of a poly(A) tail to its 3′ end by poly(A) polymerase. In addition to controlling mRNA stability, the poly(A) tail also supports the translation of the matured mRNA molecule by ribosomes in the cytoplasm. However, some mRNA is suggested to be transcribed by the RNA polymerase II machinery lacking the poly(A) tail (Yang et al. 2011). Analysis of the mechanism of the stabilization of these mRNAs lacking a poly(A) revealed the important role of ncDNA-induced triplex formation in maintaining the stability of mRNA and its translational efficiency. MEN β (multiple endocrine neoplasia β) is a ncRNA that plays important roles in maintaining the structural integrity of nuclear paraspeckles that control gene expression. This ncRNA and MALAT1 are examples of RNA pol II-transcribed RNA that lacks a poly(A) tail. Recent studies have identified the presence of highly conserved A- and U-rich motifs at the 3′ end of these ncRNAs, which are required for the stability of the ncRNAs (Wilusz et al. 2012). The studies demonstrated a possible triple helix structure formation at the 3′ end of the ncRNA, where the U-rich motif 1 forms Hoogsteen hydrogen bonding with Watson–Crick base-paired U-rich motif 2 and the A-rich sequence (Fig. 2.4c). The formed triplex helps in the protection of ncRNAs from the degradation of exonucleases, thereby regulating their stability. Further, when placed downstream from an open reading frame, the triple helix motif of ncRNAs is translated efficiently in the cytoplasm, which demonstrates the importance of the triplex at the end of the transcript in improving the translational efficiency. Thus, the endogenous ncRNA-induced triple helix structures could be critical for a variety of biological functions.

2.5 Synthetic Ligands Targeting RNA Structures: Advances and Challenges

A deeper understanding of RNA secondary structure and associated mechanisms may help us to design RNA-targeting ligands with superior specificity. Several small molecules have been shown to be capable of binding to and modulating the function of RNA structures. A rational design strategy based on a Janus wedge recognition unit yielded a ligand (Fig. 2.5a) with high affinity for CUG trinucleotide repeats, and which is known as sequester muscle blind-like (MBNL) proteins, which are involved in DM1 (Arambula et al. 2009). This ligand destabilizes the toxic poly(CUG) sequences and MBNL1, even in the presence of tRNA. Miller and coworkers used the resin-bound form of dynamic combinatorial chemistry to

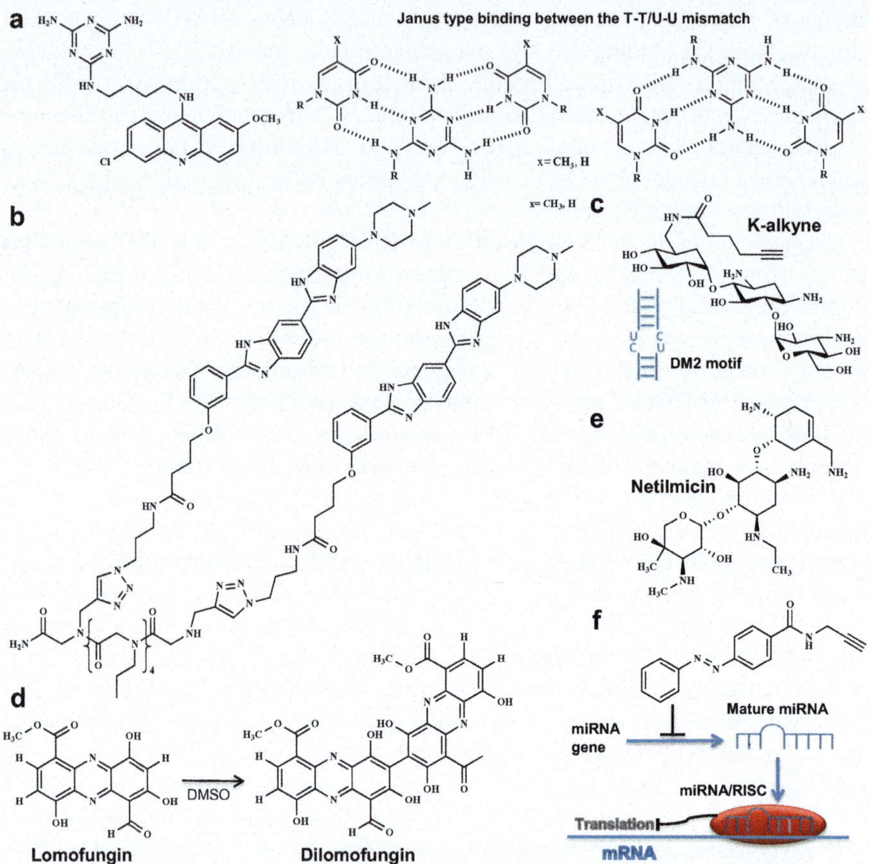

Fig. 2.5 Small molecules targeting RNA structures. (**a**) Ligand rationally designed on the basis of a Janus wedge recognition unit (Arambula et al. 2009). (**b**) Multivalent ligand 2H-4 simultaneously binding to 5′ CUG/3′ GUC motif (Childs-Disney et al. 2012). (**c**) Ligand targeting r(CCUG)^{exp} repeats in DM2 (Childs-Disney et al. 2014). (**d**) Lomofungin, dilomofungin (Hoskins et al. 2014). (**e**) Netilmicin binding TAR of HIV-1 (Stelzer et al. 2011). (**f**) miR-21-targeting small-molecule interfering with Dicer enzyme processing by binding to its precursor (Gumireddy et al. 2008)

identify compounds that inhibit (CUG)–MBNL1 and improved them by including a benzo[g]quinolone moiety (Ofori et al. 2012). Binding of *bis*-benzimidazole with the 5′CUG/3′GUC motif in r(CUG)^{exp} folds was harnessed to design multivalent ligands and identify a compound (*2H-4*) (Fig. 2.5b) capable of simultaneously binding this motif (Childs-Disney et al. 2012). The designed compounds improved the defects in alternative splicing and translation, and disrupted the nuclear foci that confer DM1. Subsequently, ligand (Fig. 2.5c) targeting the r(CCUG)^{exp} repeats improved the DM2-associated defects (Childs-Disney et al. 2014). Recently, the antimicrobial lomofungin was demonstrated to undergo spontaneous dimerization in dimethyl sulfoxide and to form dilomofungin, a potent inhibitor of MBNL1–(CUG)12 binding with distinct cellular effects (Fig. 2.5d) (Hoskins et al. 2014). An

RNA dynamic ensemble constructed by adjoining NMR spectroscopy and computational molecular dynamics and a subsequent virtual screening study with small molecules revealed a compound called netilmicin, which can bind the TAR on HIV-1 with remarkable selectivity (Stelzer et al. 2011). Netilmicin (Fig. 2.5e) also had activity under in vivo conditions by inhibiting Tat activation of the HIV-1 long terminal repeat by about 81 % in T cell lines and by perturbing HIV-1 replication.

Gene induction via transcriptional derepression can be achieved by targeting and inhibiting aberrant lncRNAs, using antisense oligonucleotides designed according to base pairing rules (Modarresi et al. 2012). Wheeler et al. (2012) successfully targeted and knocked down *Malat1* in a transgenic mouse model, using antisense oligonucleotides, and demonstrated the rectification of physiological and histopathological features of DM1. However, this lncRNA therapeutic strategy is hindered by poor cellular uptake and nonspecific immune system activation. Small molecules are mostly nonimmunogenic, and their mode of delivery and retention are easier than those of biologicals. Therefore, identifying and developing RNA-targeting small molecules is a preferred alternative strategy. Gumireddy et al. (2008) first demonstrated the utility of a small molecule (Fig. 2.5f) in modulating the miRNA miR-21, whose expression is associated with a variety of cancers. Aminoglycosides can bind to the secondary structures of RNA such as stem–loop structures with bulges observed in miRNA precursors. A screening study revealed that the aminoglycoside antibiotic streptomycin could effectively inhibit miR-21 activity by binding to its precursor and interfering with Dicer enzyme processing (Bose et al. 2012).

Recently, Disney and coworkers demonstrated a remarkable lead identification strategy termed Inforna and designed RNA-targeting small molecules based on the sequence (Velagapudi et al. 2014). Inforna integrates the following approaches: (1) RNA motif–small-molecule interaction is identified by two-dimensional combinatorial screening (2DCS); (2) the fitness of the identified interactions is determined by StARTS (structure–activity relationships through sequencing); and (3) structural information on the target DNA is derived from experimental, phylogenetic, or computational analysis (Fig. 2.6a). The identified compounds selectively inhibited miRNAs, and the designer small molecules targeting the precursors of the miRNAs (miR-96, miR-182, miR-183, and miR-210) also had significant bioactivity (Fig. 2.6b). In addition, the miR-96-targeting lead compound effectively modulated the miR-96-FoxO1 regulation pathway with high selectivity and induced apoptosis.

Programmable DNA-binding synthetic small molecules called PIPs artificially modulate gene expression in living cells in a sequence-specific manner by binding to the minor groove of DNA by following a unique DNA-binding rule. Designed PIPs can bind to their respective DNA sequence with a binding affinity similar to that of natural transcription regulators (Dervan 2001; Vaijayanthi et al. 2013). This principle of PIPs has been used as a gene-suppressing strategy against various therapeutically important genes (Syed et al. 2014). However, whether the well-established DNA-binding rule of PIPs is applicable to the chemically similar RNA molecule is yet to be determined. Because there are far fewer small molecules targeting therapeutically important RNA molecules (Guan and Disney 2013), efforts are underway to characterize the binding property of PIPs to dsRNA. Recent studies using thermal

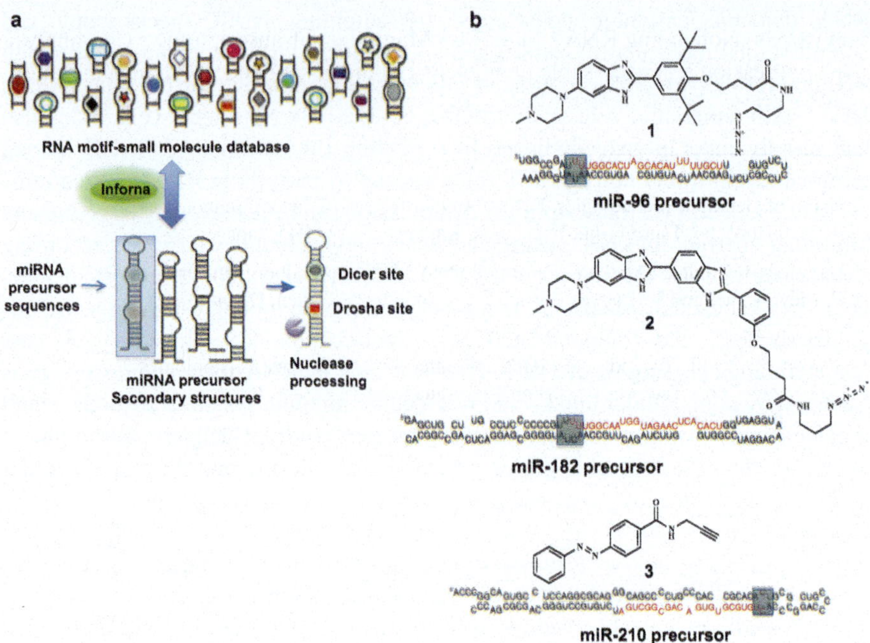

Fig. 2.6 Small molecules that target RNA from sequence applied to human miRNA precursors. (**a**) Schematic representation of Inforna approach. (**b**) Structures of small molecules 1, 2, and 3 binding to processing sites (*blue boxes*) in mir-96, mir-182, and mir-210 precursors, respectively (Modified and reproduced from Velagapudi et al. 2014)

melting temperature analysis have revealed that the PIPs display large thermal stabilization to dsDNA, but the polyamides exhibited no thermal stabilization to dsRNA (Chenoweth et al. 2013). Likewise, the binding ability of PIPs designed to target the replication site of the influenza A viral RNA sequence revealed the lower binding affinity of the designed PIPs to the viral RNA sequence than to the corresponding DNA sequence, which could be attributed to the secondary structures of the target molecule (Iguchi et al. 2013) (Fig. 2.7a–c). Recently, PIPs have been developed to have dual functions by conjugation with a chromatin-modifying histone deacetylase (HDAC) inhibitor, such as SAHA (Pandian and Sugiyama 2012, 2013; Saha et al. 2013; Pandian et al. 2014a, b, c). Screening studies to evaluate the effect of SAHA–PIPs on iPSC factors in MEFs indicated that SAHA–PIPs distinctively activate iPSC factors by triggering epigenetic marks that are associated with transcriptionally permissive chromatin (Pandian et al. 2012). In human fibroblasts, a SAHA–PIP was identified to be capable of selectively activating the endogenous expression of PIWI-interacting RNAs, which are distinctively expressed in male mammalian germ cells. It is important to note here that the PIWI-interacting RNAs that regulate the meiotic process are typically silenced in the human somatic cell (Fig. 2.7d) (Han et al. 2013). Evaluation of the effect of 32 SAHA–PIPs on the genome-wide gene expression in human fibroblasts divulged that each SAHA–PIP

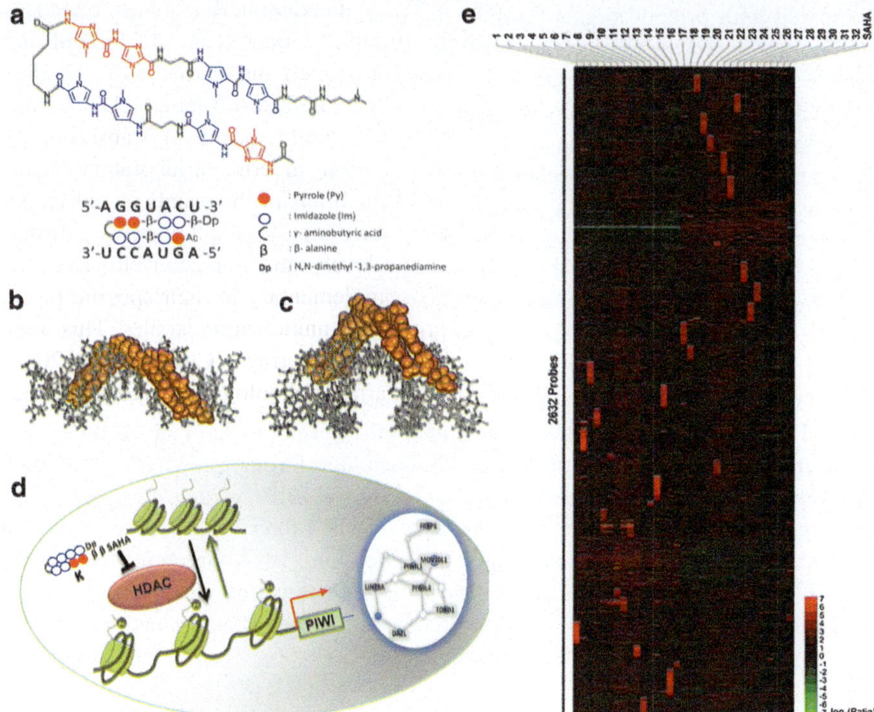

Fig. 2.7 (**a**) RNA-targeting pyrrole–imidazole polyamides (PIPs). Putative binding structures of dsDNA (**b**), and dsRNA (**c**) with TGF-β1 targeting PIP (Modified and reproduced from Iguchi et al. 2013). (**d**) A programmed PIP conjugated to the histone deacetylases inhibitor SAHA, called SAHA–PIP K, altered the heterochromatin structure to induce the PIWI-interacting RNA pathway (Han et al. 2013). (**e**) Individual SAHA–PIPs trigger transcriptional activation of distinctive noncoding RNAs (ncRNAs) (Reproduced from Pandian et al. 2014a)

activated its own unique set of therapeutically important genes and ncRNA (Pandian et al. 2014 a, c) (Fig. 2.7e). Advancing such new types of synthetic DNA-based transcriptional activators could lead to an effective strategy for targeting and modulating specific genes and RNAs of interest

2.6 Future Perspectives

Targeted modulation of RNA structures has been hampered by limitations in the understanding of RNA secondary structures. The relative expression levels of RNAs within the cell compared with the larger number of ribosomes that constitute cellular RNAs is also a major drawback because ncRNAs constitute <5 % of total cellular RNAs. However, because of the limited RNA-binding chemical scaffolds, high-throughput screening studies have been less successful in identifying RNA targets compared with protein targets because the small-molecule libraries used in such studies are biased toward binding proteins. Modern genome-wide analytical techniques and synthetic strategies aid the identification of the secondary structures of RNA and profiling of their expression in the natural cellular environment. Consequently, several screening methods have been developed to design small molecules capable of targeting repeating RNAs, miRNAs, and viral elements with high affinity and specificity. Novel strategies such as Inforna, which is based on information about RNA–ligand interactions, have obviated the notion that small molecules cannot be designed on the basis of the sequence. This approach has been shown to be more accurate in predicting the small molecules capable of targeting RNAs than the conventional medicinal chemistry approaches, such as screening and chemical similarity probing. Targeting the Dicer or Drosha processing sites could be harnessed as a general strategy to inhibit miRNA biogenesis. The development of rational and computational approaches could aid in the development of small molecules without the need for the 3D structure of RNA and high-throughput screen. Targeting transcriptional modulators such as PIPs and SAHA–PIPs, which are capable of binding and modulating specific DNA sequences and their local structures, could also be expanded to fulfill the promise of developing RNA-targeting therapeutic targets. Strategies addressing the limitations of RNA-targeting ligands, such as cell permeability and accessibility, could overcome the barriers that hamper the translation of genomics-to-patient therapeutics.

References

Agazie YM, Lee JS, Burkholder GD (1994) Characterization of a new monoclonal antibody to triplex DNA and immunofluorescent staining of mammalian chromosomes. J Biol Chem 269:7019–7023

Arambula JF, Ramisetty SR, Baranger AM, Zimmerman SC (2009) A simple ligand that selectively targets CUG trinucleotide repeats and inhibits MBNL protein binding. Proc Natl Acad Sci U S A 106:16068–16073

Barry G, Briggs JA, Vanichkina DP, Poth EM, Beveridge NJ, Ratnu VS, Nayler SP, Nones K, Hu J, Bredy TW, Nakagawa S, Rigo F, Taft RJ, Cairns MJ, Blackshaw S, Wolvetang EJ, Mattick JS (2014) The long non-coding RNA Gomafu is acutely regulated in response to neuronal activation and involved in schizophrenia-associated alternative splicing. Mol Psychiatry 19:486–494

Batista PJ, Chang HY (2013) Long noncoding RNAs: cellular address codes in development and disease. Cell 152:1298–1307

Ben-Shem A, Garreau de Loubresse N, Melnikov S, Jenner L, Yusupova G, Yusupov M (2011) The structure of the eukaryotic ribosome at 3.0 a resolution. Science 334:1524–1529

Bernhart SH, Hofacker IL, Will S, Gruber AR, Stadler PF (2008) RNAalifold: improved consensus structure prediction for RNA alignments. BMC Bioinforma 9:474

Birney E (2012) An integrated encyclopedia of DNA elements in human genome. Nature 489:57–74

Bischoff G, Hoffmann S (2002) DNA-binding of drugs used in medicinal therapies. Curr Med Chem 9:312–348

Bittencourt D, Auboeuf D (2012) Analysis of co-transcriptional RNA processing by RNA-ChIP assay. Methods Mol Biol 809:563–577

Bose D, Jayaraj G, Suryawanshi H, Agarwala P, Pore SK, Banerjee R, Maiti S (2012) The tuberculosis drug Streptomycin as a potential cancer therapeutic: inhibition of miR-21 function by directly targeting its precursor. Angew Chem Int Ed 51:1019–1023

Burkholder GD, Latimer LJ, Lee JS (1991) Immunofluorescent localization of triplex DNA in polytene chromosomes of Chironomus and Drosophila. Chromosoma 101:11–18

Buske FA, Mattick JS, Bailey TL (2011) Potential in vivo roles of nucleic acid triple-helices. RNA Biol 8:427–439

Cabianca DS, Gabellini D (2010) The cell biology of disease: FSHD: copy number variations on the theme of muscular dystrophy. J Cell Biol 191:1049–1060

Cabianca DS, Casa V, Bodega B, Xynos A, Ginelli E, Tanaka Y, Gabellini D (2012) A long ncRNA links copy number variation to a polycomb/trithorax epigenetic switch in fshd muscular dystrophy. Cell 149:819–831

Cesana M, Cacchiarelli D, Legnini I, Santini T, Sthandier O, Chinappi M, Tramontano A, Bozzoni I (2011) A long noncoding RNA controls muscle differentiation by functioning as a competing endogenous RNA. Cell 147:358–369

Chen C, Ridzon DA, Broomer AJ, Zhou Z, Lee DH, Nguyen JT, Barbisin M, Xu NL, Mahuvakar VR, Andersen MR, Lao KQ, Livak KJ, Guegler KJ (2005) Real-time quantification of microRNAs by stem-loop RT-PCR. Nucleic Acids Res 33:e179

Chenoweth DM, Meier JL, Dervan PB (2013) Pyrrole-imidazole polyamides distinguish between double-helical DNA and RNA. Angew Chem Int Ed 52:415–418

Childs-Disney JL, Hoskins J, Rzuczek SG, Thornton C, Disney MD (2012) Rationally designed small molecules targeting the RNA that causes myotonic dystrophy type 1 are potently bioactive. ACS Chem Biol 7(5):856–862

Childs-Disney JL, Yildrim I, Park H, Lohman JR, Guan L, Tran T, Sarkar P, Schatz GC, Disney MD (2014) Structure of the myotonic dystrophy type 2 RNA and designed small molecules that reduce toxicity. ACS Chem Biol 9(2):538–550

Czerwoniec A, Dunin-Horkawicz S, Purta E, Kaminska KH, Kasprzak JM, Bujnicki JM, Grosjean H, Rother K (2009) MODOMICS: a database of RNA modification pathways. 2008 update. Nucleic Acids Res 37:D118–D121

Deigan KE, Li TW, Mathews DH, Weeks KM (2009) Accurate SHAPE-directed RNA structure determination. Proc Natl Acad Sci U S A 106:97–102

Derrien T, Johnson R, Bussotti G, Tanzer A, Djebali S, Tilgner H, Guernec G, Martin D, Merkel A, Knowles DG, Lagarde J, Veeravalli L, Ruan X, Ruan Y, Lassmann T, Carninci P, Brown JB, Lipovich L, Gonzalez JM, Thomas M, Davis CA, Shiekhattar R, Gingeras TR, Hubbard TJ, Notredame R, Harrow J, Guigo R (2012) The GENCODE v7 catalog of human long noncoding RNAs: analysis of their gene structure, evolution, and expression. Genome Res 22:1775–1789

Dervan PB (2001) Molecular recognition of DNA by small molecules. Bioorg Med Chem 9:2215–2235

Disney MD, Yildirim I, Childs-Disney JL (2014) Methods to enable the design of bioactive small molecules targeting RNA. Org Biomol Chem 12:1029–1039

Du Y, Kong G, You X, Zhang S, Zhang T, Gao Y, Ye L, Zhang X (2012) Elevation of highly up-regulated in liver cancer (HULC) by hepatitis B virus X protein promotes hepatoma cell proliferation via down-regulating p18. J Biol Chem 287:26302–26311

Duca M, Vekhoff P, Oussedik K, Halby L, Arimondo PB (2008) The triple helix: 50 years later, the outcome. Nucleic Acids Res 36:5123–5138

Fedor MJ, Williamson JR (2005) The catalytic diversity of RNAs. Nat Rev Mol Cell Biol 6:399–412

Gabellini D, Green MR, Tupler R (2002) Inappropriate gene activation in FSHD: a repressor complex binds a chromosomal repeat deleted in dystrophic muscle. Cell 110:339–348

Geisler S, Coller J (2013) RNA in unexpected places: long non-coding RNA functions in diverse cellular contexts. Nat Rev Mol Cell Biol 14:699–712

Goñi JR, de la Cruz X, Orozco M (2004) Triplex-formingoligonucleotide target sequences in the human genome. Nucleic Acids Res 32:354–360

Grote P, Wittler L, Hendrix D, Koch F, Währisch S, Beisaw A, Macura K, Bläss G, Kellis M, Werber M, Herrmann BG (2013) The tissue-specific lncRNA Fendrr is an essential regulator of heart and body wall development in the mouse. Dev Cell 24:206–214

Guan L, Disney MD (2012) Recent advances in developing small molecules targeting RNA. ACS Chem Biol 7:73–86

Guan L, Disney MD (2013) Covalent small molecule-RNA complex formation enables cellular profiling of small molecule-RNA interactions. Angew Chem Int Ed 52:10010–10013

Gumireddy K, Young DD, Xiong X, Hogenesch JB, Huang Q, Deiters A (2008) Small-molecule inhibitors of microRNA miR-21 function. Angew Chem Int Ed 47:7482–7484

Gupta RA, Shah N, Wang KC, Kim J, Horlings HM, Wong DJ, Tsai M-C, Hung T, Argani P, Rin JL, Wang Y, Brzoska P, Kong B, Li R, West RB, van de Vijver MJ, Sukumar S, Chang HY (2010) Long non-coding RNA HOTAIR reprograms chromatin state to promote cancer metastasis. Nature 464:1071–1076

Han L, Pandian GN, Junetha S, Sato S, Anandhakumar C, Taniguchi J, Saha A, Bando T, Nagase H, Sugiyama H (2013) A synthetic small molecule for targeted transcriptional activation of germ cell genes in a human somatic cell. Angew Chem Int Ed 52:13410–13413

Hauptman N, Glavac D (2013) Long non-coding RNA in cancer. Int J Mol Sci 14(3):4655–4669

Hermann T, Patel DJ (2000) RNA bulges as architectural and recognition motifs. Structure 8:R47–R54

Heward JA, Lindsay MA (2014) Long non-coding RNAs in the regulation of the immune response. Trends Immunol 35:408–419

Hoskins JW, Ofori LO, Chen CZ, Kumar A, Sobczak K, Nakamori M, Southall N, Patnaik S, Marugan JJ, Zheng W, Austin CP, Disney MD, Miller BL, Thornton CA (2014) Lomofungin and dilomofungin: inhibitors of MBNL1-CUG RNA binding with distinct cellular effects. Nucleic Acids Res. doi:10.1093/nar/gku275

Iguchi A, Fukuda N, Takahashi T, Watanabe T, Matsuda H, Nagase H, Bando T, Sugiyama H, Shimizu K (2013) RNA binding properties of novel gene silencing pyrrole-imidazole polyamides. Biol Pharm Bull 36:1152–1158

Kanak M, Alseiari M, Balasubramanian P, Addanki K, Aggarwal M, Noorali S, Kalsum A, Mahalingam K, Pace G, Panasik N, Bagasra O (2010) Triplex-forming MicroRNAs form stable complexes with HIV-1 provirus and inhibit its replication. Appl Immunohistochem Mol Morphol 18:532–545

Khalil AM, Faghihi MA, Modarresi F, Brothers SP, Wahlestedt C (2008) A novel RNA transcript with antiapoptotic function is silenced in fragile X syndrome. PLoS One 3:e1486

Kilburn D, Roh JH, Guo L, Briber RM, Woodson SA (2010) Molecular crowding stabilizes folded RNA structure by the excluded volume effect. J Am Chem Soc 132:8690–8696

Kladwang W, Mann TH, Becka A, Tian S, Kim H, Yoon S, Das R (2014) Standardization of RNA chemical mapping experiments. Biochemistry 53:3063–3065

Klattenhoff CA, Scheuermann JC, Surface LE, Bradley RK, Fields PA, Steinhauser ML, Ding H, Butty VL, Torrey L, Haas S, Abo R, Tabebordbar M, Lee RT, Burge CB, Boyer LA (2013) Braveheart, a long noncoding RNA required for cardiovascular lineage commitment. Cell 152:570–583

Kretz M, Siprashvili Z, Chu C, Webster DE, Zehnder A, Qu K, Lee CS, Flockhart RJ, Groff AF, Chow J, Johnston D, Kim GE, Spitale RC, Flynn RA, Zheng GXY, Aiyer S, Raj R, Rinn JL, Howard Y, Chang HY, Paul A, Khavari PA (2013) Control of somatic tissue differentiation by the long non-coding RNA TINCR. Nature 493:231–235

Lee MM, Pushechnikov A, Disney MD (2009) Rational and modular design of potent ligands targeting the RNA that causes myotonic muscular dystrophy 2. ACS Chem Biol 4:345–355

Lubitz I, Zikich D, Kotlyar A (2010) Specific high-affinity binding of thiazole orange to triplex and g-quadruplex DNA. Biochemistry 49:3567–3574

Lucks JB, Mortimer SA, Trapnell C, Luo S, Aviran S, Schroth GP, Pachter L, Doudna JA, Arkin AP (2011) Multiplexed RNA structure characterization with selective 2′-hydroxyl acylation analyzed by primer extension sequencing (SHAPE-seq). Proc Natl Acad Sci U S A 108:11063–11068

Maass PG, Rump A, Schulz H, Stricker S, Schulze L, Platzer K, Aydin A, Tinschert S, Goldring MB, Luft FC, Bahring S (2012) A misplaced lncRNA causes brachydactyly in humans. J Clin Invest 122:3990–4002

Maass PG, Luft FC, Bahring S (2014) Long non-coding RNA in health and disease. J Mol Med 92:337–346

Martianov I, Ramadass A, Serra Barros A, Chow N, Akoulitchev A (2007) Repression of the human dihydrofolate reductase gene by a non-coding interfering transcript. Nature 445:666–670

McGinnis JL, Dunkle JA, Cate JH, Weeks KM (2012) The mechanisms of RNA SHAPE chemistry. J Am Chem Soc 134:6617–6624

Meyer ST, Hergenrother PJ (2009) Small molecule ligands for bulged RNA secondary structures. Org Lett 11:4052–4055

Modarresi F, Faghihi MA, Lopez-Toledano MA, Fatemi RP, Magistri M, Brothers SP, Brug MP, Wahlestedt C (2012) Inhibition of natural antisense transcripts *in vivo* results in gene-specific transcriptional upregulation. Nat Biotechnol 30:453–459

Morgan AR, Wells RD (1968) Specificity of the three-stranded complex formation between double-stranded DNA and single-stranded RNA containing repeating nucleotide sequences. J Mol Biol 37:63–80

Mustoe AM, Brooks CL, Al-Hashimi HM (2014) Hierarchy of RNA functional dynamics. Annu Rev Biochem 83:441–466

Ng S-Y, Lin L, Soh BS, Stanton LW (2013) Long noncoding RNAs in development and disease of the central nervous system. Trends Genet 29:461–468

Novikova IV, Hennelly SP, Sanbonmatsu KY (2012) Structural architecture of the human long non-coding RNA, steroid receptor RNA activator. Nucleic Acids Res 40:5034–5051

Novikova IV, Hennelly SP, Sanbonmatsu KY (2013) Tackling structures of long noncoding RNAs. Int J Mol Sci 14:23672–23684

Ofori LO, Hoskins J, Nakamori M, Thornton CA, Miller BL (2012) From dynamic combinatorial 'hit' to lead: in vitro and in vivo activity of compounds targeting the pathogenic RNAs that cause myotonic dystrophy. Nucleic Acids Res 40:6380–6390

Orkin SH, Hochedlinger K (2011) Chromatin connections to pluripotency and cellular reprogramming. Cell 145:835–850

Ouyang Z, Snyder MP, Chang HY (2013) Seqfold: genome-scale reconstruction of RNA secondary structure integrating high-throughput sequencing data. Genome Res 23:377–387

Pan T, Sosnick T (2006) RNA folding during transcription. Annu Rev Biophys Biomol Struct 35:161–175

Pandian GN, Sugiyama H (2012) Programmable genetic switches to control transcriptional machinery of pluripotency. Biotechnol J 7(6):798–809

Pandian GN, Sugiyama H (2013) Strategies to modulate heritable epigenetic defects in cellular machinery: lessons from nature. Pharmaceuticals 6(1):1–24

Pandian GN, Nakano Y, Sato S, Morinaga H, Bando T, Nagase H, Sugiyama H (2012) A synthetic small molecule for rapid induction of multiple pluripotency genes in mouse embryonic fibroblasts. Sci Rep 2:544

Pandian GN, Taniguchi J, Junetha S, Sato S, Han L, Anandhakumar C, Bando T, Nagase H, Vaijayanthi T, Taylor RD, Sugiyama H (2014a) Distinct DNA-based epigenetic switches trigger transcriptional activation of silent genes in human dermal fibroblasts. Sci Rep 4:3843

Pandian GN, Taylor RD, Junetha S, Saha A, Anandhakumar C, Vaijayanthi T, Sugiyama H (2014b) Alteration of epigenetic program to recover memory and alleviate neurodegeneration: prospects of multi-target molecules. Biomater Sci 2:1043–1056

Pandian GN, Sato S, Anandhakumar C, Taniguchi J, Takashima K, Syed J, Han L, Saha A, Bando T, Nagase H, Sugiyama H (2014c) Identification of a small molecule that turns 'ON' the pluripotency gene circuitry in human fibroblasts. ACS Chem Biol 9:2729–2736

Pang KC, Frith MC, Mattick JS (2006) Rapid evolution of noncoding RNAs: lack of conservation does not mean lack of function. Trends Genet 22:1–5

Prensner JR, Iyer MK, Balbin OA, Dhanasekaran SM, Cao Q, Brenner JC, Laxman B, Asangani IA, Grasso CS, Kominsky HD, Cao X, Jing X, Wang X, Siddiqui J, Wei JT, Robinson D, Iyer HK, Palanisamy N, Maher CA, Chinnaiyan AM (2011) Transcriptome sequencing across a prostate cancer cohort identifies PCAT-1, an unannotated lincRNA implicated in disease progression. Nat Biotechnol 29:742–749

Prensner JR, Iyer MK, Sahu A, Asangani IA, Cao Q, Patel L, Vergara IA, Davicioni E, Erho N, Ghadessi M, Jenkins RB, Triche TJ, Malik R, Bedenis R, McGregor N, Ma T, Chen W, Han S, Jing X, Cao X, Wang X, Chandler B, Yan J, Siddiqui J, Kunju LP, Dhanasekaran SM, Pienta KJ, Feng FY, Chinnaiyan AM (2013) The long noncoding RNA SChLAP1 promotes aggressive prostate cancer and antagonizes the SWI/ SNF complex. Nat Genet 45:1392–1398

Quagliata L, Matter MS, Piscuoglio S, Arabi L, Ruiz C, Procino A, Kovac M, Moretti F, Makowska Z, Boldanova T, Anderson JB, Hammerie M, Tornillo L, Heim MH, Diedericha S, Cillo C, Terracciano L (2013) Long noncoding RNA HOTTIP/HOXA13 expression is associated with disease progression and predicts outcome in hepatocellular carcinoma patients. Hepatology. doi:10.1002/hep.26740

Rask-Andersen M, Almen MS, Schioth HB (2011) Trends in the exploitation of novel drug targets. Nat Rev Drug Discov 10:579–590

Reuter JS, Mathews DH (2010) RNAstructure: software for RNA secondary structure prediction and analysis. BMC Bioinforma 11:129

Saha A, Pandian GN, Sato S, Taniguchi J, Hashiya K, Bando T, Sugiyama H (2013) Synthesis and biological evaluation of a targeted DNA-binding transcriptional activator with HDAC8 inhibitory activity. Bioorg Med Chem 21(14):4201–4209

Schmidt LH, Spieker T, Koschmieder S, Humberg J, Jungen D, Bulk E, Hascher A, Wittmer D, Marra A, Hillejan L, Wiebe K, Berdel WE, Wiewrodt R, Müller-Tidow C (2011) The long noncoding MALAT-1 RNA indicates a poor prognosis in non-small cell lung cancer and induces migration and tumor growth. J Thorac Oncol 6:1984–1992

Schmitz KM, Mayer C, Postepska A, Grummt I (2010) Interaction of noncoding RNA with the rDNA promoter mediates recruitment of DNMT3b and silencing of rRNA genes. Genes Dev 24:2264–2269

Shappell SB (2008) Clinical utility of prostate carcinoma molecular diagnostic tests. Rev Urol 10:44–69

Spitale RC, Crisalli P, Flynn RA, Torre EA, Kool ET, Chang HY (2013) RNA shape analysis in living cells. Nat Chem Biol 9:18–20

Stelzer AC, Frank AT, Kratz JD, Swanson MD, Gonzalez-Hernandez MJ, Lee J, Andricioaei I, Markovitz DM, Al-Hashimi HM (2011) Discovery of selective bioactive small molecules by targeting an RNA dynamic ensemble. Nat Chem Biol 7(8):553–559

Syed J, Pandian GN, Sato S, Taniguchi J, Chandran A, Hashiya K, Bando T, Sugiyama H (2014) Targeted suppression of EVI1 oncogene expression by sequence-specific pyrrole-imidazole polyamide. Chem Biol. doi:10.1016/j.chembiol.2014.07.019

Takahashi K, Yan I, Haga H, Patel T (2014) Long noncoding RNA in Liver diseases. Hepatology 60:744–753

Takayama K, Horie-Inoue K, Katayama S, Suzuki T, Tsutsumi S, Ikeda K, Urano T, Fujimura T, Takagi K, Takahashi S, Homma Y, Ouchi Y, Aburatani H, Hayashizaki Y, Inoue S (2013) Androgen-responsive long noncoding RNA CTBP1-AS promotes prostate cancer. EMBO J 32:1665–1680

Ulitsky I, Shkumatava A, Jan CH, Sive H, Bartel DP (2011) Conserved function of lincRNAs in vertebrate embryonic development despite rapid sequence evolution. Cell 147:1537–1550

Underwood JG, Uzilov AV, Katzman S, Onodera CS, Mainzer JE, Mathews DH, Lowe TM, Salama SR, Haussler D (2010) FragSeq: transcriptome-wide RNA structure probing using high-throughput sequencing. Nat Methods 7(12):995–1001

Vaijayanthi T, Bando T, Hashiya K, Pandian GN, Sugiyama H (2013) Design of a new fluorescent probe: pyrrole/imidazole hairpin polyamides with pyrene conjugation at their γ-turn. Bioorg Med Chem 21:852–855

Van Dijk M, Thulluru HK, Oudejans CBM (2012) HELLP babies link a novel lincRNA to the trophoblast cell cycle. J Clin Invest 122:4003–4011

Velagapudi SP, Gallo SM, Disney MD (2014) Sequence-based design of bioactive small molecules that target precursor microRNAs. Nat Chem Biol 10:291–297

Wang KC, Chang HY (2011) Molecular mechanisms of long noncoding RNAs. Mol Cell 43(6):904–914

Washietl S, Kellis M, Garber M (2014) Evolutionary dynamics and tissue specificity of human long noncoding RNAs in six mammals. Genome Res 24:616–628

Weeks KM (2010) Advances in RNA structure analysis by chemical probing. Curr Opin Struct Biol 20:295–304

Wells SE, Hughes JM, Igel AH, Ares M Jr (2000) Use of dimethyl sulfate to probe RNA structure in vivo. Methods Enzymol 318:479–493

Wheeler TM, Leger AJ, Pandey SK, Macleod AR, Nakamori M, Cheng SH, Wentworth BM, Bennett CF, Thornton CA (2012) Targeting nuclear RNA for in vivo correction of myotonic dystrophy. Nature 488:111–115

Wilkinson KA, Merino EJ, Weeks KM (2006) Selective 2′-hydroxyl acylation analyzed by primer extension (SHAPE): quantitative RNA structure analysis at single nucleotide resolution. Nat Protoc 1:1610–1616

Wilusz JE, JnBaptiste CK, Lu LY, Kuhn CD, Joshua-Tor L, Sharp PA (2012) A triple helix stabilizes the 3′ ends of long noncoding RNAs that lack poly(A) tails. Genes Dev 26:2392–2407

Wong TN, Sosnick TR, Pan T (2007) Folding of noncoding RNAs during transcription facilitated by pausing-induced nonnative structures. Proc Natl Acad Sci U S A 104:17995–18000

Yang L, Duff MO, Graveley BR, Carmichael GG, Chen LL (2011) Genome wide characterization of non-polyadenylated RNAs. Genome Biol 12:R16

Yildirim E, Kirby JE, Brown DE, Mercier FE, Sadreyev RI, Scadden DT, Lee JT (2013) Xist RNA is a potent suppressor of hematologic cancer in mice. Cell 152:727–742

Zengeya T, Li M, Rozners E (2011) PNA containing isocytidine nucleobase: synthesis and recognition of double helical RNA. Bioorg Med Chem Lett 21:2121–2124

Zhao J, Ohsumi TK, Kung JT, Ogawa Y, Grau DJ, Sarma K, Song JJ, Kingston RE, Borowsky M, Lee JT (2010) Genome-wide identification of polycomb-associated RNAs by RIP-seq. Mol Cell 40:939–953

Part II
Atomic and Molecular Structures of lncRNAs

Chapter 3
Structure and Interaction with Protein of Noncoding RNA: A Case for an RNA Aptamer Against Prion Protein

Masato Katahira

Abstract An RNA aptamer is RNA that strongly and specifically binds to a certain molecule, such as a small molecule or a protein. An RNA aptamer is involved in a riboswitch, one kind of noncoding RNA, as a crucial domain to sense the state of cells. We discovered that a short RNA fragment, r(GGAGGAGGAGGA) (R12), functions as an RNA aptamer against bovine prion protein (bPrP), an abnormal form of which causes bovine spongiform encephalopathy (BSE). Two portions of bPrP were found as binding sites of R12. We determined the structure of R12 in complex with binding peptides of bPrP. R12 forms a unique quadruplex structure and dimerizes. Each monomer of the R12 dimer binds to the partial peptide. It was concluded that high affinity and specificity of R12 to bPrP originates from simultaneous dual binding of each monomer of the R12 dimer to the two binding sites of a single bPrP. Furthermore, the cell-based assay demonstrated that R12 reduces the amount of the abnormal form of prion protein and thus exerts anti-prion activity. The R12 aptamer may be utilized to develop drugs against prion diseases and Alzheimer's disease.

Keywords Aptamer • Prion • Structure • Interaction • NMR • Prion disease • Alzheimer's disease • Drug

3.1 Introduction

The riboswitch was first discovered in 2002 (Winkler et al. 2002; Mironov et al. 2002). The riboswitch is one kind of noncoding RNAs that senses a level of a certain metabolite in cells and regulates the expression level of related genes. The riboswitch consists of an aptamer domain and an expression platform domain. The aptamer domain strongly and specifically binds a certain metabolite. This binding induces the structural change of the platform domain, which will have an effect on

M. Katahira (✉)
Institute of Advanced Energy, Kyoto University, Gokasho, Uji, Kyoto 611-0011, Japan
e-mail: katahira@iae.kyoto-u.ac.jp

R. Kurokawa (ed.), *Long Noncoding RNAs*, DOI 10.1007/978-4-431-55576-6_3

either translation or transcription, resulting in a change in the level of a certain gene (Wittmann and Suess 2012; Serganov and Patel 2012). An RNA aptamer can also be artificially developed with an in vitro selection method (Ellington and Szostak 1990; Tuerk and Gold 1990). As an example of the study of structure and interaction with protein of noncoding RNA, the case of an RNA aptamer against prion protein (PrP) will be discussed.

Prions are infectious particles and are composed exclusively of misfolded proteins, being devoid of nucleic acids. PrP is almost ubiquitously expressed and highly conserved in mammals, being anchored on the surface of cells. PrP exhibits two alternative forms; a normal cellular form (PrP^{C}), which is a soluble α-helix-rich isoform, and an abnormal form (PrP^{Sc}), which is an insoluble β-sheet-rich isoform and is resistant to cleavage by proteinase K (Prusiner 1998; Pan et al. 1993). The conformational change from PrP^{C} to PrP^{Sc} is thought to be crucial in prion pathogenesis, causing diseases such as Creutzfeldt-Jacob disease (CJD) in humans, bovine spongiform encephalopathy (BSE) in cattle, and scrapie in sheep (Pan et al. 1993; Prusiner 2004; Huang et al. 1994). It is supposed that the contact of PrP^{C} with PrP^{Sc} induces the conversion of PrP^{C} to PrP^{Sc}. The detailed mechanism of the conformational conversion remains unknown.

An RNA aptamer that tightly binds to PrP^{C} is expected to stabilize PrP^{C} and thus block the conversion to PrP^{Sc}. Therefore, such an RNA aptamer may prevent prion diseases. However, attempts in this context have been very limited, and a structural basis of the binding of an RNA aptamer to PrP^{C}, which would facilitate such an application, has not been available.

We discovered RNA aptamers that tightly and specifically bind to bovine PrP^{C} ($bPrP^{C}$) (Murakami et al. 2008; Nishikawa et al. 2009). Then, it was noted that a short RNA fragment, r(GGAGGAGGAGGA) (R12), functions as an RNA aptamer. We also identified two binding sites for R12 in the N-terminal half of the prion protein (Mashima et al. 2009). Then, we determined the structure of R12 both in a free form (Mashima et al. 2009) and in a complex form with the binding sites of bPrP (Mashima et al. 2013). R12 in complex with the N-terminal half of PrP, which involves the two binding sites, was also analyzed. These structural analyses elucidated the mode of interaction of R12 with PrP and the mechanism by means of which R12 exhibits high affinity for PrP. Furthermore, the anti-prion activity of R12 was demonstrated by the assay using cells that persistently express PrP^{Sc} (Mashima et al. 2013). This suggests that the R12 aptamer has the potential to be utilized as an anti-prion drug.

3.2 Identification of R12 as an RNA Aptamer Against Bovine Prion Protein (bPrP)

RNA aptamers against a bovine prion protein (bPrP) were obtained by means of an in vitro selection method (Ellington and Szostak 1990; Tuerk and Gold 1990) from RNA pools containing a 55-nucleotide randomized region (Murakami et al. 2008).

The obtained RNA aptamers showed high affinity for both $bPrP^C$ and its amyloidogenic β isoform (bPrP-β). It is well established that bPrP-β resembles $bPrP^{Sc}$ in terms of structural and biochemical properties (Lührs et al. 2006). It was demonstrated that the RNA aptamers can specifically detect bPrP in a bovine brain homogenate on a northwestern blotting assay (Murakami et al. 2008).

It was noticed that a short RNA fragment of 12 nucleotides, R12, is present in most obtained RNA aptamers. Then, it was surprisingly revealed that R12 alone can function as an RNA aptamer. R12 binds $bPrP^C$ and bPrP-β with high affinity, the dissociation constants being 8.5×10^{-9} and 2.8×10^{-7} M, respectively (Murakami et al. 2008).

3.3 Identification of Two Binding Sites for R12 in the N-Terminal Intrinsically Disordered Region of bPrP

The N-terminal half of bPrP is intrinsically disordered, while the C-terminal half is folded (Garcia et al. 2000). First, we revealed that R12 binds to the N-terminal half of bPrP (Murakami et al. 2008). Then, we synthesized a series of partial peptides of the N-terminal half of bPrP with a length of 12 amino acid residues, and the binding of R12 to these peptides was examined by microchip electrophoresis (Mashima et al. 2009). Two binding sites were identified; residues 25–35 of bPrP, P1, and residues 108–119 of bPrP, P16. Their amino acid sequences are P1: SKKRPKPGGGWN and P16: GQWNKPSKPKTN. Each binding site contains a lysine cluster and a tryptophan residue, in common. The dissociation constant was determined to be 1×10^{-5} M for both of the peptides.

3.4 Structure of an R12 Aptamer in a Free Form

The structure of R12 in a free form under physiological conditions (100 mM KCl and pH 6.2) was determined by NMR (Mashima et al. 2009). The numbering of R12 is as follows: G1G2A3G4G5A6G7G8A9G10G11A12. Two unique architectures were found in R12; a G2:G5:G8:G11 tetrad plane and a G1(:A3):G4:G7(:A9):G10 hexad plane (Fig. 3.1a, b). These architectures are formed through the tight network of hydrogen bonds. The backbone of R12 frequently changes its direction and, as a result, R12 forms a parallel-type quadruplex structure (Fig. 3.1c). Four GG stretches are aligned parallel to each other. A remarkable point is that two R12 monomers dimerize to make a dimer structure (Fig. 3.1c). R12 exists as a stable dimer in solution. The dimer structure is supposed to be stabilized by a stacking interaction between the two hexad planes of each monomer. The dimer structure may also be stabilized by possible coordination of a potassium ion between the two hexad planes.

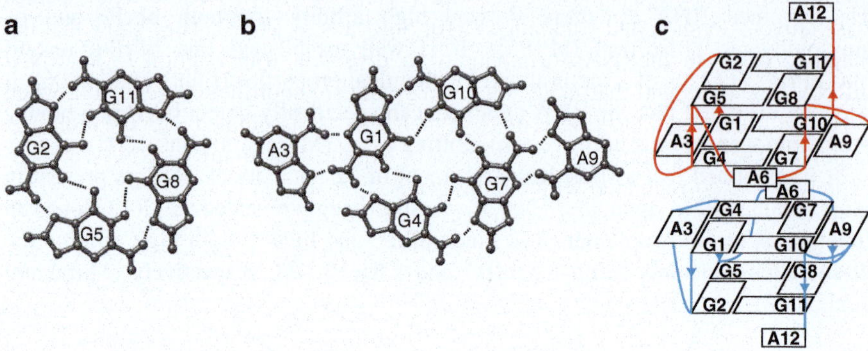

Fig. 3.1 Structure of R12 in a free form under physiological conditions (100 mM KCl, 10 mM K-phosphate buffer (pH 6.2)). Tetrad (**a**) and hexad (**b**) planes, and overall structure of the R12 dimer (**c**)

3.5 Structure of an R12 Aptamer in Complex with a P16 Binding Peptide

Two binding sites for R12, P1, and P16 were identified in bPrP, as described above. Either the P1 or P16 peptide was added step by step to the R12 solution, and almost identical chemical shift perturbations were observed for R12 resonances (Mashima et al. 2013). This indicates that the mode of interactions with R12 is basically the same for P1 and P16 peptides. Therefore, a P16 peptide was taken as a representative, and the structure of R12 in complex with the P16 peptide was determined by NMR (Mashima et al. 2013) (Fig. 3.2). The numbering of P16 is as follows: G1Q2W3N4K5P6S7K8P9K10T11N12.

The structure of R12 found in a free form remains in a complex as well. One R12 monomer folds into a parallel-type quadruplex with tetrad and hexad planes, and a dimer structure is formed (Fig. 3.2).

Two P16 peptides are bound to each R12 monomer. Three lysine residues of each P16 peptide make electrostatic interactions with a distinct phosphate group of each R12 monomer; K5, K8, and K10 of P16 interact with G8, G11, and A9 of R12, respectively (Fig. 3.2). Additionally, a tryptophan residue of each P16, W3, makes a stacking interaction with a guanine base of G8 of each R12 (Fig. 3.2). Thus, three electrostatic interactions and one stacking interaction contribute to the stabilization of the complex at each P16 peptide–R12 aptamer interface. This conclusion was supported by the mutational analysis that the replacement of either K5, K8, K10, or W3 of P16 by an alanine residue drastically reduced the affinity to R12 (Mashima et al. 2013).

Computational analysis of the structure of R12 in complex with P16 gave further insight into the driving force of binding (Hayashi et al. 2014). The importance of the contribution of water entropy to the formation of the complex was deduced.

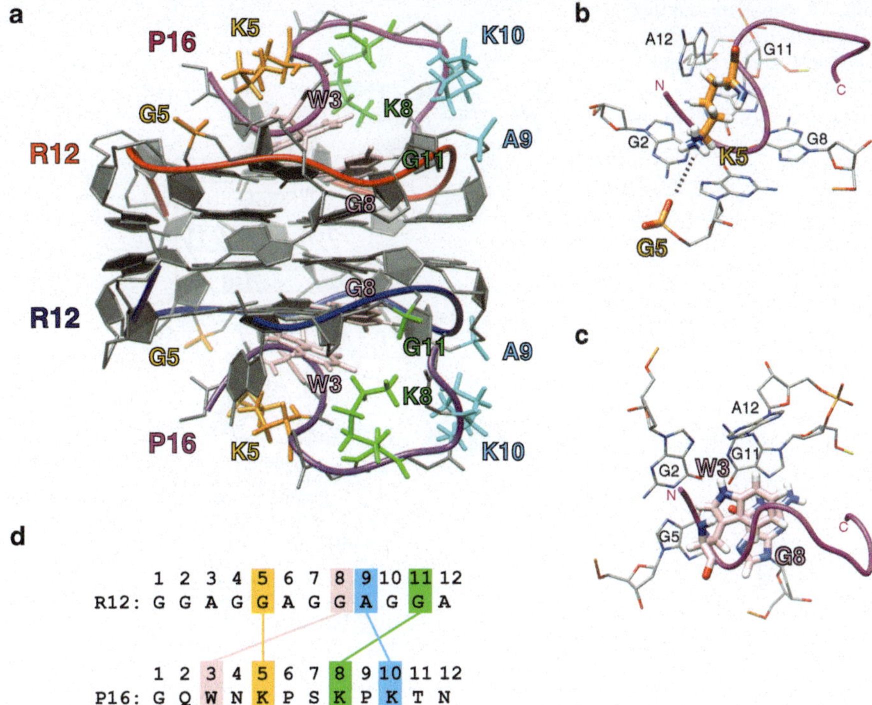

Fig. 3.2 Structure of R12 in complex with the P16 peptide. R12 dimerizes, and two P16 peptides are bound to each R12 monomer (**a**). (**b**) Electrostatic interaction between a side chain of a K5 residue of the P16 peptide and a phosphate group of a G5 residue of the R12 aptamer. (**c**) Stacking interaction between a ring of W3 of P16 and a base of G8 of R12. (**d**) Summary of electrostatic and stacking interactions found for each R12-P16 unit

3.6 Binding Mode of R12 with bPrP

In a case of the R12:P16 complex, two P16 peptides are bound to each monomer of the R12 dimer. bPrP possesses two binding sites, P1 and P16, and it was revealed that P1 interacts with R12 almost in the same way as P16. Therefore, it is quite likely in the case of the R12:bPrP complex that P1 of bPrP interacts with one monomer of the R12 dimer and that P16 of bPrP interacts with another monomer of the R12 dimer (Fig. 3.3).

In order to confirm this idea, the stoichiometry of the R12:N-terminal half of bPrP (bPrP-N) complex was determined on the basis of chemical shift perturbation. It was found that the R12 monomer:bPrP-N monomer = 1:0.5. When it is taken into account that R12 exists as a dimer in complex, this stoichiometry indicates that one R12 dimer binds to one bPrP-N, which is consistent with the idea described above (Mashima et al. 2013).

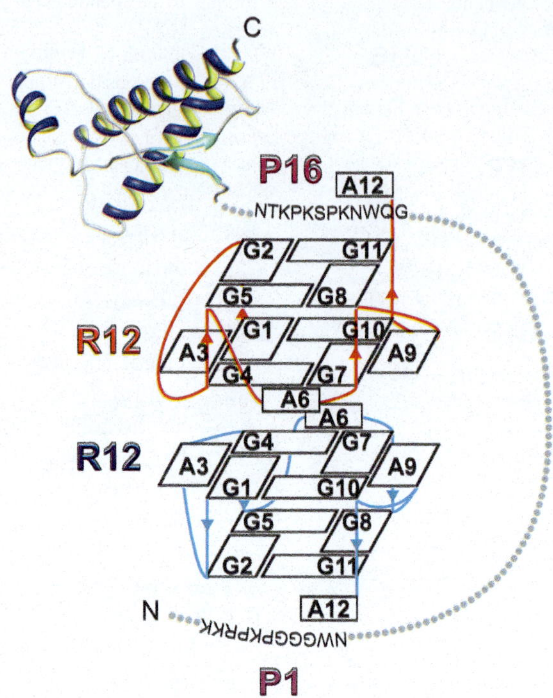

Fig. 3.3 Simultaneous dual binding mode of R12 with bPrP. R12 dimerizes, and each R12 monomer simultaneously binds to one of the two binding sites of a single bPrP

3.7 Origin of High Affinity and Specificity of R12 to bPrP

The deduced simultaneous dual binding of the R12 dimer with two binding sites of a single bPrP molecule is supposed to be the origin of the high affinity of R12 to PrP. The dual binding could ideally increase the affinity by as much as the square of the binding constant if each interaction is perfectly achieved (Campisi et al. 2001). The dissociation constants for the R12-P1 and R12-P16 complexes are both approximately 1×10^{-5} M. Simultaneous dual binding at the P1 and P16 sites of bPrP could maximally result in the dissociation constant of 1×10^{-10} M. The real dissociation constant, 1×10^{-8} M, is not as small as 1×10^{-10} M. Nonetheless, the high affinity of R12 for bPrP can be rationally interpreted on the basis of simultaneous dual binding.

The protein targeted by R12 should have a binding site that contains three lysine residues for electrostatic interaction and one tryptophan residue for stacking interaction. In addition, for tight binding, the target protein should have two binding sites with this character. The two sites should be adequately separated by linker residues to simultaneously interact with each monomer of the R12 dimer. Finally, the two sites should not be buried in an inner region of a protein but should be exposed to the solvent for the interaction with R12. The protein that satisfies these requirements is quite limited. This may rationalize why R12 specifically binds to bPrP, as experimentally revealed (Murakami et al. 2008).

3.8 Anti-prion Activity of R12

R12 binds more strongly to $bPrP^C$ than to bPrP-β, which resembles $bPrP^{Sc}$. Therefore, it is expected that PrP^C is stabilized through binding of R12 and that conversion of PrP^C to PrP^{Sc} can thus be inhibited by R12. This idea was examined with the cell-based assay (Mashima et al. 2013). Mouse neuronal cells designated GT+FK are persistently infected with the human TSE agent, Fukuoka-1 strain, and constantly express PrP^{Sc} (Nishida et al. 2000). R12 was added to the culture of GT+FK to the final concentration of 10 μM (Fig. 3.4). After 72 h of the treatment, cells were collected and lysed. The sample was digested with proteinase K, by which PrP^C is digested, while PrP^{Sc} remains undigested. Then, the amount of PrP^{Sc} was quantified with western blotting. In a case of control, just a buffer solution was added to the culture. It was found that R12 reduces the amount of PrP^{Sc} to 65.0 % of that for the control (Fig. 3.4 and Table 3.1). When the RNase inhibitor, RNasin, was added to avoid the possible degradation of R12 in the culture, R12 further reduced the amount of PrP^{Sc} to 49.4 % (Fig. 3.4 and Table 3.1). U12, which consists of 12 uridine residues, did not reduce the amount of PrP^{Sc} at all (Fig. 3.4 and Table 3.1). So, the reduction of PrP^{Sc} is specific to R12. Thus, it has been demonstrated that R12 actually exhibits anti-prion activity, as expected.

The amino acid identity between bPrP and mouse PrP is 89 %. In particular, concerning the two binding sites of bPrP for R12, mouse PrP possesses almost the same amino acid sequence. Therefore, it is supposed that R12 selected against bPrP also functions against mouse PrP in the same way.

Previously, we determined the structure of a DNA version of R12, d(GGAGGAGGAGGA), D12 (Matsugami et al. 2001). The overall structure of D12 is similar to that of R12, with some differences. Therefore, it is expected that D12 binds to bPrP in the similar way as R12 does and that D12 also exhibits anti-prion

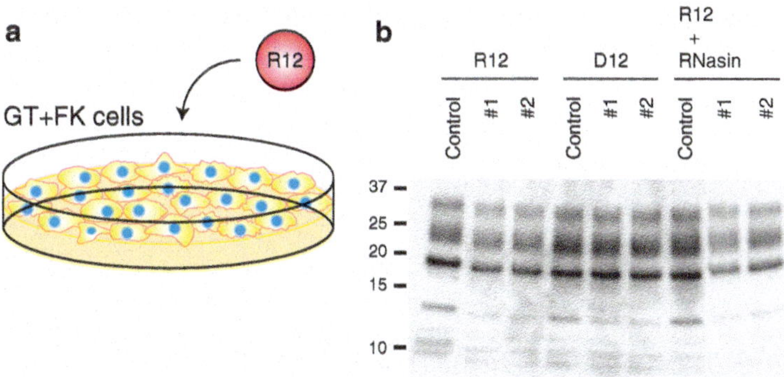

Fig. 3.4 Cell-based assay of anti-prion activity of R12. (**a**) R12 was added to the culture of GT+FK cells expressing PrP^{Sc}. After the treatment, cells were incubated for 72 h, collected and lysed. (**b**) Western blotting of PrP^{Sc} in GT+FK cells after the treatment with 10 μM of either R12 or D12 is shown for two independent experiments, #1 and #2. Each assay was repeated eight times, in fact. The effect of the addition of the RNase inhibitor, RNasin, is also examined

Table 3.1 Anti-prion activities of R12 and D12

Treatment	Relative PrP^{Sc} level (%)
Control	100
R12	65.0±9.3
R12 + RNasin	49.4±17.4
U12 + RNasin	98.8±5.8
D12	75.5±17.0

activity. It was revealed that D12 also reduces the amount of PrP^{Sc} to 75.5 % (Fig. 3.4 and Table 3.1). Thus, D12 actually exhibits anti-prion activity, the activity being weaker than R12.

3.9 Possible Therapeutic Application

Several compounds that reduce the amount of PrP^{Sc} have been developed (Kuwata et al. 2007; Vogtherr et al. 2003; Taubner et al. 2010; Schütz et al. 2011). The binding sites and the mode of the interaction of R12 with PrP differ from those of these compounds. Therefore, the mechanism for R12 to exert anti-prion activity is supposed to be distinct from that for these compounds. Thus, R12 may be utilized to develop a new anti-prion drug.

Recently, it was suggested that PrP^C may be a receptor of amyloid-β-oligomer, which is supposed to be related to Alzheimer's disease (Laurén et al. 2009). When binding of amyloid-β-oligomer to PrP^C was inhibited by anti-PrP antibodies, synaptic plasticity in hippocampal slices was rescued (Laurén et al. 2009). As R12 binds tightly to PrP^C as the antibodies did, R12 may also inhibit the binding of amyloid-β-oligomer to PrP^C and thus rescue synaptic plasticity. In this context, it is remarkable that the amyloid-β-oligomer binds to the region very close to P16 where R12 binds. Therefore, it can really be expected that R12 inhibits the binding of the amyloid-β-oligomer to PrP^C. Then, R12 may be utilized to develop an anti-Alzheimer's disease drug.

3.10 Conclusions

A riboswitch is one of the typical examples of functional noncoding RNAs. An RNA aptamer domain of the riboswitch plays a crucial role to sense the state of cells. An RNA aptamer can also be artificially developed with an in vitro selection method. We found that a short RNA of just 12 residues, R12, functions as an RNA aptamer. We determined the structure of R12 in complex with the binding peptide of bPrP and elucidated the mechanism by means of which R12 tightly and specifically binds to bPrP. Simultaneous dual binding of each monomer of the R12 dimer to the two binding sites of bPrP turned out to be a key point. It was demonstrated by the cell-based assay that R12 can reduce the amount of PrP^{Sc}. This anti-prion activity of R12 may be utilized for the development of drugs against prion and Alzheimer's diseases.

Acknowledgments This work was accomplished in collaboration with Dr. T. Mashima and Prof. T. Nagata of Kyoto University, Dr. F. Nishikawa and Dr. S. Nishikawa of AIST, and Dr. Y. O. Kamatari and Prof. K. Kuwata of Gifu University. This work was supported by Grants-in-Aid for Scientific Research from the Ministry of Education, Culture, Sports, Science and Technology of Japan (25115507, 25291013, 26104520, and 26650014).

References

Campisi DM, Calabro V, Frankel AD (2001) Structure-based design of a dimeric RNA-peptide complex. EMBO J 20:178–186

Ellington A, Szostak JW (1990) In vitro selection of RNA molecules that bind specific ligands. Nature 346:818–822

García FL, Zahn R, Riek R, Wüthrich K (2000) NMR structure of the bovine prion protein. Proc Natl Acad Sci U S A 97:8334–8339

Hayashi T, Oshima H, Mashima T, Nagata T, Katahira M, Kinoshita M (2014) Binding of an RNA aptamer and a partial peptide of a prion protein: crucial importance of water entropy in molecular recognition. Nucleic Acids Res 42:6861–6875

Huang Z, Gabriel JM, Baldwin MA, Fletterick RJ, Prusiner SB, Cohen FE (1994) Proposed three-dimensional structure for the cellular prion protein. Proc Natl Acad Sci U S A 91:7139–7143

Kuwata K, Nishida N, Matsumoto T, Kamatari YO, Hosokawa-Muto J, Kodama K, Nakamura HK, Kimura K, Kawasaki M, Takakura Y et al (2007) Hot spots in prion protein for pathogenic conversion. Proc Natl Acad Sci U S A 104:11921–11926

Laurén J, Gimbel DA, Nygaard HB, Gilbert JW, Strittmatter SM (2009) Cellular prion protein mediates impairment of synaptic plasticity by amyloid-β oligomers. Nature 457:1128–1132

Lührs T, Zahn R, Wüthrich K (2006) Amyloid formation by recombinant full-length prion proteins in phospholipid bicelle solutions. J Mol Biol 357:833–841

Mashima T, Matsugami A, Nishikawa F, Nishikawa S, Katahira M (2009) Unique quadruplex structure and interaction of an RNA aptamer against bovine prion protein. Nucleic Acids Res 37:6249–6258

Mashima T, Nishikawa F, Kamatari YO, Fujiwara F, Saimura M, Nagata T, Kodaki T, Nishikawa S, Kuwata K, Katahira M (2013) Anti-prion activity of an RNA aptamer and its structural basis. Nucleic Acids Res 41:1355–1362

Matsugami A, Ouhashi K, Kanagawa M, Liu H, Kanagawa S, Uesugi S, Katahira M (2001) An intramolecular quadruplex of $(GGA)_4$ triplet repeat DNA with a G:G:G:G tetrad and a G(:A):G(:A):G(:A):G heptad, and its dimeric interaction. J Mol Biol 313:255–269

Mironov AS, Gusarov I, Rafikov R, Lopetz LE, Shatalin K, Kreneva RA, Perumov DA, Nudler R (2002) Sensing small molecules by nascent RNA: a mechanism to control transcription in bacteria. Cell 111:747–756

Murakami K, Nishikawa F, Noda K, Yokoyama T, Nishikawa S (2008) Anti-bovine prion protein RNA aptamer containing tandem GGA repeat interacts both with recombinant prion protein and its β isoform with high affinity. Prion 2:73–78

Nishida N, Harris DA, Vilette D, Laude H, Frobert Y, Grassi J, Casanova D, Milhavet O, Lehmann S (2000) Successful transmission of three mouse-adapted scrapie strains to murine neuroblastoma cell lines overexpressing wild-type mouse prion protein. J Virol 74:320–325

Nishikawa F, Murakami K, Matsugami A, Katahira M, Nishikawa S (2009) Structural studies of an RNA aptamer containing GGA repeats under ionic conditions using microchip electrophoresis, circular dichroism, and 1D-NMR. Oligonucleotides 19:179–190

Pan KM, Baldwin M, Nguyen J, Gasset M, Serban A, Groth D, Mehlhorn I, Huang Z, Fletterick RJ, Cohen FE et al (1993) Conversion of alpha-helices into beta-sheets features in the formation of the scrapie prion proteins. Proc Natl Acad Sci U S A 90:10962–10966

Prusiner SB (1998) Prions. Proc Natl Acad Sci U S A 95:13363–13383

Prusiner SD (2004) Novel proteinaceous infectious particles cause scrapie. Science 216:136–144

Schütz AK, Soragni A, Hornemann S, Aguzzi A, Ernst M, Böckmann A, Meier BH (2011) The amyloid-Congo red interface at atomic resolution. Angew Chem Int Ed 50:5956–5960

Serganov A, Patel DJ (2012) Molecular recognition and function of riboswitches. Curr Opin Struct Biol 22:279–286

Taubner LM, Bienkiewicz EA, Copie V, Caughey B (2010) Structure of the flexible amino-terminal domain of prion protein bound to a sulfated glycan. J Mol Biol 395:475–490

Tuerk C, Gold L (1990) Systematic evolution of ligands by exponential enrichment: RNA ligands to bacteriophage T4 DNA polymerase. Science 249:505–510

Vogtherr M, Grimme S, Elshorst B, Jacobs DM, Fiebig K, Griesinger C, Zahn R (2003) Antimalarial drug quinacrine binds to C-terminal helix of cellular prion protein. J Med Chem 46:3563–3564

Winkler W, Nahvi A, Breaker RR (2002) Thiamine derivatives bind messenger RNAs directly to regulate bacterial gene expression. Nature 419:952–956

Wittmann A, Suess B (2012) Engineered riboswitches: expanding researchers' toolbox with synthetic RNA regulators. FEBS Lett 586:2076–2083

Chapter 4
Characterization of G-Quadruplex DNA- and RNA-Binding Protein

Takanori Oyoshi

Abstract Mammalian telomeres containing TTAGG repeats are bound by a multiprotein complex with a telomeric repeat-containing RNA (TERRA) containing UUAGGG repeats, which is a long noncoding RNA transcribed from the telomeres. Telomere DNA and TERRA form a G-quadruplex in vitro. The functions of the G-quadruplex structures in the telomere, however, are not clear, because little is known about G-quadruplex specific binding proteins and G-quadruplex RNA-binding molecules without binding to G-quadruplex telomere DNA. We have reported that the Arg-Glu-Gly motif in Translocated in Liposarcoma (TLS) forms G-quadruplex telomere DNA and TERRA simultaneously in vitro. Furthermore, TLS promotes the methylation of hinstone H4 and H3 at lysine and regulates telomere length. These finding suggest that the G-quadruplex functions as a scaffold for telomere-binding protein, TLS. Moreover, we have shown that substitution of Tyr for Phe in the RGG motif of TLS converts its binding specificity solely toward G-quadruplex TERRA. This molecule binds to loops within the G-quadruplexes of TERRA by recognizing the 2′-OH of the riboses. It will be useful for investigating biological roles of the G-quadruplex in long noncoding RNA.

Keywords G-quadruplex • Telomere • Telomeric repeat-containing RNA • G-quadruplex binding protein • RGG motif • Histone modification • Heterochromatin

4.1 Introduction

Mammalian telomeres cap chromosome termini to prevent chromosome loss and are complexed with telomere-binding proteins and a telomeric repeat-containing RNA (TERRA), which is a long noncoding RNA and transcribed from the telomere region (de Lange 2005; Luke and Lingner 2009; Azzalin et al. 2007). Mammalian telomere DNA comprises tandems of 5′-TTAGGG-3′ sequences,

T. Oyoshi (✉)
Department of Chemistry, Faculty of Science, Shizuoka University, 836 Ohya Suruga, Shizuoka 422-8529, Japan
e-mail: stohyos@ipc.shizuoka.ac.jp

R. Kurokawa (ed.), *Long Noncoding RNAs*, DOI 10.1007/978-4-431-55576-6_4

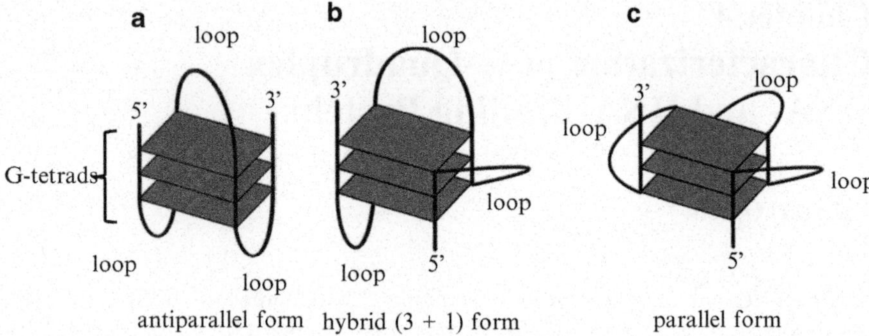

Fig. 4.1 Schematic representation of G-quadruplex structures. (**a**) Basket-type antiparallel form. (**b**) Hybrid (3 + 1) form. (**c**) Parallel-stranded form. G-quadruplex consists of G-tetrads and loops

and TERRA contains tandem arrays of 5′-UUAGGG-3′ (Luu et al. 2006; Matadinate and Phan 2009). Mammalian telomere DNA forms an equilibrium G-quadruplex hybrid (3 + 1) forms together with a parallel-stranded form and a basket-type antiparallel form, and TERRA forms a parallel-stranded form, but the biological significance of their G-quadruplex formation is unclear (Fig. 4.1) (Luu et al. 2006; Matadinate and Phan 2009; Xu et al. 2010). Guanine-rich sequences are able to fold into the G-quadruplex structure, which consists of cyclic Hoogsteen base pairs of four guanine bases and is stabilized by the presence of monovalent cations.

G-quadruplex DNA-binding proteins have been reported to interact with the G-quadruplex, which forms at a single-stranded 3′-overhang in mammalian telomeres, and to relate to telomere maintenance. A component of the telomere shelterin complex, Protection of Telomerase 1 (POT1), binds to the single-stranded 3′-overhang and disrupts the G-quadruplex (Kelleher et al. 2005). POT1 inhibits telomerase by blocking the overhang but also forms a complex with TPP1 to increase telomerase activity (Kelleher et al. 2005; Wang et al. 2007). The heterogeneous nuclear ribonucleoprotein (hnRNP) family represents proteins involved in the unfolding of the telomeric G-quadruplex and interacts with telomerase in vivo (Wang et al. 2012). A splicing variant of mammalian hnRNP A2 was recently identified as a telomeric DNA-binding protein in mammalian cells. It unfolds telomeric G-quadruplex DNA and leads to telomere elongation by enhancing the catalytic activity of telomerase. G-quadruplex binding proteins in telomeres might regulate the structure and role of the G-quadruplex, but their function in mammalian telomeres is not clear. This chapter describes the mechanism of mammalian telomere maintenance regulated by G-quadruplex DNA- and RNA-binding protein, Translocated in Liposarcoma (TLS). Moreover, the substitution of Tyr for Phe in the nucleic acid binding domain of TLS converts its binding specificity solely toward G-quadruplex RNA. This engineered molecule will be useful for investigating the biological roles of the G-quadruplex in long noncoding RNA.

4.2 Roles of the G-Quadruplex Binding Protein, TLS/FUS, in Telomere

G-quadruplex structures might exist not only in a single-stranded 3′-overhang but also in double-stranded telomere DNA, which is a scaffold for telomere-binding proteins. Visualization of G-quadruplexes in human cells using anti-G-quadruplex antibody revealed that G-quadruplex structures exist in double-stranded DNA (Biffi et al. 2013). In vitro, G-quadruplex formation is promoted under physiologic concentrations of K^+ and molecular crowding conditions, which is typical of the aqueous environment in living cells (Zheng et al. 2010). In particular, a stable G-quadruplex in long double-stranded DNA is formed during transcription under these conditions. The G-quadruplex binding protein, TLS, also termed FUS, binds to telomere DNA in a double-stranded region in mammalian cells (Takahama et al. 2013). The G-quadruplex binding molecule, tetra-(N-methyl-4-pyridyl)porphine (TMPyP4), inhibits TLS binding in a telomere double strand, but not TRF1, which is identified as a component of shelterin that binds to telomere double strands and distinguish telomeres from sites of DNA damage (Fig. 4.2). Moreover, Ewing's sarcoma, which is related to TLS as a subgroup within the RNA-binding protein family, induces G-quadruplex formation in vitro (Takahama et al. 2011), suggesting that TLS binds to the G-quadruplex in a telomere double strand and might induce G-quadruplex folding (Fig. 4.2). On the other hand, DNA helicases RTEL1 (Regulator of Telomere Length) and BLM (Bloom Syndrome Protein) facilitate telomere replication by unfolding the G-quadruplex (Vannier et al. 2012), which suggests that the formation of G-quadruplex structures in double-stranded telomere DNA is regulated by G-quadruplex binding proteins and DNA helicases during transcription and DNA replication (Fig. 4.2).

G-quadruplex binding proteins might regulate telomere histone modifications. The C-terminal RGG (Arg-Gly-Gly) motif in TLS specifically targets a fold in the G-quadruplex telomere DNA and TERRA (Takahama et al. 2013). TLS can simultaneously bind to G-quadruplex telomere DNA and TERRA in vitro. The functions of TERRA are thought to be important for telomerase activity and histone H3 trimethylation in telomeres (Deng et al. 2009; Arnoult et al. 2012; Cusanelli et al. 2013). Recent studies have reported that TERRA is associated with several heterochromatin markers, including the origin recognition complex, trimethylated histone

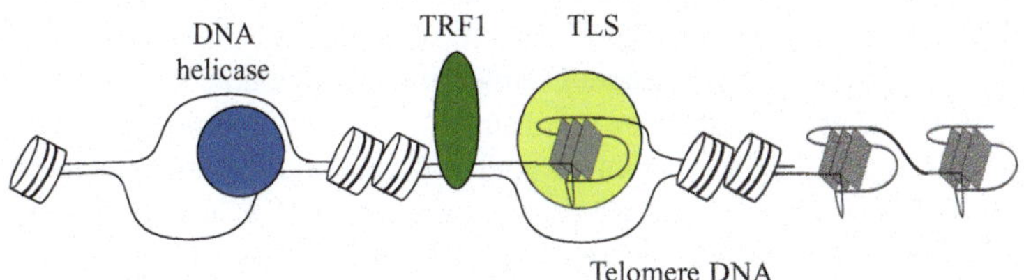

Fig. 4.2 Model of TLS, TRF1, and DNA helicase at telomere

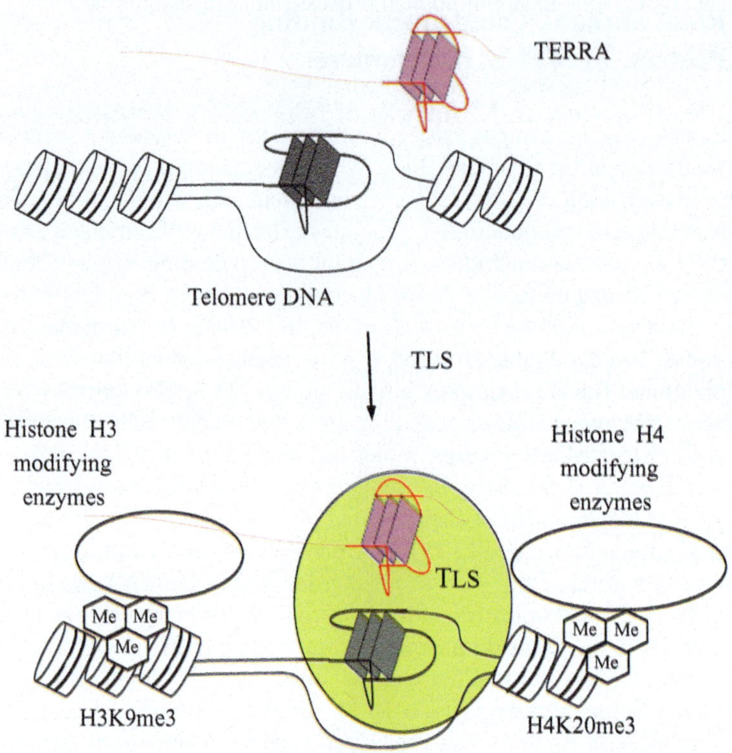

Fig. 4.3 Model of the roles of TLS at a telomere double strand. Recruitment of histone H4-modifying enzymes to the telomere by TLS, which binds to G-quadruplex TERRA and DNA in a telomere double strand, results in an increase in H4K20 and H3K9 trimethylations

H3 (H3me3), and HP1 (heterochromatin protein 1) isoforms, which accumulate at constitutive heterochromatin. Trimethylated histone H4 (H4me3) is induced sequentially by H3me3 at heterochromatin. How TERRA localizes to the telomere is not clear. TLS might regulate histone modification due to G-quadruplex telomere DNA and TERRA binding (Fig. 4.3). We reported that TLS overexpression results in forming H3me3 and H4me3 (Takahama et al. 2013). TLS interacts with the enzyme Suv4-20h2, resulting in H4me3 at telomeres. TRF2, a component of the telomere shelterin complex, has G-quadruplex TERRA and double strand telomere DNA-binding ability and regulates histone H3 modifications (Deng et al. 2009; Arnoult et al. 2012). This suggests that G-quadruplex binding proteins, such as TLS and TRF2, tether TERRA to the telomere G-quadruplex and double-stranded DNA, and indirectly form heterochromatin by recruiting the histone trimethylation enzyme. In addition, a G-quadruplex at a telomere double strand might contribute to the localization of TERRA at the telomere by forming a DNA/RNA hybrid. The transcription of G-rich coding regions produces a G-quadruplex and C-rich DNA/G-rich RNA hybrid (Duquette et al. 2004). In fact, TERRA/DNA hybrid formation in *Saccharomyces cerevisiae* is resolved by overexpressed RNase H, which digests

RNA at a specific RNA/DNA hybrid structure (Luke et al. 2008). Overexpressed RNase H digests and shortens TERRA in vivo. Further studies are needed to confirm whether a TERRA/DNA hybrid exists in double-stranded human telomere DNA (Htelo). Moreover, if a TERRA/DNA hybrid exists in telomeres, the effect of the G-quadruplex in a double strand to form this hybrid should be investigated.

Normal cells progressively lose telomeres during cell division because of incomplete DNA-end replication, and this shortening of the telomeres limits the lifespan of the cell. Cancer cells, however, have at least two main mechanisms to maintain telomere length: the addition of telomeric repeats by telomerase and an alternative lengthening of telomeres that relies on homologous recombination between telomeric sequences (Blackburn et al. 2006; Cesare and Reddel 2010). G-quadruplex binding proteins might regulate recombination at telomere double strands and telomere length. A previous study suggested that H4me3 in telomeres affects alternative lengthening of telomeres (Benetti et al. 2007). In fact, TLS overexpression results in telomere shortening (Takahama et al. 2013). In contrast, overexpression of mutated TLS, which is a deleted C-terminal RGG motif that cannot bind the G-quadruplex, does not affect telomere length or histone modification in telomeres. These findings suggest that TLS binding to the G-quadruplex in the telomere double-stranded region and TERRA with G-quadruplex structure specificity results in in vivo telomere shortening.

4.3 Engineering of the G-Quadruplex RNA-Binding Protein

Previously, the functions of G-quadruplex DNA in gene promoters, telomeres, and genomes have been elucidated using G-quadrulex DNA-binding molecules (Blasubramanian et al. 2011, Rodriguez et al. 2012, Neidle 2010). G-quadruplex RNA-binding molecules will be useful for elucidating TERRA functions, but little is known about the molecules that bind to G-quadruplex TERRA without binding to G-quadruplex telomere DNA. We recently reported that an engineered RGG motif, translocated in liposarcoma (TLS), specifically binds to G-quadruplex TERRA (Takahama and Oyoshi 2013).

The RGG motif of the C-terminal in TLS can simultaneously bind to G-quadruplex telomere DNA and TERRA in vitro, and binding stoichiometry was determined as one RGG motif per one Htelo, and one RGG motif per one TERRA. The RGG motif in TLS contains three Tyr in addition to two Phe as aromatic amino acids. To evaluate the roles of Tyr and Phe for G-quadruplex DNA and RNA recognition, we performed simultaneous substitution of two Phe by Tyr within an RGG motif in TLS (RGGY). Tyr substitution dramatically reduced Htelo binding. Binding stoichiometries were determined as one RGGY per two TERRA. These indicate that Tyr substitution increased the TERRA-binding capacity of the RGG motif in TLS, in addition to reducing Htelo binding (Fig. 4.4).

In order to investigate whether RGGY recognizes the loops and/or the G-tetrad of RNA, we examined RGGY binding to several G-quadruplexes: G-quadruplex

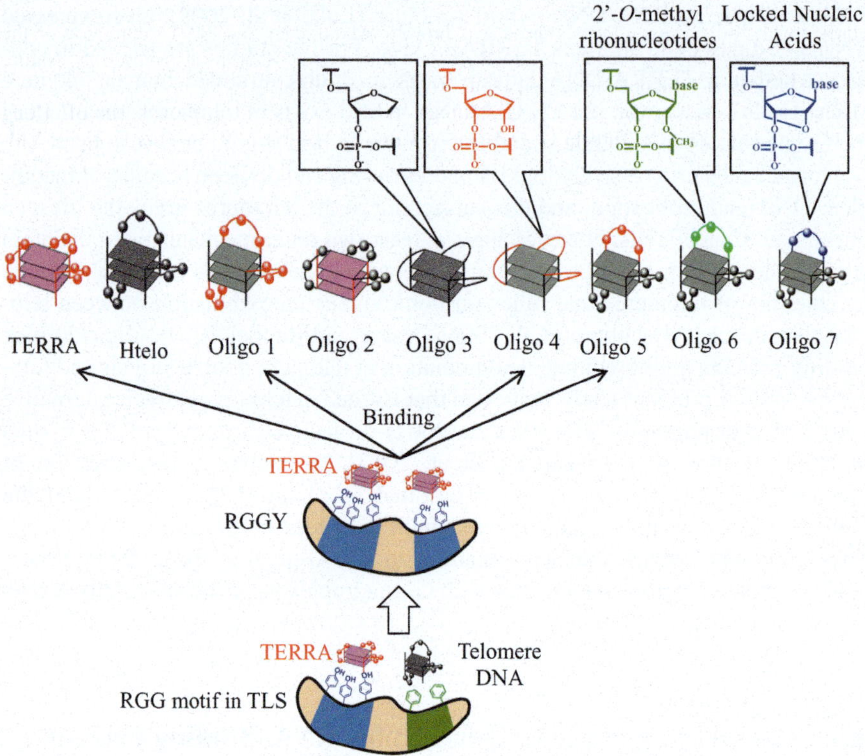

Fig. 4.4 Effect of DNA or RNA loops on the G-quadruplex binding selectivity of RGGY. The nucleic acid structures of abasic DNA, abasic RNA, 2′-*O*-methylribonucleotide, and locked nucleic acid are indicated. *Red*, *gray*, *green*, and *blue* in the cartoon show, respectively, RNA, DNA, 2′-*O*-methylribonucleotide, and locked nucleic acid

comprising DNA tetrads and three r(UUU) loops (oligo 1), RNA tetrads and three d(TTA) loops (oligo 2), DNA tetrads and three DNA abasic loops (oligo 3), and DNA tetrads and three RNA abasic loops (oligo 4) (Fig. 4.4). This showed that the G-quadruplexes of oligo 1 and oligo 4 are favorable for binding, like TERRA, while the G-quadruplexes of oligo 2 and oligo 3 were unfavorable. The fact that RGGY recognizes RNA loops suggests that 2′-OH in the loops is responsible for the recognition. To identify which part of the nucleotides on the loop is recognized by RGGY, we investigated the role of the base and ribose on the loop in the G-quadruplex for RGGY binding to G-quadruplexes comprising DNA tetrads with several DNA/RNA loops; G-quadruplex DNA with a single RNA loop (oligo 5), a single loop containing an artificial 2′-modified nucleic acid loop, 2′-*O*-methylribonucleotides (oligo 6), and locked nucleic acids (oligo 7). This showed that oligo 6 and oligo 7 were weakly bound, compared with oligo 5 (Fig. 4.4). It indicates the preferential recognition of the 2′-OH of the loop in the G-quadruplex by RGG.

4.4 Conclusions

G-quadruplexes that form at a double-stranded region and a single-stranded 3′ overhang in telomeres relate to telomere maintenance via G-quadruplex binding proteins with telomerase and histone-modifying enzymes. To our knowledge, TLS is the first known molecule that binds to G-quadruplex telomere DNA and upregulates the level of heterochromatin. TLS binds to the G-quadruplex telomere DNA and TERRA depending on the RGG motif, which contains Arg-Gly-Gly repeats. Many RNA-binding proteins with an RGG motif are members of the hnRNP family, TET family proteins containing TLS, Ewing's sarcoma and TATA-binding protein-associated factor 15, Fragile X Mental Retardation Protein, and others (Wang et al. 2012; Takahama et al. 2011, 2013; Phan et al. 2011). For example, the RGG motif of Fragile X Mental Retardation Protein recognizes G-quadruplexes and surrounding RNA sequences. The simple Arg-Gly-Gly repeat polypeptide, however, is not able to specifically bind the G-quadruplex. Not only RGG repeats, but also Pro-rich and Arg-rich sequences of the RGG motif in Ewing's sarcoma are required for specific G-quadruplex binding (Takahama et al. 2011). Further studies are required to identify the mechanism of RGG motif recognition of G-quadruplex structures and the possible function of G-quadruplex binding proteins in vivo.

TERRA performs various cellular functions, such as telomere length regulation, telomeric heterochromatin formation, and telomere protection (Takahama et al. 2013; Deng et al. 2009). TLS and TRF2 might tether TERRA to telomeres depending on the G-quadruplex of TERRA. The G-quadruplex-dependent functions of TERRA in telomere maintenance in human cells, however, are not clear. The functions of G-quadruplex DNA in gene promoters, telomeres, and genomes have been elucidated using G-quadruplex DNA-binding molecules (Blasubramanian et al. 2011; Rodriguez et al. 2012; Neidle 2010). Small molecules targeting the G-quadruplex in the 5′-untranslated regions of mRNA that regulate translation initiation could modulate translational activity by stabilizing or destabilizing the G-quadruplex structures (Bugaut and Balasubramanian 2012). G-quadruplex RNA-binding molecules will be useful for elucidating TERRA functions, but little is known about the molecules that bind to G-quadruplex TERRA without binding to G-quadruplex telomere DNA. To our knowledge, RGGY is the first known molecule that specifically recognizes the 2′-OH of the ribose of loops in the G-quadruplex. RGGY will be useful for investigating the role of the G-quadruplex form of TERRA without affecting G-quadruplex telomere DNA functions.

The G-quadruplex structure is thought to be widely present in promoter regions, telomeres, certain minisatellites, etc. (Zhang et al. 2013; Huppert and Balasubramanian 2007; Chiarella et al. 2013; Amrane et al. 2012). Recent computational studies in humans have indicated the possibility of as many as 2,394 G-quadruplex hits in long noncoding RNA (Jayaraj et al 2012). These potential quadruplex motif peaks, ranging in size from 200 to 300 bases, would be expected to be stable structures with mainly single and dinucleotide loops. These findings suggest that G-quadruplex binding proteins modify chromatin via long noncoding G-quadruplex RNA, such as TERRA, at promoter or other regions.

Several of the above-discussed biologic functions of the G-quadruplex and G-quadruplex binding proteins are still speculative. Further analysis of the mechanisms of the G-quadruplex in the telomere and genome, the timing of G-quadruplex formation, and how G-quadruplex binding proteins regulate the structures of telomere and genome, will be important to provide a better understanding of the functions of the G-quadruplex and G-quadruplex binding proteins. Visualization of the G-quadruplex in human cells, using the anti-G-quadruplex antibody, revealed that G-quadruplex structures are modulated during the cell cycle. G-quadruplex formation is maximal during S phase in cell cycles (Biffi et al. 2013). G-quadruplex formation might be spontaneously regulated by G-quadruplex binding proteins to perform its function at a specific site and time.

References

Amrane S et al (2012) Formation of pearl-necklace monomorphic G-quadruplexes in the human CEB25 minisatellite. J Am Chem Soc 134:5807–5816

Arnoult N et al (2012) Telomere length regulates TERRA levels through increased trimethylation of telomeric H3K9 and HI1α. Nat Struct Mol Biol 19:948–956

Azzalin CM et al (2007) Telomeric repeat containing RNA and RNA surveillance factors at mammalian chromosome ends. Science 318:798–801

Benetti R et al (2007) Suv4-20 h deficiency results in telomere elongation and derepression of telomere recombination. J Cell Biol 178:925–936

Biffi G et al (2013) Quantitative visualization of DNA G-quadruplex structures in human cells. Nat Chem 5:182–186

Blackburn EH et al (2006) Telomeres and telomerase: the path from maize, Tetrahymena and yeast to human cancer and aging. Nat Med 12:1133–1138

Blasubramanian S et al (2011) Targeting G-quadruplexes in gene promoters: a novel anticancer strategy? Nat Rev Drug Discov 10:261–275

Bugaut A, Balasubramanian S (2012) 5′-UTR RNA G-quadruplexes: translation regulation and targeting. Nucleic Acid Res 40:4727–4741

Cesare AJ, Reddel RR (2010) Alternative lengthening of telomeres: models, mechanisms and implications. Nat Rev Genet 11:319–330

Chiarella S et al (2013) Nucleophosmin mutations alter its nucleolar localization by impairing G-quadruplex binding at ribosomal DNA. Nucleic Acid Res 41:3228–3239

Cusanelli E et al (2013) Telomeric noncoding RNA TERRA is induced by telomere shorting to nucleate telomerase molecules at short telomeres. Mol Cell 51:780–791

de Lange T (2005) Shelterin: the protein complex that shapes and safeguards human telomeres. Genes Dev 19:2100–2110

Deng Z et al (2009) TERRA RNA binding to TRF2 facilitates heterochromatin formation and ORC recruitment at telomeres. Mol Cell 35:403–413

Duquette ML et al (2004) Intracellular transcription of G-rich DNAs induces formation of G-loops, novel structures containing G4 DNA. Genes Dev 18:1618–1629

Huppert JL, Balasubramanian S (2007) G-quadruplexes in promoters throughout the human genome. Nucleic Acid Res 35:406–413

Jayaraj GG et al (2012) Potential G-quadruplexes in the human long non-coding transcriptome. RNA Biol 9:81–86

Kelleher C et al (2005) Human protection of telomeres 1 (POT1) is a negative regulator of telomerase activity in vitro. Mol Cell Biol 25:808–818

Luke B, Lingner J (2009) TERRA: telomeric repeat-containing RNA. EMBO J 28:2503–2510

Luke B et al (2008) The Rat1p 5′ to 3′ exonuclease degrades telomeric repeat-containing RNA and promotes telomere elongation in Saccharomyces cerevisiae. Mol Cell 32:465–477

Luu KN et al (2006) Structure of the human telomere in K^+ solution: an intramolecular (3 + 1) G-quadruplex scaffold. J Am Chem Soc 128:9963–9970

Matadinate H, Phan AT (2009) Structure of propeller-type parallel-etranded RNA G-quadruplexed, formed by human telomeric RNA sequences in K^+ solution. J Am Chem Soc 131:2570–2578

Neidle S (2010) Human telomeric G-quadruplex: the current status of telomeric G-quadruplexes as therapeutic targets in human cancer. FEBS J 277:1118–1125

Phan AT et al (2011) Structure-function studies of FMRP RGG peptide recognition of an RNA duplex-quadruplex junction. Nat Struct Mol Biol 18:796–804

Rodriguez R et al (2012) Small-molecule-induced DNA damage identifies alternative DNA structures in human genes. Nat Chem Biol 8:301–310

Takahama K, Oyoshi T (2013) Specific binding of modified RGG domain in TLS/FUS to G-quadruplex RNA: tyrosines in RGG domain recognize 2′-OH of the ribose of loops in G-quadruplex. J Am Chem Soc 135:18016–18019

Takahama K et al (2011) Loop lengths of G-quadruplex structures affect the G-quadruplex DNA binding selectivity of the RGG motif in Ewing's sarcoma. Biochemistry 50:5369–5378

Takahama K et al (2013) Regulation of telomere length by G-quadruplex telomere DNA- and TERRA-binding protein TLS/FUS. Chem Biol 20:341–350

Vannier JB et al (2012) RTEL1 dismantles T loops and counteracts telomeric G4-DNA to maintain telomere integrity. Cell 149:795–806

Wang F et al (2007) The POT1-TPP1 telomere complex is a telomerase processivity factor. Nature 445:506–510

Wang F et al (2012) Telomere- and telomerase-interacting protein that unfold telomere G-quadruplex and promotes telomere. Proc Natl Acad Sci U S A 109:20413–20418

Xu Y et al (2010) Telomeric repeat-containing RNA structure in living cells. Proc Natl Acad Sci U S A 107:14579–14584

Zhang C et al (2013) DNA G-quadruplex formation in response to remote downstream transcription activity: long-range sensing and signal transducing in DNA double helix. Nucleic Acid Res 41:7144–7152

Zheng KW et al (2010) Molecular crowding creates an essential environment for the formation of stable G-quadruplexes in long double-stranded DNA. Nucleic Acid Res 38:327–338

Part III
Molecular Functions of lncRNAs

Chapter 5
Initiation of Transcription Generates Divergence of Long Noncoding RNAs

Riki Kurokawa

Abstract Global analyses have revealed that the majority of the human genome should be transcribed into RNAs, although less than 5 % of the genome possesses information on amino acid sequences and consists of coding regions. Most noncoding regions produce long noncoding RNAs (lncRNAs), and some of them have been found to play biological roles in living cells. These sequences bearing lncRNAs display extensive diversity because of their origins from intergenic regions, pseudogenes, and repetitive sequences. The quest for a divergent origin of lncRNAs is an elusive problem. We focus on transcription as the only driving force generating these lncRNAs. For covering such kinds of diversity of lncRNA, the transcription would need to be initiated all over the genome. Actually, pervasive transcriptions have been reported around many promoter regions of active genes. This tells us that the transcription of lncRNAs should be initiated from non-canonical initiation structures like TATA and initiators, suggesting an unrevealed principle for transcriptional initiation of lncRNAs. This chapter provides an overview of recent and older publications regarding transcription and especially focuses on the initiation of the transcription of lncRNA. There is still no clear presentation regarding the precise mechanism of transcription of lncRNA. Here, we discuss a possible hypothesis of transcription of lncRNAs that produces diversity of lncRNAs, and also their biological significance.

Keywords Transcription • RNA polymerase II • Long noncoding RNA • Transcription factor • General transcription factor • Chromosome • Divergent transcription • Coactivator • Corepressor

5.1 Introduction

Tremendous numbers of transcripts have been documented with next-generation sequence analyses (Khalil et al. 2009; Derrien et al. 2012; Djebali et al. 2012; Necsulea et al. 2014; Lipovich et al. 2014). The majority are long noncoding RNAs

---

R. Kurokawa (✉)
Division of Gene Structure and Function, Research Center for Genomic Medicine, Saitama Medical University, 1397-1, Yamane, Hidaka-shi, Saitama-ken 350-1241, Japan
e-mail: rkurokaw@saitama-med.ac.jp

R. Kurokawa (ed.), *Long Noncoding RNAs*, DOI 10.1007/978-4-431-55576-6_5

whose nucleotide length is more than 100 (Carninci et al. 2005; Kapranov et al. 2007a, b). Cell-type-specific expression of lncRNAs has also been reported in mammalian systems, while developmental stage-dependent expression of lncRNAs has been observed (Cabili et al. 2011; Djebali et al. 2012; Ravasi et al. 2006; Lipovich et al. 2014). These data indicate that the expression of lncRNAs should have been regulated with some specific transcriptional apparatus including DNA-binding transcription factors and general transcription factors like TFIID and TFIIB, nucleated with RNA polymerase II (Pol II). There have been reports to show that transcription of lncRNA should be conducted mainly with Pol II (Goodrich and Kugel 2006; Martianov et al. 2007; Wang et al. 2008). Thus, this review focuses on the Pol II-dependent transcription of lncRNAs.

The transcriptional machinery of eukaryotic cells has been extensively investigated over a half century (Roeder 2003; Kadonaga 2012; Lee and Young 2013). A solid model of the transcription has been presented. Sequence-specific transcription factors recognize each binding site and recruit cognate coactivator leading to mediator linking to basic core machinery around the transcription start sites (TSS) (Malik and Roeder 2010). Then, multiple factor complexes of basic transcription factors, TATA-binding protein (TBP) of TFIID, TFIIA, TFIIB, polII/TFIIF, TFIIE, and TFIIH, and the preinitiation complex (PIC) run off from the TSS to generate nascent transcript. Now, the process of assembly of PIC has been well examined with crystallographic analysis and the precise order of the assembly has been documented (He et al. 2013; Cianfrocco et al. 2013; Murakami et al. 2013). These data help to show how transcription of lncRNAs works in living cells.

Divergent transcription of mammalian genes has been widely observed (Core et al. 2008; Preker et al. 2008; Seila et al. 2008; Almada et al. 2013). Most lncRNA turns out to be transcribed from active gene loci (Sigova et al. 2013). Especially, the active promoter of genes is a major site of initiation of lncRNAs. The transcription frequently starts both sense and antisense directions from TSS. Therefore, antisense transcripts of the promoter regions of the transcription-active genes are major groups of lncRNAs. Actually, various promoter-associated lncRNAs have been shown. Wang et al. presented, however, that both sense and antisense transcripts of cyclin D1 promoter are elaborated upon DNA-damaging stimulation (Wang et al. 2008). The implication of the antisense strand of the promoter-associated lncRNAs remains an unresolved question. The enhancer of genes is also found to be frequently initiation of the lncRNA transcription and gives rise to enhancer RNA that is also a sort of lncRNA (Kim et al. 2010; De Santa et al. 2010; Natoli and Andrau 2012).

The molecular mechanism of divergent transcription remains elusive. The majority of the lncRNA transcription does not relate to any sequence-specific transcription factor. Thus, a major player of lncRNA transcription is supposed to be general transcription factors. The function of general transcription factors is to assemble PIC around TSS. The first step of formation is recognition of TFIID with the core promoter around TSS (He et al. 2013; Juven-Gershon and Kadonaga 2010). TFIID is a multi-subunit complex formed with TBP associated with a dozen TBP-associated factors (TAFs). The cooperative actions of TAFs and TBP bind a specific core promoter and start to form PIC there. Therefore, TFIID has a central role in the initial step of transcription of lncRNAs. In this chapter, we dissect the mechanism of transcriptional initiation through analysis of general transcription factors.

There are two categories of transcription initiation of coding genes. One is focused transcription initiation. This is mainly employed with stringently regulated genes (Kadonaga 2012; Juven-Gershon and Kadonaga 2010). Another is dispersed transcription initiation, which is utilized with constitutively expressed genes like housekeeping genes. In focused transcription, there is either a single major transcription start site or short regions of transcriptional start. The focused transcription is the predominant mode of transcription in simpler organisms. In dispersed transcription, there are several weak TSSs over a broad region of 50–100 nucleotides. For example, dispersed transcription is observed in two thirds of human genes. In vertebrates, focused transcription tends to be associated with regulated promoters, whereas dispersed transcription is typically observed in constitutive promoters in CpG islands (Illingworth et al. 2010; Maunakea et al. 2010; Sleutels et al. 2002). The dispersed transcription should be one of the major forces to drive the lncRNA transcripts.

Divergence of sequences of the lncRNAs is one of the central questions in modern biology (Cech and Steitz 2014; Wu and Sharp 2013). In the process of biological evolution, vast numbers of lncRNAs survive after stringent selections. This suggests that some benefits of the lncRNA should add to the kingdoms of organisms. A large population of lncRNAs provides a pool of RNA sequences to be selected for biologically versatile functions. The lncRNA pool in living cells is a possible supplier for the functional RNAs to regulate divergent biological programs. Then, only transcription is a generating process for the kinds of lncRNAs. This chapter presents a discussion of the origination of lncRNAs thorough the eukaryotic transcriptional machinery with the hope of yielding a fruitful conceptual outcome.

5.2 Cell-Type-Specific Expression of Long Noncoding RNAs and Their Possible Functions

It has been reported that there are more than 30,000 lncRNAs in the human genome, and most of their functions have not been documented yet. The GENCODE v7 catalog of human lncRNAs has been published recently, giving an extensive examination of their gene structure, evolution, and expression pattern (Derrien et al. 2012). The lncRNA catalog shows a subset of the manually annotated GENCODE human gene annotation catalog containing 15,512 transcripts clustered into 9,640 gene loci. The GENCODE catalog demonstrates that lncRNAs have canonical gene structures and histone modifications as equivalent to protein-coding genes. These lncRNAs are more likely to be under weaker evolutionary pressure and to be expressed at lower levels than coding genes. Generally, lncRNAs are localized in the chromatin and nucleus of the living cells. The GENCODE catalog stimulates a series of substantial discussions.

The lncRNA annotation is confronted with a tough obstacle with low expression levels of these RNAs, which accidentally may lead to fragmentary annotation and poor definition of the transcript boundaries. One of the major issues is doubt as to whether lncRNAs are independent transcripts or whether they are simply unidentified extensions of neighboring protein-coding transcripts. To eliminate the problems, the GENCODE project employed various high-throughput sequencing data in the context of the ENCODE project to search for evidence of the position of lncRNA transcripts in the genome and neighboring protein-coding genes. With all these endeavors, the GENCODE project has obtained data showing that the majority of lncRNAs are unlikely to represent unannotated extensions of neighboring protein-coding genes. Most lncRNAs have been confirmed to have independent transcriptional units. This is a principal viewpoint of the chapter regarding the origination of lncRNAs.

Natural selection of genomic sequences represents solid evidence for biological functionality of relevant lncRNAs. Therefore, the GENCODE project tried to assess whether lncRNAs experienced this kind of selection. These assessments were performed with precomputed, nucleotide-level calculations of evolutionary selection provided by the phastCons algorithm (Siepel et al. 2005), based on a phylogenetic hidden Markov model. By this analysis, the lncRNA exons are significantly more conserved than corresponding ancestral repeat sequences, although at lower expression levels than protein-coding genes. These data agree with findings from previous publications of lncRNAs (Guttman et al. 2009; Marques and Ponting 2009; Orom et al. 2010). The GENCODE project inspected the sequence conservation of different regions of lncRNA genes consisting of promoters, exons, and introns. Actually, the promoters of lncRNAs are, on average, more conserved than their exons and are almost as conserved as protein-coding gene promoters. The promoters of lncRNAs should be a key component of the lncRNA "coding" DNAs.

The lncRNAs have been shown to have lower and more tissue-specific expression than protein-coding genes in the GENCODE catalog. The expression patterns of lncRNAs in a wide range of human organs and cell lines was analyzed using available RNA-seq data as well as a custom lncRNA microarray. The analysis was executed with particular interest in understanding the magnitude of lncRNA expression, as well as its degree of tissue specificity. Using RNA-seq data obtained from the Illumina Human Body Map Project (HBM; www.illumina.com; Array Express ID:E-MTAB-513), computation of the distribution of expression of lncRNAs and protein-coding genes was done across the 16 tissues profiled in the HBM Project. As a result, lncRNAs exhibit lower expression in all tissues, compared with mRNAs, although lncRNAs display relatively high expression in the testis. It was revealed that the lncRNAs also display more tissue-specific patterns compared with protein-coding genes, although this might be a result of their lower expression levels and resultant false negative detection in some tissues when applying a strict cutoff of expression. Sixty-five percent of protein-coding genes were detected in all HBM tissues, compared with 11 % of lncRNAs. In agreement with this observation, the lncRNAs were shown to have higher expression variability, detected as the coefficient of variation across cell lines and tissues tested, than protein-coding genes.

The GENCODE catalog presents intriguing points regarding dissection of lncRNA transcription. It authenticates the entity of lncRNAs in the human genome rather than supporting skepticism about their biological significance. The lncRNAs are firmly expressed in living cells, but the level of expression is much lower than that of protein-coding genes and is also labile (Flynn et al. 2011; Ntini et al. 2013; Preker et al. 2008), although distinctive tissue-specific expression of them is validated. These data suggest that transcription of lncRNAs is distinct from regular protein-coding genes. Then, we review previously achieved data regarding eukaryotic transcription. This should be a clue to know how lncRNAs are generated in the human genome.

5.3 Transcriptional Machinery of Eukaryotic Cells

Transcription alone has potency in generating ribonucleic acid and also lncRNAs as well in living cells. In this section, we inspect transcription as a generation process of lncRNAs. Initiation of transcription is a starting event for every RNA synthesis. Before transcriptional initiation, there is a robust barrier against the initiation of transcription, with a chromatin structure consisting of nucleosomes with histones and other proteins. The whole genomic DNA is wrapped into nucleosome structures and forced into the inactive state of transcription. Then, some signals from outside the cells induce activity of transcription factors with protein modification like phosphorylation and also an allosteric effect by ligand binding to bind to enhancers located upstream of TSS and unwind the well-wrapped nucleosome structure with recruitment of histone acetyltransferase in coactivator protein CBP/p300 (Fig. 5.1). The acetylation of histones stimulates methyltransferase activity of histones at the sites, and in turn causes exposure of promoter regions containing TSS to form PIC. Formation of PIC is presumably the most critical step to initiate transcription of lncRNAs over the human genome. It is logical that many of the lncRNA transcriptions initiate from regions without known transcription factor binding sites, because the initiation occurs from numerous regions of the genome. For elucidation of transcription initiation of lncRNAs, we need to know more about general transcription machinery. Thus, we provide an overview of the eukaryotic transcription machinery.

Robert G. Roeder has outlined the eukaryotic basic transcriptional machinery consisting of eukaryotic general/basal transcriptional factors (Roeder 2003). Dozens of polypeptides are involved in the process of initiation of transcripts. Precise and prescribed initiation of the gene transcription represents a major step in gene regulation, requiring the coordinated activity of a large number of proteins and protein complexes (He et al. 2013; Murakami et al. 2013). The basal transcriptional machinery includes Pol II along with a series of general transcription factors (GTFs; TFIIA, TFIIB, TFIID, TFIIE, and TFIIH) that assemble into a 2 megaDalton (MDa) complex on the core promoter DNA sequence. This PIC is essential to direct accurate TSS selection, promoter melting and Pol II promoter escape. Despite recent structural advances on Pol II and subcomplexes of the PIC, the molecular assembly details of this essential complex remain elusive.

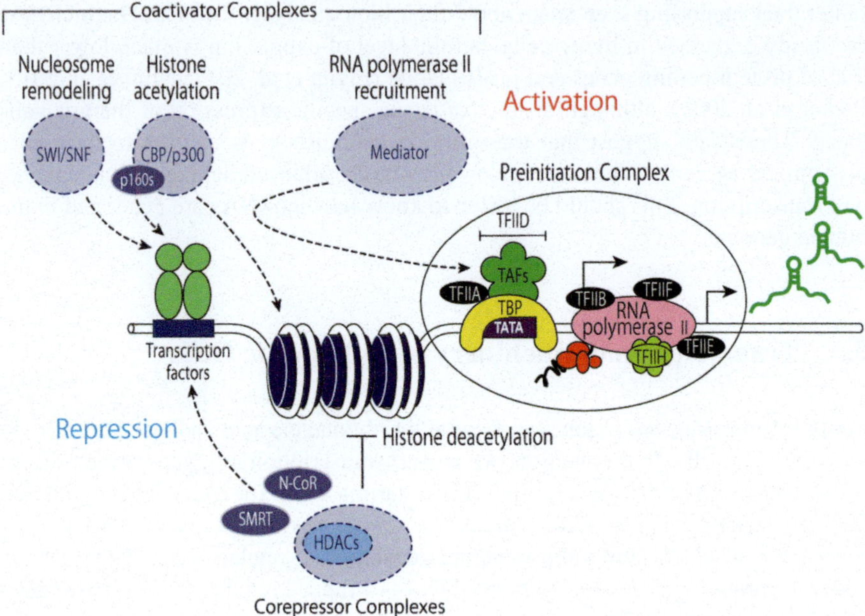

Fig. 5.1 Eukaryotic transcriptional machinery. TBP forms TFIID with TAFs. TFIID and several general transcription factors are shown with RNA polymerase II forming the preinitiation complex (*PIC*). PIC plays a central role in initiation of transcription from the core promoter

In vitro reconstitution of this process has provided a model for the sequential assembly pathway of transcription initiation. TFIID is the initial factor specifically recruited to the promoter. TFIID, a megaDalton complex, includes TBP, which is alone sufficient for forming PIC on TATA box-containing prompters. TFIIA and TFIIB are then recruited, further stabilizing the interaction between TBP (or TFIID) and promoter DNA. Next, Pol II, probably in association with TFIIF, adds to the growing PIC. Finally, TFIIE and TFIIH, which is required for DNA melting, are recruited to form the transcriptionally competent PIC. Therefore, TFIID is the first one to be built into the PIC and plays a pivotal role in recognizing the specific promoter sequence. This step is especially determinant for initiation of lncRNA transcription.

Structural characterization of PIC assemblies is challenging and has been limited to a small number of electron microscopy (ME) studies. Crystallographic structures of individual components, combined with biochemical data, have led to a number of structural models for PIC subcomplexes, in either a closed or open-promoter conformation. Overviewing the PIC assembly, recruitment of TFIID to the promoter is the first step for the transcription of lncRNA (Cianfrocco et al. 2013). TFIID is a putative major player in generating the diversity of lncRNA. We discuss the functional contribution of TFIID in transcriptional initiation later in this chapter.

5.4 Divergent Transcription from the Mammalian Genome

The majority of the human genome is transcribed into numerous species of transcripts. However, most intergenic transcription activity generates short and unstable noncoding transcripts, the amounts of which are usually far lower than those from standard protein-coding genes (Wu and Sharp 2013; Guttman and Rinn 2012; Ponting et al. 2009). It is uncertain if most intergenic transcripts have biological significance or cellular function. Recent data have shown that most intergenic transcription occurs near active gene loci, such as the promoter and gene body (Sigova et al. 2013). Especially, the promoters are major DNA sequences in generating lncRNAs (Fig. 5.2). The majority of mammalian promoters have transcription toward both sides, a well-known profile of RNA biosynthesis as divergent transcription (Core et al. 2008; Preker et al. 2008; Seila et al. 2008) (Sigova et al. 2013). Divergent transcription generates upstream antisense RNAs (uaRNAs) near the 5′ end of genes that are typically short (50–2,000 nucleotides) and relatively labile (Preker et al. 2008; Flynn et al. 2011). Similar divergent transcription also occurs at distal enhancer regions, providing RNAs termed enhancer RNAs (Kim et al. 2010). In mouse and human embryonic stem (ES) cells, most long noncoding RNAs (lncRNAs, longer than 100 nucleotides) are associated with protein-coding genes, including ~50 % as uaRNAs and ~20 % as eRNAs. These observations suggest that divergent transcription from promoters and enhancers of protein-coding genes is the major source of intergenic transcription in ES cells.

It is an established model of the eukaryotic promoter that the directionality is set by the arrangement of an upstream cis-element region followed by a core promoter typically consisting of BREu (upstream TFIIB recognition element), TATA box, Inr

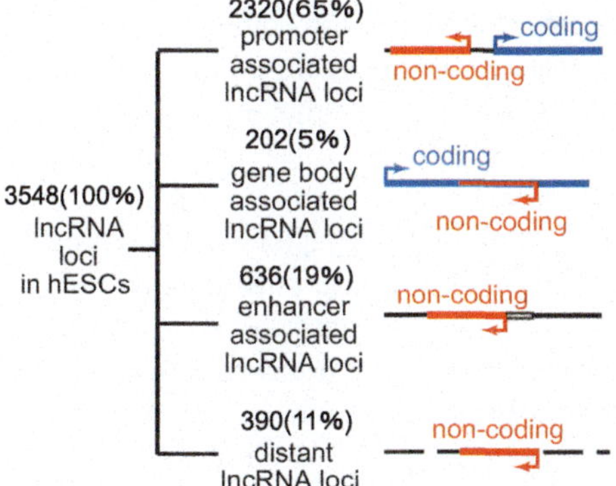

Fig. 5.2 Most lncRNAs are associated with active protein-coding genes in human ES cells. Summary of various types and numbers of lncRNA loci in hES cells. The diagrams on the right depict lncRNA loci as *red lines*, protein-coding genes as *blue lines*, and an enhancer as an *open box*. An *arrow* indicates the direction of the transcription initiation. Enhancer-associated lncRNAs overlap or originate at genomic regions enriched in nucleosomes with histone H3 acetylated at lysine 27 (H3K27Ac) (This figure is reproduced from Figure 1B of the reference Sigova et al. (2013) upon permission from the PNAS office)

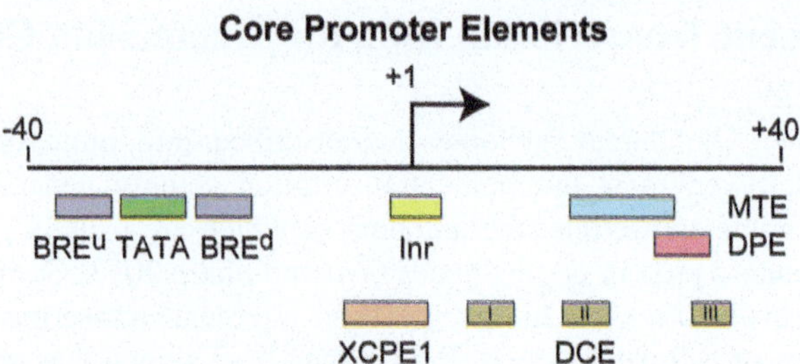

Fig. 5.3 The core promoter elements. It is an established model of the eukaryotic promoter that the directionality is set by the arrangement of an upstream cis-element region followed by a core promoter typically consisting of BREu (upstream TFIIB recognition element), TATA box, Inr (initiator), MTE (motif ten element), DPE (downstream promoter element), DCE (downstream core element), and XCPE1 (the X gene core promoter element 1)

(initiator), MTE (motif ten element), DPE (downstream promoter element), DCE (downstream core element), and XCPE1 (the X gene core promoter element 1) (Fig. 5.3). The cis-elements are bound by DNA-binding transcription factors, whereas the core promoter is bound by TBP (or forming as TFIID complex) and other factors that recruit the basic transcriptional machinery. Most mammalian promoters do not have a TATA element (TATA-less) and just have CpG-rich regions (Sandelin et al. 2007). For these promoters, TBP is recruited through transcription factors such as Sp1, which binds CpG-rich sequences and components of the TFIID complex that have less sequence specificity, although TFIID is capable of recognizing TSS regions as a full TFIID complex. Thus, in the absence of strong TATA elements such as CpG island promoters, TFIID is supposed to be recruited on both sides of the transcription factors to form PICs in both orientations (Fig. 5.4). This model is supported by the observation that divergent transcription occurs at most promoters that are associated with CpG islands in mammals and worms are rather associated with unidirectional transcription (Core et al. 2008; Kruesi et al. 2013). Moreover, divergent transcription is less common in Drosophila where CpG islands are rare (Core et al. 2014). Since transcription factors with chromatin remodeling potency and transcription activation domains also bind at enhancer sites, it is not surprising that these are also sites of divergent transcription. In fact, promoters and enhancers have many properties in common, and it has been shown recently that many intergenic enhancers can act as alternative promoters producing tissue-specific lncRNAs.

Intriguing data have been reported regarding the orientation of transcription of lncRNAs with ES cells (Sigova et al. 2013). ES cells have been widely employed as a model system to study transcriptional regulation of cell states during early development, yet there is no comprehensive catalog of lncRNAs in human ESCs, and it is not clear how lncRNAs are regulated in these cells. Some catalogs of lncRNAs have been recently described in various murine and human cell types, but the majority were limited to spliced lncRNA species and those distant from protein-coding genes (Guttman et al. 2009, 2011; Khalil et al. 2009; Cabili et al. 2011). Because

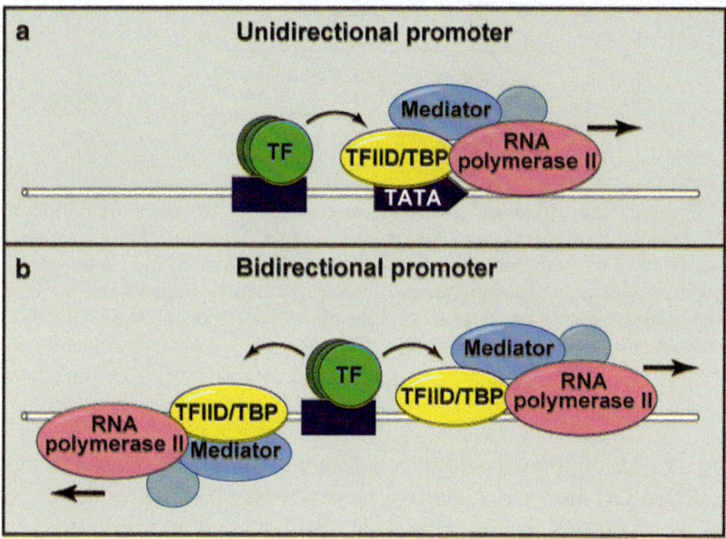

Fig. 5.4 TBP (TFIID) is a major molecule in generating divergent transcription. (**a**) Transcription factor (*TF*) binding assists to recruit TBP and associated factors (*TFIID*), which bind the directional TATA element in the DNA and orientate RNA polymerase II to transcribe downstream DNA. (**b**) In the absence of canonical TATA elements, a common feature of CpG island promoters, TF-recruited TBP and associated factors (TFIID) bind to low-specificity sequences and form initiation complexes at similar frequencies in both directions

lncRNAs tend to be cell-type specific, these catalogs likely contains only a very small fraction of lncRNAs expressed in hES cells. Sigova et al. compiled a catalog of lncRNA species expressed in hES cells that originated from 3,548 non-redundant loci. The sizes of these lncRNAs in the hES cell catalog range from 105 to 687,089 nucleotides and have a median size of 1,831 nts. The amount of these lncRNAs is, on average, ten times less than that of mRNAs in the cells. Half of the lncRNA loci have spliced transcripts. The majority of the lncRNAs in the catalog have not been previously identified. Examination of the genomic positions of lncRNA loci revealed that 89 % are associated with the genic regions, the promoters, enhancers, and bodies of coding genes. Most lncRNAs were found to originate within a 2 kb region surrounding the TSS of coding genes (65 %), and others originate from antisense transcription of coding genes (5 %), enhancers (19 %), and other more distant (more than 2 kb) sites from coding genes (11 %) (Fig. 5.2). The catalogs show human and murine ES cell lncRNAs and the genomic regions from which these RNA species arise (Juven-Gershon and Kadonaga 2010; Sigova et al. 2013). They show that the majority of these lncRNAs originate from divergent transcription of lncRNA/mRNA gene pairs and that many such gene pairs are coordinately regulated when ES cells differentiate. Strikingly, 65 % of lncRNAs are associated with the promoter regions of active coding genes. This could explain that many lncRNAs are co-expressed and co-regulated with coding genes.

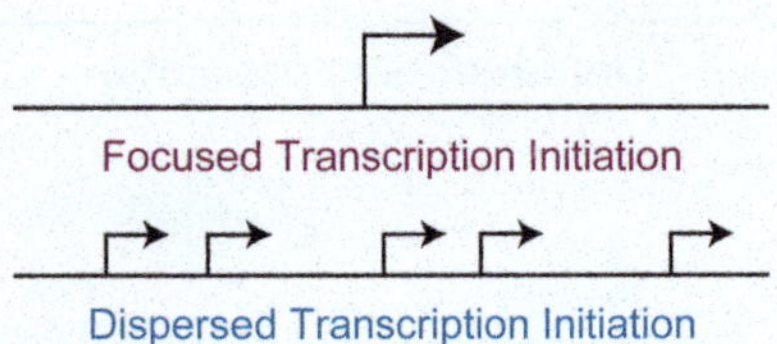

Fig. 5.5 "Focused" and "dispersed" transcription initiation. In "focused" transcription, there is either a single major transcription start site or several start sites within a narrow region of several nucleotides. Focused transcription is the predominant mode of transcription in simpler organisms and is typically found in regulated promoters. In "dispersed" transcription, there are several weak transcription start sites over a broad region of about 50–100 nucleotides. Dispersed transcription is the most common mode of transcription in vertebrates and is commonly found in constitutive promoters

Divergent transcription should have initiations from many places in the human genome. Juven-Gershon and Kadonaga present a classification of the transcription initiation as "focused" versus "dispersed" initiation of transcription (Fig. 5.5). Inspection of the features of transcription initiation reveals two distinctive modes of transcription initiation, focused and dispersed initiations (Smale and Kadonaga 2003; Carninci et al. 2006; Juven-Gershon and Kadonaga 2010). In the focused initiation, transcription starts at a single nucleotide or within narrow regions of several nucleotides, whereas in dispersed initiation, there are multiple weak start sites over a broad region of about 50–100 nucleotides. Focused transcription initiation occurs in all organisms and appears to be the predominant or exclusive mode of transcription in simpler organisms. In vertebrates, however, around 70 % of genes have dispersed promoters, which are typically found in CpG islands. It generally appears that focused promoters are associated with regulated genes, whereas dispersed promoters are used in constitutive genes. From a teleological standpoint, this arrangement is consistent with the notion that it would be easier to regulate the transcription of a gene with a single transcription start site than one with multiple start sites. Conversely, variations in the expression of a constitutive gene would be minimized by the use of multiple start sites.

Most previous studies of RNA polymerase II transcription have been carried out with focused promoters. Although focused promoters constitute a minority of all promoters in vertebrates, there is an extravagant amount of labor devoted to focused promoters relative to dispersed promoters because of the biological significance of the regulated genes with which the focused promoters are associated. The analysis of focused core promoters has led to the discovery of sequence motifs in core promoters such as BREu, TATA box, Inr, MTE, DPE, DCE, and XCPE1. In contrast, dispersed promoters generally lack BRE, TATA, DPE, and MTE (Carninci et al. 2006; Sandelin et al. 2007). It is likely that there are fundamental differences in the mechanisms of transcription from focused versus dispersed promoters, although the mechanism of transcription from dispersed promoters remains elusive. The dispersed initiation might be one of driving force to generate diverse species of lncRNAs.

5.5 Molecular Mechanism of Divergent Transcription

Until recently, it was a consensus that a universal and highly conserved RNA polymerase II core promoter recognition apparatus, PIC, initiated transcription in all eukaryotic cells. The core promoter is the region of a gene locus to which RNA polymerase II and the general transcription factors bind to initiate transcription. Core promoters encompass from around 40 base pairs upstream to 40 base pairs downstream of the transcription start site and are composed of DNA elements like TATA box, where subunits of TFIID or TFIIB bind (Fig. 5.3). Central components of the PIC such as TFIID, a protein complex of TBP and TAFs, were generally considered essential but passive partners that were designed to follow the regulatory instructions provided by DNA-binding transcription factors. This model came in part from studying a limited set of cell types—for example, yeast, Drosophila melanogaster S2 cells, and human HeLa cells, which divide rapidly and were preferred for practical reasons such as large-scale production for biochemical analysis or ease of genetic manipulation. Furthermore, many experiments in the transcription field have used recombinant model genes and promoters and artificial regulators. More recent studies have shifted toward an analysis of endogenous genes and physiologically relevant regulators observed in the context of nearly homogeneous populations of a single, specific, differentiated cell type like ES and iPS cells and in distinct cell cycle stages. These studies have revealed the requirement for a number of non-typical core promoter recognition factors for transcription, including cell-type-specific TAFs and TBP-related factors (TRFs). Furthermore, new functions of the prototypical core promoter recognition machinery have been identified. These TRFs have been identified to regulate specific sets of genes during somatic and germ cell development.

The core promoter recognition is the first step in the mechanism of transcription initiation. The major general transcription factor in core promoter recognition for protein-coding genes is TFIID, which binds multiple core promoter elements to begin the process of forming PIC containing RNA polymerase II (Fig. 5.6). The RNA polymerase II core promoters in higher eukaryotes are highly diverse and the core promoters of many genes do not contain any known core promoter elements. The most recognizable core promoter element is the TATA box, but TATA-containing promoters are actually in a minority compared with the group of TATA-less promoters (Kadonaga 2012). It now seems unlikely that one can simply classify promoters into TATA-containing versus TATA-less, as there seem to be many potentially diverse TATA-less classes of promoters. However, the TFIID complex works at most of the RNA polymerase II promoters in eukaryotic cells. This suggests that TFIID has the potential to recognize divergent and "hidden" promoter sequences to initiate transcription.

TFIID, which consists of TBP and 13 or 14 TAF subunits, binds core promoter DNA through multiple subunits (for example, TBP, TAF1, TAF2, TAF6, and TAF9) (Fig. 5.6). The TBP subunit of TFIID binds TATA boxes, which, when present in promoters, are centered approximately 27 bp upstream of TSS. Several TAFs also bind the promoter elements downstream of the TATA box. TAF1 and TAF2 bind the initiator element, which spans the transcription start site; TAF6 and TAF9 bind

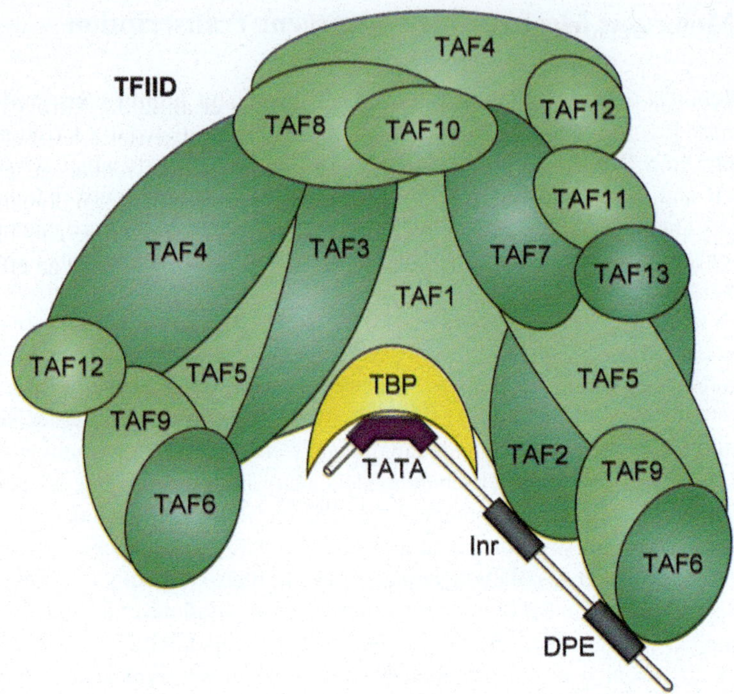

Fig. 5.6 Core promoter recognition by TFIID. Multiple subunits of TFIID complex bind core promoter elements. TBP binds TATA boxes. TBP-associated factor 1 (*TAF1*) and TAF2 bind the initiator element (*Inr*). TAF6 and TAF9 bind the downstream promoter element (*DPE*) (This figure is reproduced from the reference Goodrich and Tjian (2010) upon the license of the Nature Publishing Group (no. 3547410222285))

DPE; and TAF1 is in close proximity to the downstream core element when TFIID is bound to promoters. Some of the TAF subunits are also the targets of transcriptional activators, thereby allowing TFIID to integrate signals from activators to the core promoter. General transcription factor TFIIA assists the function of TFIID in binding core promoters, after which the remaining general transcription machinery can associate, including TFIIB, PolII, TFIIF, TFIIE and TFIIH, as well as the mediator coactivator complex. Once formed, PICs are competent to initiate transcription. In living cells, as opposed to in vitro experiments with purified components, this process is far more complicated, primarily because of the DNA being associated with nucleosomes as part of chromatin plus the additional requirement for transcription activators, coactivators, chromatin-modifying factors, and transcription elongation factors. However, although these additional factors are important for transcription in living cells, they are not considered to be core promoter recognition factors. Recently, subunits of TFIID have been found to show activities that were not expected on the basis of the dictatorial viewpoint of core promoter recognition and PIC formation (Goodrich and Tjian 2010). Proteins related in sequence to TBP and several TAFs have been discovered that seem to have unique functions during development, differentiation, and cell proliferation.

The Nogales laboratory presented data showing that TFIID interacts with diverse promoter architectures through the rearranged conformation, using cryo-electron microscopy technology. Because the majority of promoters within the Drosophila and human genomes do not contain all four of the core promoter motifs, for efficient in vitro analysis a synthetic promoter sequence was engineered into the super core promoter (SCP) which consists of TATA, Inr, MTE, and DPE (Juven-Gershon et al. 2006) (Fig. 5.7). Cianfrocco et al. analyzed promoter architectures with SCP to investigate the relevance of the rearranged TFIID-TFIIA-SCP structure (Fig. 5.7) (Cianfrocco et al. 2013). The "rearranged" TFIID is stimulated by TFIIA with SCP to form a ternary complex consisting of TFIID, TFIIA, and SCP. To this end, they compared TFIID interactions with wild-type versus mutant versions of the SCP. These data suggest that TFIID interacts with TATA-Inr, TATA-MTE/DPE, and Inr-MTE/DPE promoters in the rearranged conformation. With both the wild-type SCP and the three mutant versions of the SCP (mTATA, mInr, and mMTE/DPE), they observed that TFIIA stimulates the binding of TFIID to the

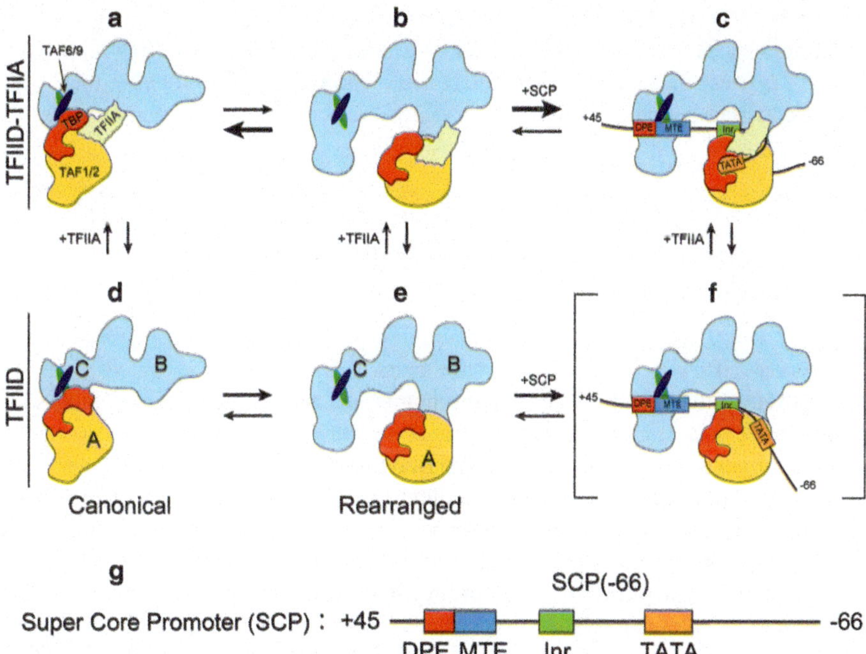

Fig. 5.7 TFIID interaction with core promoter DNA in a conformation- and TFIIA-dependent manner. (**a–c**) Conformations adopted by TFIID-TFIIA. (**d–f** Conformations adopted by TFIID alone. (**a, b** TFIIA stabilizes TFIID in the canonical conformation. (**c**) The addition of SCP DNA stabilizes the rearranged conformation for the ternary complex TFIID-TFIIA-SCP. (**d, e**) In contrast to TFIID-TFIIA (**a, b**), TFIID adopts a conformational landscape that populates canonical and rearranged states equally. (**f**) Upon addition, super core promoter (*SCP*) DNA is bound by TFIID in the rearranged conformation. *Brackets* ([ ]) denote that (**f**) was observed only through biochemical footprinting. (**g**) SCP structure (This figure is reproduced from the reference Cianfrocco et al. (2013) upon the license of Elsevier (no. 3512961163404))

TATA box (Thomas and Chiang 2006). With the mTATA promoter, the primary interaction of TFIID with the DNA is via the Inr, MTE, and DPE motifs, and a weak stimulation by TFIIA of the binding of TFIID to the mutant TATA box region is also observed. With the mInr and mMTE/DPE promoters, it seems likely that TFIIA stimulates the binding of TBP to the TATA box, and that the remainder of the TFIID complex then interacts with the Inr through the DPE region of the core promoter, irrespective of the presence of consensus Inr or MTE/DPE elements. These findings may be analogous to the previously observed stimulation of the partially purified TFIID to the downstream promoter region of the adenovirus major late promoter (which lacks MTE/DPE motifs) by the upstream stimulatory factor (Sawadogo and Roeder 1985; Van Dyke et al. 1988). In this light, it is possible that other sequence-specific activators, as well as coactivators, may function in a related manner to stabilize TFIID on promoter DNA and thus promote the formation of the rearranged conformation.

These data on the conformational circumstance of TFIID provide a conceptual outline for understanding the molecular interactions that occur between TFIIA and TFIID on the core promoter sequence. In particular, the dynamics of conformational alteration of TFIID have regulatory roles within living cells by providing specific structural targets that can be recognized by transcriptional activators and repressors (Fig. 5.7). These data suggest that TFIID could enforce transcriptional initiation all over the human genome, leading to pervasive transcription of divergent species of lncRNAs.

5.6 "Chance and Necessity" in Origination of Diverse lncRNAs in the Human Genome

Studies on eukaryotic transcription have been enthusiastically executed for more than half a century and have provided a marvelous achievement in providing substantial understanding of the transcriptional machinery. The comprehension of the transcription has mostly concerned protein-coding genes, for historical reasons, and later their importance or needs for medical outcomes. However, the progress of sequencing technology has been boosting deep analysis of numerous transcripts from the human genome, presenting numerous species of lncRNAs.

We have been working on an emerging interrogation about the origination of diversity of lncRNAs in the human genome. Then, there is a straightforward resolution for it. It should be from diversity of initiation of transcription. For generating many species of lncRNAs, initiation of transcription requires numerous start sites over the human genome. Recent analysis of the basic transcription machinery indicates that the primary event of the initiation is guided with the TFIID subunit of the basic transcription machinery. It has been shown that TFIID recognizes putative promoters even lacking TATA box, Inr, and other components of the prototypical promoter possibly with unidentified associated proteins. These data suggest that transcriptional initiation elicited with TFIID plays a central role in making unanticipated numbers of species of lncRNAs. This potency of TFIID is a driving force toward generation of divergent lncRNAs. From the teleological point of view, it is a central question why the human genome generates so many lncRNAs. It is, however, just by "chance", not by "necessity". The human genome has intrinsic potential to make divergent lncRNAs and provide a huge pool of species of lncRNAs (Fig. 5.8) (Reinberg and Roeder 1987; Shenkin and Burdon 1966). This pool should work for the SELEX (systematic evolution of ligands by exponential enrichment) technology of biochemistry in which a functional RNA is selected out from a pool of randomized RNA oligos to provide an RNA aptamer with specific binding ability (Tuerk and Gold 1990; Kurokawa 2011). During the evolutionary process, the SELEX-like reaction would occur in living cells and make selection of lncRNAs with biologically significant ability from the pool of divergent lncRNAs (Fig. 5.8). The whole process selects valuable lncRNAs in biological events in cells.

Divergent transcription is a major property of generating the diverse species of lncRNAs from the human genome. The diversity of transcription is allowed with potency of TFIID recognition of degenerative sequences of putative promoters. Therefore, the lncRNA diversity can be attributed to the function of the basic transcriptional machinery centered with TFIID.

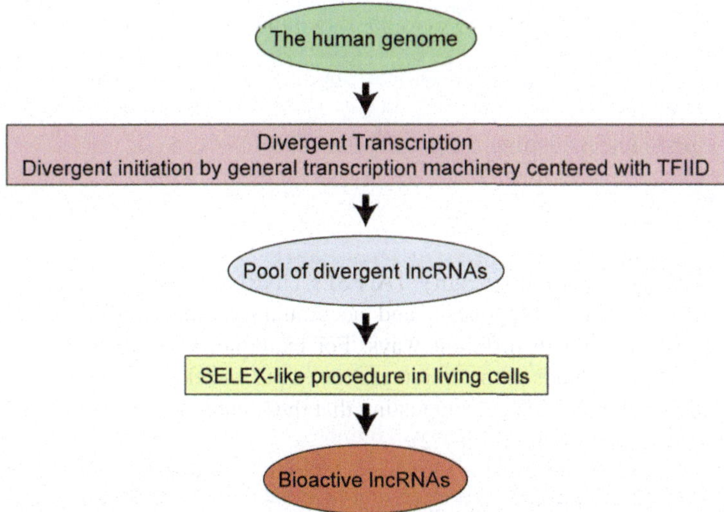

Fig. 5.8 A hypothesis of evolution of lncRNAs by divergent transcription. Divergent transcription drives generation of a pool of lncRNA molecules. SELEX (Systematic evolution of ligands by exponential enrichment)-like chemical procedures select biologically functional lncRNAs

5.7 Taxonomy of lncRNAs

The penultimate section of this chapter focuses on classification of lncRNAs. Attempts to provide a taxonomy of lncRNAs have given rise to an idea regarding the origination of lncRNAs. It has been identified that there are divergent species of lncRNAs. There are at least two classes of lncRNAs. One is a class of noncoding RNA that is spliced and poly adenylated. These are supposed to arise from promoter-like regions that are probably similar to those that give rise to mRNAs. These lncRNAs are likely to be more stable and to have the potential to reside in both the nucleus and cytoplasm, and for nuclear forms to exert possible trans-effects on gene expression.

A second class of lncRNAs emerges from enhancers. For instance, they are so-called enhancer RNAs. The great majority of these RNAs are not spliced or polyadenylated and they are most likely unstable, are confined to the nucleus, and exert local effects on gene expression if they have any function. It has been shown that analysis of intergenic RNA transcripts indicates many similarities of enhancer-directed transcripts with transcripts initiating from mRNA promoters (Glass CK, personal communication 2015). Similar conclusions were reported by the Riken Consortium and John Lis recently (Forrest et al. 2014; Core et al. 2014). However, to our knowledge, there is no clear understanding of why some transcripts get polyadenylated and others do not. This is highly correlated with splicing, but some unspliced transcripts are polyadenylated. It has been suggested that a splice site should be the most distinguishing feature of initiation sites that produce polyadenylated mRNAs. Therefore, it is possible to classify lncRNAs into two categories. One is promoter-associated lncRNAs, while the other is enhancer lncRNAs. The promoter-associated lncRNAs have splicing and polyadenylation, and are more akin to mRNA molecules. The enhancer lncRNAs are transcribed away from mRNA promoters and are also less related to coding regions. These might open the door to another clue to understanding the origination of lncRNAs.

5.8 Conclusions and Future Prospects

There are various species of lncRNAs expressed in living cells, although the expression levels are approximately ten times less than those of protein-coding genes. The lower expression levels of lncRNAs are one of the clues to understanding why so many lncRNAs are expressed from the human genome. Actually, transcription initiation of lncRNAs is not as efficient as protein-coding genes. This is because lncRNA transcription initiates through non-canonical core promoters without BRE, TATA box, and MTE. For initiation of lncRNAs, TFIID is required to recognize less specific core promoters mainly with CpG-rich regions, and forms PIC with TFIIB, TFIIE, TFIIF, and TFIIH to run off the transcription. There has not been extensive study of the transcriptional mechanism of the lncRNAs. Here, we have explored the literature regarding the diversity of lncRNAs and grasped the thought that a main force for generation of numerous lncRNAs is divergent transcription of lncRNAs, especially initiation of the transcription. Taken together, divergent or less stringent initiation of transcription is essential for generating numerous lncRNAs. The transcription of lncRNAs is less efficient than that of coding genes and causes lower expression of lncRNAs.

These observations raise the question as to why such a kind of unstable and incomplete transcription has been developed for accommodating expression of lncRNAs. One possibility is that divergent transcription of lncRNAs was able to induce rapid evolution of lncRNAs and their rapid evolution should have some merit for organisms bearing lncRNAs. For instance, rapid evolution of lncRNAs might have played a role in originating modern human beings from other primates like chimpanzees (Beniaminov et al. 2008). It has been reported that there are distinctive sequences between chimpanzees and humans in brain noncoding RNA (Beniaminov et al. 2008). Human accelerated region 1 (HAR1) is a short DNA region identified to have evolved rapidly among highly conserved regions since divergence from our common ancestor with chimpanzees. It is transcribed as part of a lncRNA termed HAR1 RNA specifically expressed in the developing human neocortex. Analysis with enzymatic and chemical probes proposed fairly different structures of HAR1 RNA between humans and chimpanzees (Fig. 5.9). Intriguingly, the substitutions between the chimpanzee and human sequences led the human HAR1 RNA to embrace a cloverleaf-like structure instead of an extended and

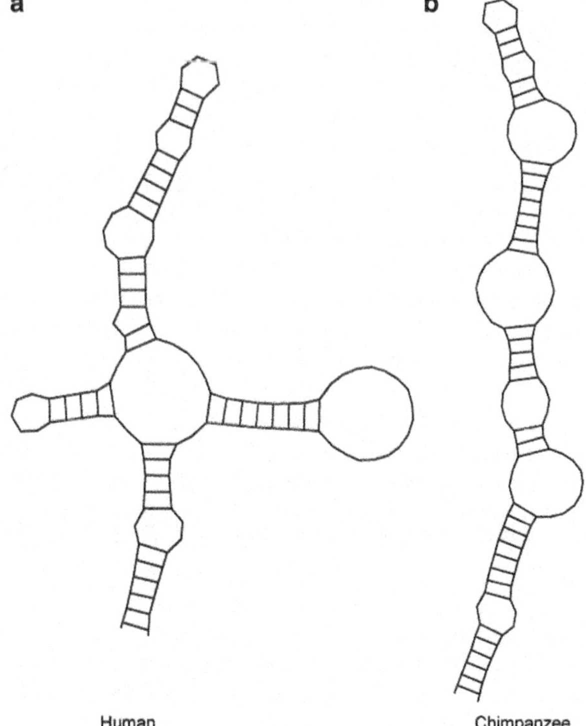

Fig. 5.9 Distinctive molecular shapes of HAR1 lncRNAs of humans and chimpanzees. (**a**) The cloverleaf-like model of the human. (**b**) The chimpanzee HAR1 RNA adopts a hairpin structure

unstable hairpin in the chimpanzee sequence (Fig. 5.9). Thus, the rapid evolution of HAR1 RNA resulted in a profound rearrangement of the HAR1 RNA structure and presumably modulates its function. More recently, developmental specific expression of lncRNAs has been reported in surgical samples of human cerebral cortex tissues (Lipovich et al. 2014). Microarray analysis of the surgically resected human neocortical samples from 36 patients (ages: 0.9–47 years olds) detected nearly 6,000 lncRNAs and identified eight lncRNA loci with distinct developmental expression patterns. These lncRNAs contained anthropoid-specific exons and splice sites and polyadenylation signals in primate-specific sequences. These observations provide solid evidence showing that these lncRNAs have potential biological functions in the human brain. Together, all of these data provide a hypothesis that rapid evolution of the brain-specific lncRNA might have prompted an ancestor to evolve into *Homo sapiens*. Therefore, evolution of lncRNAs might contribute to species formation of *Homo sapiens*. These data provide a firm motivation to drive investigation of lncRNAs over the human genome.

Acknowledgments The author thanks Ms. R. Tanji for her excellent assistance in preparing a set of figures and the manuscript. This study was supported by Grants-in-Aid for Scientific Research (B: no. 22390057; no. 25293073). This work was also supported in part by a Grant-in-Aid for "Support Project of Strategic Research Center in Private Universities" from the Ministry of Education, Culture, Sports, Science and Technology (MEXT) to Saitama Medical University Research Center for Genomic Medicine.

References

Almada AE, Wu X, Kriz AJ, Burge CB, Sharp PA (2013) Promoter directionality is controlled by U1 snRNP and polyadenylation signals. Nature 499(7458):360–363. doi:10.1038/nature12349

Beniaminov A, Westhof E, Krol A (2008) Distinctive structures between chimpanzee and human in a brain noncoding RNA. RNA 14(7):1270–1275. doi:10.1261/rna.1054608

Cabili MN, Trapnell C, Goff L, Koziol M, Tazon-Vega B, Regev A, Rinn JL (2011) Integrative annotation of human large intergenic noncoding RNAs reveals global properties and specific subclasses. Genes Dev 25(18):1915–1927. doi:10.1101/gad.17446611

Carninci P, Kasukawa T, Katayama S, Gough J, Frith MC, Maeda N, Oyama R, Ravasi T, Lenhard B, Wells C, Kodzius R, Shimokawa K, Bajic VB, Brenner SE, Batalov S, Forrest AR, Zavolan M, Davis MJ, Wilming LG, Aidinis V, Allen JE, Ambesi-Impiombato A, Apweiler R, Aturaliya RN, Bailey TL, Bansal M, Baxter L, Beisel KW, Bersano T, Bono H, Chalk AM, Chiu KP, Choudhary V, Christoffels A, Clutterbuck DR, Crowe ML, Dalla E, Dalrymple BP, de Bono B, Della Gatta G, di Bernardo D, Down T, Engstrom P, Fagiolini M, Faulkner G, Fletcher CF, Fukushima T, Furuno M, Futaki S, Gariboldi M, Georgii-Hemming P, Gingeras TR, Gojobori T, Green RE, Gustincich S, Harbers M, Hayashi Y, Hensch TK, Hirokawa N, Hill D, Huminiecki L, Iacono M, Ikeo K, Iwama A, Ishikawa T, Jakt M, Kanapin A, Katoh M, Kawasawa Y, Kelso J, Kitamura H, Kitano H, Kollias G, Krishnan SP, Kruger A, Kummerfeld SK, Kurochkin IV, Lareau LF, Lazarevic D, Lipovich L, Liu J, Liuni S, McWilliam S, Madan Babu M, Madera M, Marchionni L, Matsuda H, Matsuzawa S, Miki H, Mignone F, Miyake S, Morris K, Mottagui-Tabar S, Mulder N, Nakano N, Nakauchi H, Ng P, Nilsson R, Nishiguchi S, Nishikawa S, Nori F, Ohara O, Okazaki Y, Orlando V, Pang KC, Pavan WJ, Pavesi G, Pesole G, Petrovsky N,

Piazza S, Reed J, Reid JF, Ring BZ, Ringwald M, Rost B, Ruan Y, Salzberg SL, Sandelin A, Schneider C, Schonbach C, Sekiguchi K, Semple CA, Seno S, Sessa L, Sheng Y, Shibata Y, Shimada H, Shimada K, Silva D, Sinclair B, Sperling S, Stupka E, Sugiura K, Sultana R, Takenaka Y, Taki K, Tammoja K, Tan SL, Tang S, Taylor MS, Tegner J, Teichmann SA, Ueda HR, van Nimwegen E, Verardo R, Wei CL, Yagi K, Yamanishi H, Zabarovsky E, Zhu S, Zimmer A, Hide W, Bult C, Grimmond SM, Teasdale RD, Liu ET, Brusic V, Quackenbush J, Wahlestedt C, Mattick JS, Hume DA, Kai C, Sasaki D, Tomaru Y, Fukuda S, Kanamori-Katayama M, Suzuki M, Aoki J, Arakawa T, Iida J, Imamura K, Itoh M, Kato T, Kawaji H, Kawagashira N, Kawashima T, Kojima M, Kondo S, Konno H, Nakano K, Ninomiya N, Nishio T, Okada M, Plessy C, Shibata K, Shiraki T, Suzuki S, Tagami M, Waki K, Watahiki A, Okamura-Oho Y, Suzuki H, Kawai J, Hayashizaki Y (2005) The transcriptional landscape of the mammalian genome. Science 309(5740):1559–1563, doi:309/5740/1559 [pii] 10.1126/science.1112014

Carninci P, Sandelin A, Lenhard B, Katayama S, Shimokawa K, Ponjavic J, Semple CA, Taylor MS, Engstrom PG, Frith MC, Forrest AR, Alkema WB, Tan SL, Plessy C, Kodzius R, Ravasi T, Kasukawa T, Fukuda S, Kanamori-Katayama M, Kitazume Y, Kawaji H, Kai C, Nakamura M, Konno H, Nakano K, Mottagui-Tabar S, Arner P, Chesi A, Gustincich S, Persichetti F, Suzuki H, Grimmond SM, Wells CA, Orlando V, Wahlestedt C, Liu ET, Harbers M, Kawai J, Bajic VB, Hume DA, Hayashizaki Y (2006) Genome-wide analysis of mammalian promoter architecture and evolution. Nat Genet 38(6):626–635, doi:ng1789 [pii] 10.1038/ng1789

Cech TR, Steitz JA (2014) The noncoding RNA revolution-trashing old rules to forge new ones. Cell 157(1):77–94. doi:10.1016/j.cell.2014.03.008

Cianfrocco MA, Kassavetis GA, Grob P, Fang J, Juven-Gershon T, Kadonaga JT, Nogales E (2013) Human TFIID binds to core promoter DNA in a reorganized structural state. Cell 152(1–2):120–131. doi:10.1016/j.cell.2012.12.005

Core LJ, Waterfall JJ, Lis JT (2008) Nascent RNA sequencing reveals widespread pausing and divergent initiation at human promoters. Science 322(5909):1845–1848. doi:10.1126/science.1162228

Core LJ, Martins AL, Danko CG, Waters CT, Siepel A, Lis JT (2014) Analysis of nascent RNA identifies a unified architecture of initiation regions at mammalian promoters and enhancers. Nat Genet 46(12):1311–1320. doi:10.1038/ng.3142

De Santa F, Barozzi I, Mietton F, Ghisletti S, Polletti S, Tusi BK, Muller H, Ragoussis J, Wei CL, Natoli G (2010) A large fraction of extragenic RNA Pol II transcription sites overlap enhancers. PLoS Biol 8(5):e1000384. doi:10.1371/journal.pbio.1000384

Derrien T, Johnson R, Bussotti G, Tanzer A, Djebali S, Tilgner H, Guernec G, Martin D, Merkel A, Knowles DG, Lagarde J, Veeravalli L, Ruan X, Ruan Y, Lassmann T, Carninci P, Brown JB, Lipovich L, Gonzalez JM, Thomas M, Davis CA, Shiekhattar R, Gingeras TR, Hubbard TJ, Notredame C, Harrow J, Guigo R (2012) The GENCODE v7 catalog of human long noncoding RNAs: analysis of their gene structure, evolution, and expression. Genome Res 22(9):1775–1789. doi:10.1101/gr.132159.111

Djebali S, Davis CA, Merkel A, Dobin A, Lassmann T, Mortazavi A, Tanzer A, Lagarde J, Lin W, Schlesinger F, Xue C, Marinov GK, Khatun J, Williams BA, Zaleski C, Rozowsky J, Roder M, Kokocinski F, Abdelhamid RF, Alioto T, Antoshechkin I, Baer MT, Bar NS, Batut P, Bell K, Bell I, Chakrabortty S, Chen X, Chrast J, Curado J, Derrien T, Drenkow J, Dumais E, Dumais J, Duttagupta R, Falconnet E, Fastuca M, Fejes-Toth K, Ferreira P, Foissac S, Fullwood MJ, Gao H, Gonzalez D, Gordon A, Gunawardena H, Howald C, Jha S, Johnson R, Kapranov P, King B, Kingswood C, Luo OJ, Park E, Persaud K, Preall JB, Ribeca P, Risk B, Robyr D, Sammeth M, Schaffer L, See LH, Shahab A, Skancke J, Suzuki AM, Takahashi H, Tilgner H, Trout D, Walters N, Wang H, Wrobel J, Yu Y, Ruan X, Hayashizaki Y, Harrow J, Gerstein M, Hubbard T, Reymond A, Antonarakis SE, Hannon G, Giddings MC, Ruan Y, Wold B, Carninci P, Guigo R, Gingeras TR (2012) Landscape of transcription in human cells. Nature 489(7414):101–108. doi:10.1038/nature11233

Flynn RA, Almada AE, Zamudio JR, Sharp PA (2011) Antisense RNA polymerase II divergent transcripts are P-TEFb dependent and substrates for the RNA exosome. Proc Natl Acad Sci U S A 108(26):10460–10465. doi:10.1073/pnas.1106630108

Forrest AR, Kawaji H, Rehli M, Baillie JK, de Hoon MJ, Haberle V, Lassman T, Kulakovskiy IV, Lizio M, Itoh M, Andersson R, Mungall CJ, Meehan TF, Schmeier S, Bertin N, Jorgensen M, Dimont E, Arner E, Schmidl C, Schaefer U, Medvedeva YA, Plessy C, Vitezic M, Severin J, Semple C, Ishizu Y, Young RS, Francescatto M, Alam I, Albanese D, Altschuler GM, Arakawa T, Archer JA, Arner P, Babina M, Rennie S, Balwierz PJ, Beckhouse AG, Pradhan-Bhatt S, Blake JA, Blumenthal A, Bodega B, Bonetti A, Briggs J, Brombacher F, Burroughs AM, Califano A, Cannistraci CV, Carbajo D, Chen Y, Chierici M, Ciani Y, Clevers HC, Dalla E, Davis CA, Detmar M, Diehl AD, Dohi T, Drablos F, Edge AS, Edinger M, Ekwall K, Endoh M, Enomoto H, Fagiolini M, Fairbairn L, Fang H, Farach-Carson MC, Faulkner GJ, Favorov AV, Fisher ME, Frith MC, Fujita R, Fukuda S, Furlanello C, Furino M, Furusawa J, Geijtenbeek TB, Gibson AP, Gingeras T, Goldowitz D, Gough J, Guhl S, Guler R, Gustincich S, Ha TJ, Hamaguchi M, Hara M, Harbers M, Harshbarger J, Hasegawa A, Hasegawa Y, Hashimoto T, Herlyn M, Hitchens KJ, Ho Sui SJ, Hofmann OM, Hoof I, Hori F, Huminiecki L, Iida K, Ikawa T, Jankovic BR, Jia H, Joshi A, Jurman G, Kaczkowski B, Kai C, Kaida K, Kaiho A, Kajiyama K, Kanamori-Katayama M, Kasianov AS, Kasukawa T, Katayama S, Kato S, Kawaguchi S, Kawamoto H, Kawamura YI, Kawashima T, Kempfle JS, Kenna TJ, Kere J, Khachigian LM, Kitamura T, Klinken SP, Knox AJ, Kojima M, Kojima S, Kondo N, Koseki H, Koyasu S, Krampitz S, Kubosaki A, Kwon AT, Laros JF, Lee W, Lennartsson A, Li K, Lilje B, Lipovich L, Mackay-Sim A, Manabe R, Mar JC, Marchand B, Mathelier A, Mejhert N, Meynert A, Mizuno Y, de Lima Morais DA, Morikawa H, Morimoto M, Moro K, Motakis E, Motohashi H, Mummery CL, Murata M, Nagao-Sato S, Nakachi Y, Nakahara F, Nakamura T, Nakamura Y, Nakazato K, van Nimwegen E, Ninomiya N, Nishiyori H, Noma S, Noazaki T, Ogishima S, Ohkura N, Ohimiya H, Ohno H, Ohshima M, Okada-Hatakeyama M, Okazaki Y, Orlando V, Ovchinnikov DA, Pain A, Passier R, Patrikakis M, Persson H, Piazza S, Prendergast JG, Rackham OJ, Ramilowski JA, Rashid M, Ravasi T, Rizzu P, Roncador M, Roy S, Rye MB, Saijyo E, Sajantila A, Saka A, Sakaguchi S, Sakai M, Sato H, Savvi S, Saxena A, Schneider C, Schultes EA, Schulze-Tanzil GG, Schwegmann A, Sengstag T, Sheng G, Shimoji H, Shimoni Y, Shin JW, Simon C, Sugiyama D, Sugiyama T, Suzuki M, Suzuki N, Swoboda RK, 't Hoen PA, Tagami M, Takahashi N, Takai J, Tanaka H, Tatsukawa H, Tatum Z, Thompson M, Toyodo H, Toyoda T, Valen E, van de Wetering M, van den Berg LM, Verado R, Vijayan D, Vorontsov IE, Wasserman WW, Watanabe S, Wells CA, Winteringham LN, Wolvetang E, Wood EJ, Yamaguchi Y, Yamamoto M, Yoneda M, Yonekura Y, Yoshida S, Zabierowski SE, Zhang PG, Zhao X, Zucchelli S, Summers KM, Suzuki H, Daub CO, Kawai J, Heutink P, Hide W, Freeman TC, Lenhard B, Bajic VB, Taylor MS, Makeev VJ, Sandelin A, Hume DA, Carninci P, Hayashizaki Y (2014) A promoter-level mammalian expression atlas. Nature 507(7493):462–470. doi:10.1038/nature13182

Goodrich JA, Kugel JF (2006) Non-coding-RNA regulators of RNA polymerase II transcription. Nat Rev Mol Cell Biol 7(8):612–616. doi:10.1038/nrm1946

Goodrich JA, Tjian R (2010) Unexpected roles for core promoter recognition factors in cell-type-specific transcription and gene regulation. Nat Rev Genet 11(8):549–558. doi:10.1038/nrg2847

Guttman M, Rinn JL (2012) Modular regulatory principles of large non-coding RNAs. Nature 482(7385):339–346

Guttman M, Amit I, Garber M, French C, Lin MF, Feldser D, Huarte M, Zuk O, Carey BW, Cassady JP, Cabili MN, Jaenisch R, Mikkelsen TS, Jacks T, Hacohen N, Bernstein BE, Kellis M, Regev A, Rinn JL, Lander ES (2009) Chromatin signature reveals over a thousand highly conserved large non-coding RNAs in mammals. Nature 458(7235):223–227, doi:nature07672 [pii] 10.1038/nature07672

Guttman M, Donaghey J, Carey BW, Garber M, Grenier JK, Munson G, Young G, Lucas AB, Ach R, Bruhn L, Yang X, Amit I, Meissner A, Regev A, Rinn JL, Root DE, Lander ES (2011) lincRNAs act in the circuitry controlling pluripotency and differentiation. Nature 477(7364):295–300. doi:10.1038/nature10398

He Y, Fang J, Taatjes DJ, Nogales E (2013) Structural visualization of key steps in human transcription initiation. Nature 495(7442):481–486. doi:10.1038/nature11991

Illingworth RS, Gruenewald-Schneider U, Webb S, Kerr AR, James KD, Turner DJ, Smith C, Harrison DJ, Andrews R, Bird AP (2010) Orphan CpG islands identify numerous conserved promoters in the mammalian genome. PLoS Genet 6(9):e1001134. doi:10.1371/journal.pgen.1001134

Juven-Gershon T, Kadonaga JT (2010) Regulation of gene expression via the core promoter and the basal transcriptional machinery. Dev Biol 339(2):225–229. doi:10.1016/j.ydbio.2009.08.009

Juven-Gershon T, Cheng S, Kadonaga JT (2006) Rational design of a super core promoter that enhances gene expression. Nat Methods 3(11):917–922

Kadonaga JT (2012) Perspectives on the RNA polymerase II core promoter. Wiley Interdiscip Rev Dev biol 1(1):40–51. doi:10.1002/wdev.21

Kapranov P, Cheng J, Dike S, Nix DA, Duttagupta R, Willingham AT, Stadler PF, Hertel J, Hackermuller J, Hofacker IL, Bell I, Cheung E, Drenkow J, Dumais E, Patel S, Helt G, Ganesh M, Ghosh S, Piccolboni A, Sementchenko V, Tammana H, Gingeras TR (2007a) RNA maps reveal new RNA classes and a possible function for pervasive transcription. Science 316(5830):1484–1488, doi:1138341 [pii] 10.1126/science.1138341

Kapranov P, Willingham AT, Gingeras TR (2007b) Genome-wide transcription and the implications for genomic organization. Nat Rev Genet 8(6):413–423, doi:nrg2083 [pii] 10.1038/nrg2083

Khalil AM, Guttman M, Huarte M, Garber M, Raj A, Rivea Morales D, Thomas K, Presser A, Bernstein BE, van Oudenaarden A, Regev A, Lander ES, Rinn JL (2009) Many human large intergenic noncoding RNAs associate with chromatin-modifying complexes and affect gene expression. Proc Natl Acad Sci U S A 106(28):11667–11672, doi:0904715106 [pii] 10.1073/pnas.0904715106

Kim TK, Hemberg M, Gray JM, Costa AM, Bear DM, Wu J, Harmin DA, Laptewicz M, Barbara-Haley K, Kuersten S, Markenscoff-Papadimitriou E, Kuhl D, Bito H, Worley PF, Kreiman G, Greenberg ME (2010) Widespread transcription at neuronal activity-regulated enhancers. Nature 465(7295):182–187, doi:nature09033 [pii] 10.1038/nature09033

Kruesi WS, Core LJ, Waters CT, Lis JT, Meyer BJ (2013) Condensin controls recruitment of RNA polymerase II to achieve nematode X-chromosome dosage compensation. Elife 2:e00808. doi:10.7554/eLife.00808

Kurokawa R (2011) Long noncoding RNA as a regulator for transcription. Prog Mol Subcell Biol 51:29–41. doi:10.1007/978-3-642-16502-3_2

Lee TI, Young RA (2013) Transcriptional regulation and its misregulation in disease. Cell 152(6):1237–1251. doi:10.1016/j.cell.2013.02.014

Lipovich L, Tarca AL, Cai J, Jia H, Chugani HT, Sterner KN, Grossman LI, Uddin M, Hof PR, Sherwood CC, Kuzawa CW, Goodman M, Wildman DE (2014) Developmental changes in the transcriptome of human cerebral cortex tissue: long noncoding RNA transcripts. Cereb Cortex 24(6):1451–1459. doi:10.1093/cercor/bhs414

Malik S, Roeder RG (2010) The metazoan Mediator co-activator complex as an integrative hub for transcriptional regulation. Nat Rev Genet 11(11):761–772. doi:10.1038/nrg2901

Marques AC, Ponting CP (2009) Catalogues of mammalian long noncoding RNAs: modest conservation and incompleteness. Genome Biol 10(11):R124, doi:gb-2009-10-11-r124 [pii] 10.1186/gb-2009-10-11-r124

Martianov I, Ramadass A, Serra Barros A, Chow N, Akoulitchev A (2007) Repression of the human dihydrofolate reductase gene by a non-coding interfering transcript. Nature 445(7128):666–670, doi:nature05519 [pii] 10.1038/nature05519

Maunakea AK, Nagarajan RP, Bilenky M, Ballinger TJ, D'Souza C, Fouse SD, Johnson BE, Hong C, Nielsen C, Zhao Y, Turecki G, Delaney A, Varhol R, Thiessen N, Shchors K, Heine VM,

Rowitch DH, Xing X, Fiore C, Schillebeeckx M, Jones SJ, Haussler D, Marra MA, Hirst M, Wang T, Costello JF (2010) Conserved role of intragenic DNA methylation in regulating alternative promoters. Nature 466(7303):253–257. doi:10.1038/nature09165

Murakami K, Elmlund H, Kalisman N, Bushnell DA, Adams CM, Azubel M, Elmlund D, Levi-Kalisman Y, Liu X, Gibbons BJ, Levitt M, Kornberg RD (2013) Architecture of an RNA polymerase II transcription pre-initiation complex. Science 342(6159):1238724. doi:10.1126/science.1238724

Natoli G, Andrau JC (2012) Noncoding transcription at enhancers: general principles and functional models. Annu Rev Genet 46:1–19. doi:10.1146/annurev-genet-110711-155459

Necsulea A, Soumillon M, Warnefors M, Liechti A, Daish T, Zeller U, Baker JC, Grutzner F, Kaessmann H (2014) The evolution of lncRNA repertoires and expression patterns in tetrapods. Nature 505(7485):635–640. doi:10.1038/nature12943

Ntini E, Jarvelin AI, Bornholdt J, Chen Y, Boyd M, Jorgensen M, Andersson R, Hoof I, Schein A, Andersen PR, Andersen PK, Preker P, Valen E, Zhao X, Pelechano V, Steinmetz LM, Sandelin A, Jensen TH (2013) Polyadenylation site-induced decay of upstream transcripts enforces promoter directionality. Nat Struct Mol Biol 20(8):923–928. doi:10.1038/nsmb.2640

Orom UA, Derrien T, Beringer M, Gumireddy K, Gardini A, Bussotti G, Lai F, Zytnicki M, Notredame C, Huang Q, Guigo R, Shiekhattar R (2010) Long noncoding RNAs with enhancer-like function in human cells. Cell 143(1):46–58, doi:S0092-8674(10)01011-1 [pii] 10.1016/j.cell.2010.09.001

Ponting CP, Oliver PL, Reik W (2009) Evolution and functions of long noncoding RNAs. Cell 136(4):629–641, doi:S0092-8674(09)00142-1 [pii] 10.1016/j.cell.2009.02.006

Preker P, Nielsen J, Kammler S, Lykke-Andersen S, Christensen MS, Mapendano CK, Schierup MH, Jensen TH (2008) RNA exosome depletion reveals transcription upstream of active human promoters. Science 322(5909):1851–1854, doi:1164096 [pii] 10.1126/science.1164096

Ravasi T, Suzuki H, Pang KC, Katayama S, Furuno M, Okunishi R, Fukuda S, Ru K, Frith MC, Gongora MM, Grimmond SM, Hume DA, Hayashizaki Y, Mattick JS (2006) Experimental validation of the regulated expression of large numbers of non-coding RNAs from the mouse genome. Genome Res 16(1):11–19. doi:10.1101/gr.4200206

Reinberg D, Roeder RG (1987) Factors involved in specific transcription by mammalian RNA polymerase II. Purification and functional analysis of initiation factors IIB and IIE. J Biol Chem 262(7):3310–3321

Roeder RG (2003) Lasker basic medical research award. The eukaryotic transcriptional machinery: complexities and mechanisms unforeseen. Nat Med 9(10):1239–1244. doi:10.1038/nm938

Sandelin A, Carninci P, Lenhard B, Ponjavic J, Hayashizaki Y, Hume DA (2007) Mammalian RNA polymerase II core promoters: insights from genome-wide studies. Nat Rev Genet 8(6):424–436. doi:10.1038/nrg2026

Sawadogo M, Roeder RG (1985) Interaction of a gene-specific transcription factor with the adenovirus major late promoter upstream of the TATA box region. Cell 43(1):165–175

Seila AC, Calabrese JM, Levine SS, Yeo GW, Rahl PB, Flynn RA, Young RA, Sharp PA (2008) Divergent transcription from active promoters. Science 322(5909):1849–1851, doi:1162253 [pii] 10.1126/science.1162253

Shenkin A, Burdon RH (1966) Asymmetric transcription of deoxyribonucleic acid by deoxyribonucleic acid-dependent ribonucleic acid polymerase of Krebs II ascites-tumour cells. Biochem J 98(1):5C–7C

Siepel A, Bejerano G, Pedersen JS, Hinrichs AS, Hou M, Rosenbloom K, Clawson H, Spieth J, Hillier LW, Richards S, Weinstock GM, Wilson RK, Gibbs RA, Kent WJ, Miller W, Haussler D (2005) Evolutionarily conserved elements in vertebrate, insect, worm, and yeast genomes. Genome Res 15(8):1034–1050, doi:gr.3715005 [pii] 10.1101/gr.3715005

Sigova AA, Mullen AC, Molinie B, Gupta S, Orlando DA, Guenther MG, Almada AE, Lin C, Sharp PA, Giallourakis CC, Young RA (2013) Divergent transcription of long noncoding RNA/mRNA gene pairs in embryonic stem cells. Proc Natl Acad Sci U S A 110(8):2876–2881. doi:10.1073/pnas.1221904110

Sleutels F, Zwart R, Barlow DP (2002) The non-coding air RNA is required for silencing autosomal imprinted genes. Nature 415(6873):810–813. doi:10.1038/415810a

Smale ST, Kadonaga JT (2003) The RNA polymerase II core promoter. Annu Rev Biochem 72:449–479. doi:10.1146/annurev.biochem.72.121801.161520

Thomas MC, Chiang CM (2006) The general transcription machinery and general cofactors. Crit Rev Biochem Mol Biol 41(3):105–178. doi:10.1080/10409230600648736

Tuerk C, Gold L (1990) Systematic evolution of ligands by exponential enrichment: RNA ligands to bacteriophage T4 DNA polymerase. Science 249(4968):505–510

Van Dyke MW, Roeder RG, Sawadogo M (1988) Physical analysis of transcription preinitiation complex assembly on a class II gene promoter. Science 241(4871):1335–1338

Wang X, Arai S, Song X, Reichart D, Du K, Pascual G, Tempst P, Rosenfeld MG, Glass CK, Kurokawa R (2008) Induced ncRNAs allosterically modify RNA-binding proteins in cis to inhibit transcription. Nature 454(7200):126–130, doi:nature06992 [pii] 10.1038/nature06992

Wu X, Sharp PA (2013) Divergent transcription: a driving force for new gene origination? Cell 155(5):990–996. doi:10.1016/j.cell.2013.10.048

Chapter 6
Beneath the Veil of Biological Complexity There Lies Long Noncoding RNA: Diverse Utilization of lncRNA in Yeast Genomes

Tomohiro Kumon and Kunihiro Ohta

Abstract Biological complexity may be partly attributable to the diversity of the noncoding genome, where a plethora of noncoding RNAs (ncRNAs) are actively transcribed. Although yeasts have relatively small portion of noncoding genome, they produces some functional ncRNAs. In this chapter, we overview recent studies on the function of yeast long ncRNAs in epigenetic regulation, cell-cycle control, centromere/telomere functions, and responses to environmental stimuli.

Keywords CUTs • SUTs • XUTs • Exosome • Epigenetics • Heterochromatin • Centromere • RNAi • Telomere • TERRA • mlonRNA

6.1 Introduction

What defines biological complexity has been a perennial question. Neither the chromosome size nor the number of genes appeared directly related to the intuitive perception of biological complexity (Claverie 2001). It was consistent with the notion that biological sophistication evolved through the development of elaborate gene regulation mechanisms, rather than a sheer increase in the number of protein-coding genes (Claverie 2001). On the other hand, the ratio of noncoding to protein-coding DNA rises as the biological complexity increases. Prokaryotes have less than 25 % noncoding DNA, simple eukaryotes (e.g. yeasts) 25–50 %, and higher eukaryotes more than 50 %, reaching approximately 98.5 % in humans (Mattick 2004). This suggests an intriguing possibility that transcripts from noncoding DNA increase the repertoires of gene regulation units, thereby providing precise control of complex gene network of eukaryotes.

Among such transcripts, *l*ong *n*on*c*oding RNAs (lncRNAs) biochemically resemble mRNAs and are arbitrarily defined as RNAs longer than 200 bp that do not appear to have coding potential (Rinn and Chang 2012). Both mRNAs and lncRNAs

T. Kumon • K. Ohta (✉)
Department of Life Sciences, Graduate School of Arts and Sciences, The University of Tokyo, Komaba 3-8-1, Meguro-ku, Tokyo 153-8902, Japan
e-mail: kohta@bio.c.u-tokyo.ac.jp

R. Kurokawa (ed.), *Long Noncoding RNAs*, DOI 10.1007/978-4-431-55576-6_6

possess 5′-methylguanosine caps (Neil et al. 2009) and are transcribed by RNA polymerase II (Pol II) (Rhee and Pugh 2012). Recent genome-wide transcriptome analyses revealed the widespread presence of lncRNAs from yeasts to humans. However, the functional importance of lncRNA remains poorly understood, partly because of its broad and arbitral definition that hinders distinct categorization with respect to the functions.

In this chapter, we introduce the overview of lncRNA transcription in budding yeast, *S. cerevisiae*, and fission yeast, *S. pombe*, followed by the epigenetic regulation of mRNA and lncRNA transcription. Co-transcriptional regulation of histone modifications has significant effects on transcriptional transition between mRNA and lncRNA transcription. In addition to their roles in gene regulation, lncRNAs are important regulators of chromosome integrity, and we describe centromeric and telomeric transcripts, of which functions are conceptually conserved from yeast to human. Finally, we illustrate how the cells utilize lncRNA in order to respond adequately to particular environmental stimuli, which might give an insight into the functional importance of lncRNAs in complex gene regulatory networks. Such diverse usage of lncRNA suggests that lncRNA has been an integrated component of gene regulatory networks throughout the evolution of eukaryotes, and that the increased proportion of lncRNA in higher eukaryotic genomes underlies their biological complexity.

6.2 lncRNAs of Budding and Fission Yeasts

Several studies have identified different types of lncRNAs in budding yeast. Comprehensive identification of lncRNAs in budding yeast originated from analyses of mutants in RNA degradation pathways. The absence of Rrp6p, a subunit of the nuclear-specific RNA exosome complex of an RNA degradation pathway, unveiled the "hidden transcription" that generated *c*ryptic *u*nstable *t*ranscripts (CUTs) (Davis and Ares 2006; Wyers et al. 2005). Subsequent genome-wide identification of CUTs also revealed *s*table *u*nannotated *t*ranscripts (SUTs), which were less sensitive to Rrp6p activity (Neil et al. 2009; Xu et al. 2009). The absence of Xrn1p, a cytoplasmic exoribonuclease, further unveiled *X*rn1-sensitive *u*nstable *t*ranscripts (XUTs) (van Dijk et al. 2011). As for mRNAs, RNA Pol II transcribes CUTs, SUTs, and XUTs (Jensen et al. 2013). Although XUTs were reported to be exported and degraded in the cytoplasm, the distinct subcellular localization of XUTs is somewhat controversial, for some XUTs seem to remain in the nucleus to be degraded in a nuclear RNA degradation pathway (Tuck and Tollervey 2013).

CUTs and SUTs are often generated from divergent transcription from bidirectional gene promoters (Fig. 6.1). Because of the high gene density in budding yeast, approximately 50 % of 5′ *n*ucleosome-*d*epleted *r*egions (NDRs) of genes, which are potential promoters, share 3′ regions of the upstream genes. Divergent transcription away from such promoters results in antisense transcription of the upstream gene (Jensen et al. 2013). In addition, antisense transcripts sometimes appear from 3′ NDRs, even when they do not accompany divergent promoters (Murray et al. 2012).

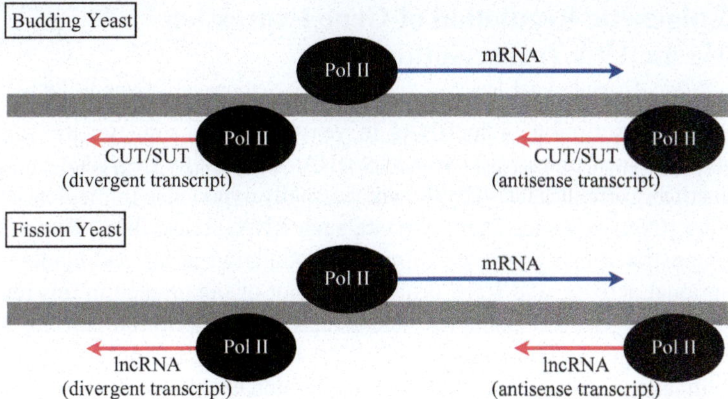

Fig. 6.1 Divergent and antisense transcription of lncRNAs in budding and fission yeasts. In budding yeast, CUTs and SUTs can be transcribed from divergent or downstream promoters in an antisense direction with respect to sense mRNA transcription. In fission yeast, lncRNAs that are similar to CUTs and SUTs in budding yeast are transcribed. In both budding and fission yeasts, mRNAs and lncRNAs are transcribed from RNA Pol II

Genome-wide studies have also identified lncRNAs in fission yeast (Fig. 6.1). These lncRNAs are reminiscent of SUTs, CUTs, and XUTs in budding yeast (Rhind et al. 2011; Wilhelm et al. 2008). Out of 1,557 lncRNAs in fission yeast, more than 80 % (1,346/1,557) are expressed in proliferating cells. Most of the expressed lncRNAs are polyadenylated and expressed with around one copy per cell. However, there is a small number of lncRNAs (38/1,557) that are nonpolyadenylated and expressed with 1–200 copies/cell. Indeed, these nonpolyadenylated lncRNAs occupy the majority of lncRNA molecules in the cell (Marguerat et al. 2012). Of note, unlike the budding yeast genome, the fission yeast genome contains large heterochromatic regions. Constitutive heterochromatin domains at centromeres, subtelomeres, and mating-type loci share *dg* and *dh* repeats, which are transcribed to silence the regions through RNAi machineries (Bühler and Moazed 2007; Grewal and Jia 2007). Other genomic regions such as meiotic gene loci also form facultative heterochromatin, depending on RNA decay systems (Zofall et al. 2012).

Most mRNAs are exported to the cytoplasm, where they are translated into proteins. In contrast, lncRNAs are predominantly localized to the nucleus, modifying epigenetic marks on chromatin and thereby regulating mRNA transcription. SUTs are more stable than CUTs that are rapidly degraded by nuclear RNA exosome. In addition, 34.4 % of SUTs are transported into the cytoplasm, compared with just 6 % of CUTs, suggesting the overlaps between SUTs and mRNAs. Indeed, SUTs and mRNAs possess stable poly(A) tails due to the presence of common sequence elements, and SUTs undergo cleavage and polyadenylation like mRNAs (Tuck and Tollervey 2013). Such "mRNA-like" lncRNAs might represent functional transcripts exported to the cytoplasm. In contrast, there are "lncRNA-like" mRNAs that are retained and degraded in the nucleus. As described later, such RNAs might represent a class of lncRNAs that possess ORF of overlapping genes but are not translated into proteins.

6.3 Epigenetic Regulation of Gene Expression via lncRNA Transcription

Widespread transcription of lncRNAs in yeast genomes suggests the functional importance of such transcripts. Some lncRNAs are rapidly degraded upon being transcribed, as exemplified by CUTs, which usually experience immediate degradation by the RNA exosome. Then, why are such RNAs transcribed? Recent studies have demonstrated that a co-transcriptional modulation of histone marks regulate transcriptional activity, and RNA products are not of importance in this regard. In this section, we will see examples where lncRNA transcription recruits histone-modifying enzymes to regulate nearby mRNA transcription, implying that epigenetic modification deposited by lncRNA transcription enables complex regulation of eukaryotic gene expression.

6.3.1 Co-transcriptional Regulation of Histone Modifications

Transcription of lncRNA can alter the expression of nearby genes by co-transcriptional regulation of histone modifications. As RNA Pol II-dependent transcription —both mRNAs and lncRNAs— recruits several histone-modifying enzymes, lncRNA transcription can affect the epigenetic landscape of chromatin, thereby regulating the transcriptional activity of nearby genes (Fig. 6.2). RNA Pol II-dependent transcription regulates histone methylation and acetylation. In brief, acetylated histones are the hallmark of actively transcribed regions, and the *h*istone *a*cetyl*t*ransferase (HAT) complex SAGA mediates increased H3 acetylation throughout the transcription unit (Govind et al. 2007). Two *h*istone *deac*etylase (HDAC) complexes, Set3 and Rpd3S, are responsible for repression of cryptic transcription from the 5′ end and 3′ end of actively transcribed regions, respectively.

In budding yeast and other eukaryotes, two *h*istone *m*ethyl*t*ransferases (HMTs) mark distinct regions of a transcription unit: Set1-mediated H3K4me3 is enriched near *t*ranscription *s*tart *s*ites (TSSs), whereas Set2-mediated H3K36 marks 3′ transcribed regions (Thornton et al. 2014; Venkatesh et al. 2012). The Rpd3S HDAC complex is co-transcriptionally recruited to 3′ transcribed regions marked with H3K36me3 (Govind et al. 2010). In between H3K4me3 and H3K36me3, that is, 5′ transcribed regions, H3K4me2 provides a binding site for the Set3 HDAC complex (Kim and Buratowski 2009; Kirmizis et al. 2007; Pokholok et al. 2005). Such a histone modification pattern is summarized in Fig. 6.2.

In budding yeast, more than 60 % of genes repressed by Set3 have overlapping lncRNAs (Kim et al. 2012). For example, *DCI1*, *FUN19*, *ATH1*, and *DUR3* are conditionally expressed genes, and they have at least two distinct transcription units (Fig. 6.3a). The distal promoter constitutively generates lncRNAs that contain ORFs that are not translated to proteins, whereas the proximal promoters are activated in specific growth conditions (Kim et al. 2012). Since depletion of *Set3* increases transcription from the proximal but not distal promoters, the upstream lncRNA

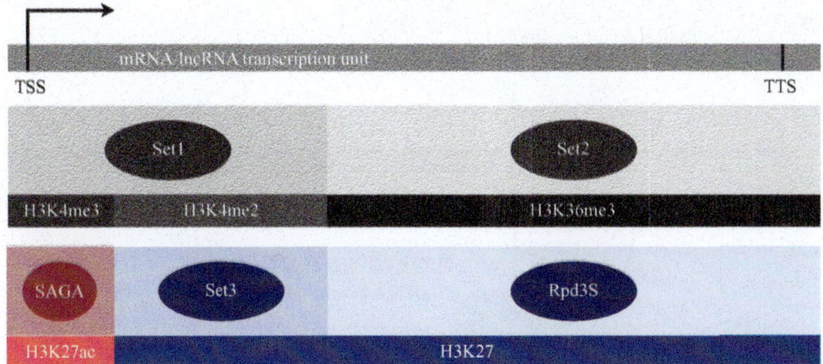

Fig. 6.2 Histone methylation and acetylation in a transcription unit of budding yeast. Set1 methylates H3K4, and Set2 H3K36. HAT complex SAGA deposits acetylation throughout the transcription unit. However, H3K4me2 and H3K36me3 recruit Set3 and Rpd3S HDAC complexes, respectively. *TSS* transcription start site, *TTS* transcription termination site

transcription units repress mRNA transcription from the downstream promoter by targeting Set3 complexes to the proximal TSSs. Transcription units that generate lncRNAs can also be in the antisense direction relative to the mRNAs (Fig. 6.3b). As an example, the antisense transcript *SUT103* overlaps the *EPL1* promoter region, and Set3 depletion leads to increased *EPL1* mRNA production (Kim et al. 2012). In both cases, lncRNA transcription recruits Set3 HDAC to promoters of nearby genes to repress transcription. Set3 depletion leads to increased transcription of mRNA but has little effect on lncRNA transcription because lncRNA transcription no longer recruits Set3 HDAC to repress transcription of mRNA. Thus, the Set3 HDAC complex represses cryptic mRNA transcription from 5′ transcribed regions of both sense and antisense lncRNAs.

6.3.2 *Transcriptional Interplay Between mRNAs and lncRNAs*

Co-transcriptional recruitment of histone modifiers partly explains why lncRNAs should be transcribed regardless of their rapid decay. Transcription of lncRNAs, but not the RNAs themselves, seems necessary to regulate conditional gene expression via co-transcriptional modulation of histone marks (Venkatesh and Workman 2013). Sporulation provides an example where lncRNA transcription modulates histone modifications that enable expression of the master transcription factor only in a specific mating genotype.

Diploid budding yeast usually proliferates through vegetative cell division. However, upon nitrogen starvation, yeast cells undergo meiosis to produce four haploid gametes called spores for sexual reproduction —the process referred to as sporulation (Honigberg and Purnapatre 2003). Nitrogen starvation causes the cells to arrest in G1 phase. For this process, fermentable sugars must be absent, whereas a non-fermentable carbon source is in turn metabolized through respiration. Such

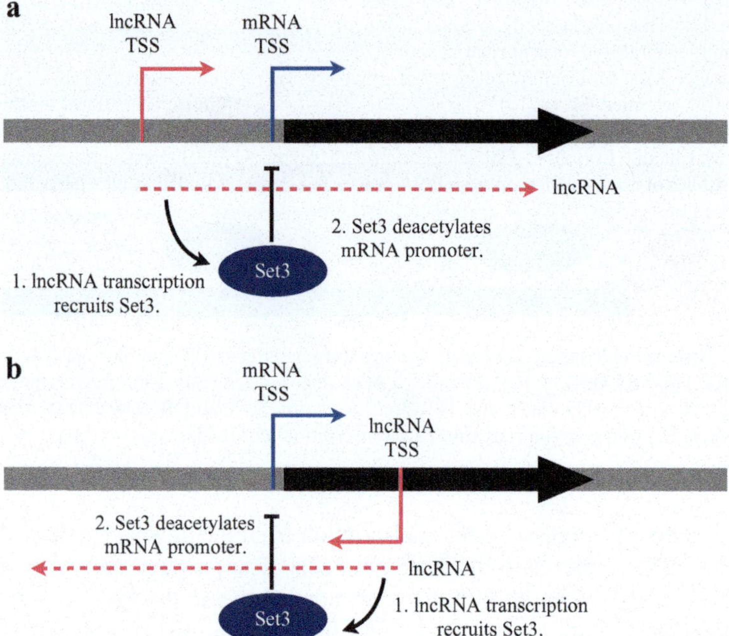

Fig. 6.3 Transcription of lncRNA recruits Set3 HDAC complex to repress transcription of mRNA. (**a**) lncRNA transcription from an upstream promoter recruits Set3 to deacetylate histones in a downstream promoter of mRNA. (**b**) Likewise, antisense lncRNA transcription from a downstream promoter recruits Set3 to an upstream promoter of mRNA. Thus, in these examples, lncRNA transcription represses cryptic transcription of mRNA. *Black* regions represent a coding region

environmental cues converge on the promoter of *IME1*, the master transcription factor of gametogenesis that initiates the sporulation program (Kassir et al. 1988; van Werven and Amon 2011).

Sporulation only occurs in *MAT***a**/α diploid cells because *IME1* expression is inhibited in *MAT***a** and *MAT*α haploid cells as well as *MAT***a**/**a** and *MAT*α/α diploid cells (Fig. 6.4). The transcription factor Rme1 binds to the *IME1* promoter and inhibits *IME1* expression (Covitz and Mitchell 1993). In *MAT***a**/α diploid cells, *RME1* is not expressed, because *MAT***a** encodes **a**1 and *MAT*α encodes α2, which together form the **a**1-α2 repressor complex that inhibits *RME1* expression (Covitz et al. 1991).

Genome-wide study revealed the noncoding transcript, named *IRT1*, in the *IME1* promoter (Xu et al. 2009; van Werven et al. 2012). In *MAT***a**/α diploid cells, *IME1* mRNA is expressed only under sporulation-induced conditions, with increased transcription during early stages of sporulation and a decreased level thereafter (van Werven et al. 2012). Interestingly, *IME1* mRNA is transiently expressed upon starvation even in *MAT***a** and *MAT*α haploid cells. However, concomitantly with *IRT1* lncRNA expression, *IME1* mRNA is repressed in these cells. The transcription

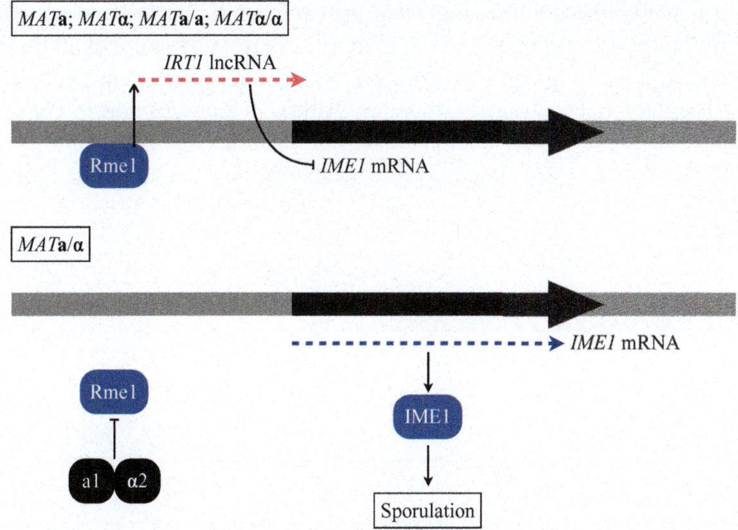

Fig. 6.4 *IRT1* lncRNA transcription inhibits *IME1* mRNA transcription. *IRT1* lncRNA is transcribed from an upstream promoter, and *IME1* mRNA transcription is repressed by the mechanism described in the previous section. In the presence of the a1-α2 repressor complex, *IRT1* lncRNA is not transcribed, and *IME1* mRNA is expressed to initiate the sporulation. *Black* regions represent a coding region of *IME1*

factor Rme1 activates *IRT1* lncRNA transcription, thereby inhibiting the *IME1* mRNA expression in *MAT***a** and *MAT*α haploid cells (Fig. 6.4).

Histone methylation-dependent HDAC activation is responsible for repression of *IME1* mRNA by *IRT1* lncRNA transcription. Indeed, after 6 h of sporulation induction, when *IRT1* lncRNA is expressed in haploid cells, both H3K4me2 and H3K36me3 are enriched in the *IME1* mRNA promoter. Deletion of *SET2* (HMT that deposits H3K36me3) and *SET3* (HDAC that is recruited by H3K4me2) allows *MAT* homozygous diploid cells to sporulate and induce haploid meiosis. Transcription of *IRT1* lncRNA deposits H3K4me2 and H3K36me3 on the *IME1* mRNA promoter and recruits Set3 and Rpd3S HDAC complexes, leading to repression of *IME1* mRNA expression. Thus, regulation of sporulation by *IRT1* lncRNA transcription exemplifies the co-transcriptional regulation of histone modifications and demonstrates the importance of lncRNA transcription rather than RNA products themselves.

6.4 Noncoding Transcripts Required for Chromosome Integrity

In a previous section, we saw an example in which lncRNA transcription recruits histone modifiers so that the overlapping gene is repressed until environmental conditions meet the requirements for gene activation. In addition, there are groups of lncRNAs that maintain chromosome integrity: centromeric and telomeric transcripts. Transcription from centromere and telomere is a conserved feature of all lineages of eukaryotes, and budding and fission yeasts utilize centromeric and telomeric transcripts to regulate chromosome structures. Although transcription in centromeres and telomeres is a conserved feature, the way of the transcript usage varies from one organism to another, suggesting divergent mechanisms in maintaining chromosome integrity.

6.4.1 Centromeric Transcripts

In higher eukaryotes, centromeres are epigenetically defined (Allshire and Karpen 2008), and centromeric transcripts contribute to the centromere integrity. In fission yeast, like higher eukaryotes, centromeric regions are associated with repetitive sequences, which are enriched with H3K9 methylation for heterochromatin formation (Hall et al. 2002; Volpe et al. 2002). In this organism, centromeric heterochromatin is established by both RNAi-dependent and -independent pathways (reviewed in Reyes-Turcu and Grewal 2012). The heterochromatic domains (outer repeats, *otr*) flank the central kinetochore domain (innermost repeat, *imr*; central core, *cnt*/*cc*). In the central core, Cnp1$^{\text{CENP-A}}$ partly replaces canonical histone H3 to form a platform for a kinetochore assembly (Folco et al. 2008). Both heterochromatin and central kinetochore domains are transcribed in an RNA Pol II-dependent manner (Fig. 6.5).

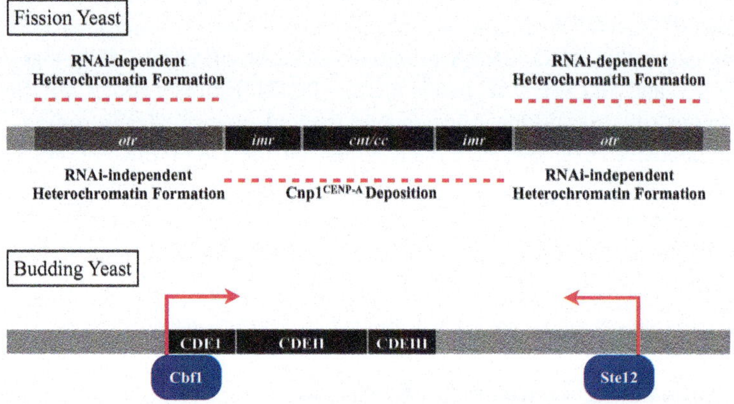

Fig. 6.5 Centromeric transcripts of budding and fission yeasts. In fission yeast, heterochromatin (*otr*) and central kinetochore (*imr-cnt*/*cc-imr*) domains produce centromeric transcripts, which contribute to RNAi-dependent heterochromatin formation and Cnp1$^{\text{CENP-A}}$ deposition, respectively. In addition, there is an RNAi-independent pathway of heterochromatin formation. In budding yeast, two transcription factors induce transcription from both centromeric and pericentromeric regions

A large proportion of the central kinetochore domain is transcribed by RNA Pol II but immediately degraded by the exosome. In fission yeast, transcription from the promoter within the centromeric domain may promote replacement of canonical H3 with $Cnp1^{CENP\text{-}A}$ by the activity of the associated chromatin remodeler $Hrp1^{Chd1}$ (Choi et al. 2011) and histone chaperones Mis16 and Mis18 (Hayashi et al. 2004, 2014). The other centromeric transcripts are generated from *otr*, and they recruit RNAi machineries to induce heterochromatin formation in subcentromeric regions. However, such transcripts are not produced until S phase, and antisense RNA of *otr* is transcribed throughout the cell cycle, even in the presence of heterochromatin (Fig. 6.6a) (Castel and Martienssen 2013; Chen et al. 2008).

When the cell enters S phase, double-stranded RNAs, which are composed of constitutive antisense RNA and S-phase-specific sense RNA, are processed by Dicer, together with the RITS (RNA-induced transcriptional gene silencing) complex (Fig. 6.6b). RdRP (RNA-dependent RNA polymerase) interacts with Dicer and RITS complex to amplify siRNAs (Fig. 6.6c) that target nascent sense-direction RNA. RNAi inhibits transcription possibly by releasing RNA Pol II (Fig. 6.6d), and

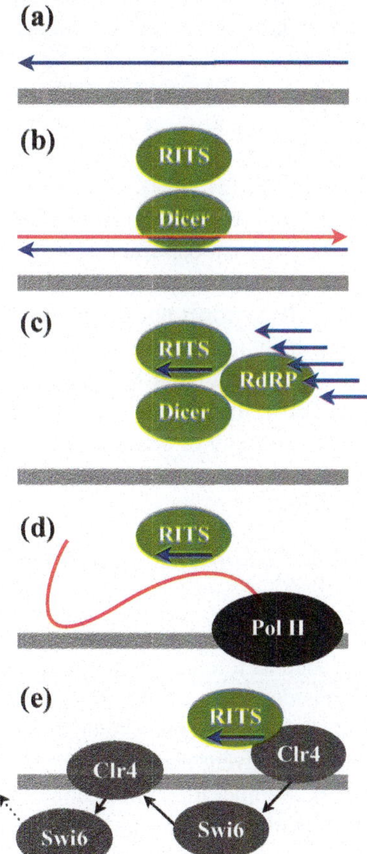

Fig. 6.6 RNAi-dependent pathway for heterochromatin formation in fission yeast. (**a**) Antisense transcripts are present throughout the cell cycle. (**b**) When the cells enter S phase, sense RNAs of otr are transcribed to form double-stranded RNAs, which are processed by Dicer. (**c**) RdRP complexes interact with Dicer and RITS to amplify siRNAs. (**d**) Processed siRNAs are loaded onto RITS, and the complex is guided to nascent RNA transcripts. (**e**) RITS recruits the CLRC, which contain $Clr4^{Suv39}$ histone methyltransferase. Methylation on H3K9 recruits $Swi6^{HP1}$, which further recruits $Clr4^{Suv39}$

RITS complexes facilitate the formation of heterochromatin. RITS complexes recruit the CLRC (cryptic loci regulator complex), which contains Clr4^{Suv39} histone methyltransferase that deposits methylation on H3K9. Once the repressive histone mark is deposited, heterochromatin can be propagated by the sequential recruitments of Swi6^{HP1} and Clr4^{Suv39} (Fig. 6.6e) (Castel and Martienssen 2013; Grewal and Jia 2007).

There is an RNAi-independent pathway for heterochromatin formation in fission yeast (Fig. 6.5). In addition to Swi6^{HP1}, H3K9 methylation can recruit yet another HP1 protein, Chp2 (Fischer et al. 2009). Both Swi6^{HP1} and Chp2 recruit SHREC, a chromatin-modifying complex that is related to mammalian NuRD. Chp2 and SHREC can form the SHREC2 complex. These complexes facilitate pericentromeric heterochromatin assembly, independently of RNAi (Motamedi et al. 2008; Sugiyama et al. 2007; Yamada et al. 2005). Interestingly, pericentromeric transcripts are also recognized by Seb1, which recruits SHREC to assist heterochromatin formation (Marina et al. 2013).

Although centromeres of budding yeast harbor three short conserved centromere DNA elements—CDEI, CDEII, and CDEIII—centromeric transcripts per se are required for proper centromere function (Ohkuni and Kitagawa 2011, 2012). Two transcription factors, Cbf1 and Ste12, contribute to the RNA Pol II-dependent production of centromeric transcripts (Fig. 6.5). Cbf1 binds to a palindromic consensus sequence at CDEI, and Ste12 binds to a consensus sequence located in the pericentromeric region. A certain amount of centromeric transcripts is required for centromere function, albeit that the exact mechanism remains elusive (Ohkuni and Kitagawa 2011, 2012).

In budding yeast, centromeric transcripts are necessary for centromere function. In fission yeast, centromeric transcripts are required for Cnp1$^{CENP\text{-}A}$ deposition and heterochromatin formation by both RNAi-dependent and -independent mechanisms. It is worth noting, however, that it is unclear to what extent the effect of RNAi-mediated pericentromeric heterochromatin formation on chromosome segregation is evolutionarily conserved beyond fission yeast (Gent and Dawe 2012). Altogether, the ways of centromeric RNA utilization are diverse, implicating mechanistic diversity of centromeric transcripts in eukaryotes.

6.4.2 *Telomeric Transcripts*

The linear nature of eukaryotic chromosomes is supposed to threaten genome integrity such as telomere erosion and fusion. Chromosome extremities, telomeres, can be misidentified as double strand breaks to induce DNA damage response (DDR) and chromosome rearrangements. Chromatin organization and binding of shelterin proteins prevent such DDR activation and rearrangements (reviewed in Ye et al. 2014). Another threat to genome integrity is the inability of the replication machinery to fully replicate the end of chromosomes (the telomere end replication problem). Thus, telomeres progressively shorten unless otherwise telomerase extends the 3′ ends of chromosomes (Ye et al. 2014).

Shelterin protein assembly at the telomere induces compaction and heterochromatin formation in budding and fission yeasts, respectively (Fig. 6.7). In budding yeast, Rap1 binds to double-stranded telomeric repeats and recruits Rif1 and Rif2 (Rap1-interacting factors) and Sir2-3-4 complexes. Compaction starts from telomeres and extends to subtelomeric regions. Rap1 recruits Sir proteins at telomeres, and similar to the case of sequential recruitment of Swi6^{HP1} and Clr4^{Suv39}, the self-organization cycle of Sir2-3-4 complex recruitment and deacetylation of histones allows compaction to spread along subtelomeric regions (Kueng et al. 2013). In fission yeast, Taz1 binds to double-stranded telomeric repeats and further recruits homologues of budding yeast Rap1 and Rif1 proteins (Blasco 2007). Heterochromatin spreading over subtelomeric regions starts from the telomere, just like Sir2-3-4-mediated compaction of budding yeast.

Despite their compacted state in budding yeast and the presence of heterochromatin in fission yeast, telomeres are transcribed into TERRA (telomeric repeat-containing

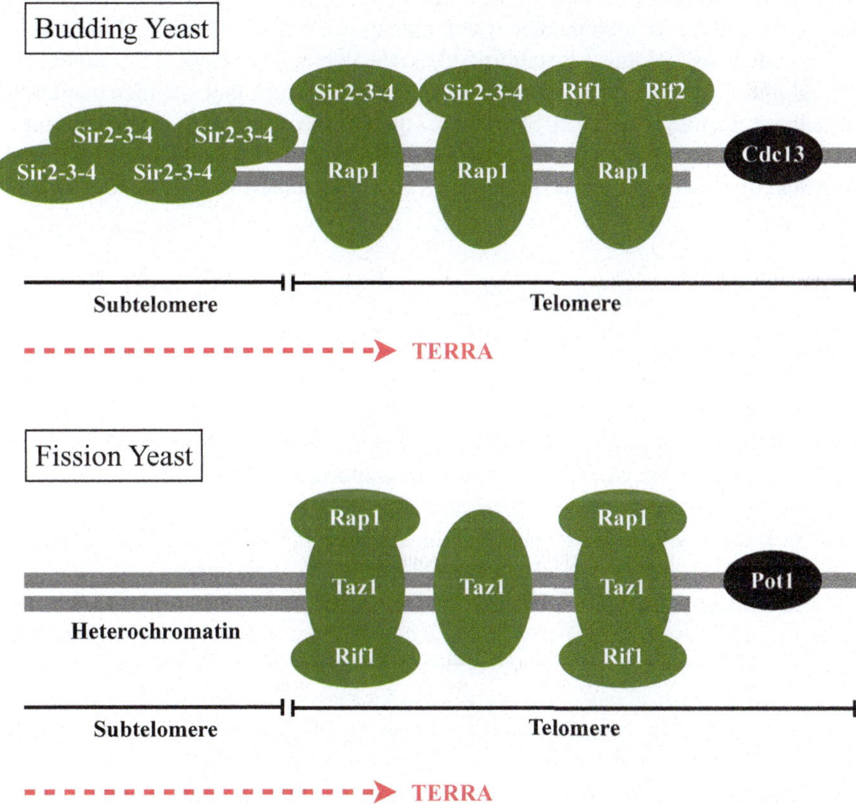

Fig. 6.7 Telomeric transcripts of budding and fission yeasts. In both budding and fission yeasts, TERRA is transcribed from subtelomeric regions. Subtelomeric regions are compacted by Sir2-3-4 complex in budding yeast and heterochromatin in fission yeast. The very ends of the telomere are capped by Cdc13 in budding yeast and Pot1 in fission yeast. Shelterin proteins are assembled in telomeric regions in both budding and fission yeasts

RNA) (Azzalin et al. 2007), and TERRA transcription is evolutionally conserved from yeast to human (reviewed in Smekalova and Baumann 2013). TERRA is heterogeneous in length and transcribed from the subtelomeric region to the repetitive telomeric region in an RNA Pol II-dependent manner (Fig. 6.7).

TERRA is implicated in the establishment of heterochromatin at telomeres and the regulation of telomere length. Since regulators of TERRA are associated with its repression, high levels of TERRA expression may have negative effects for the cells. Indeed, inhibition of TERRA degradation leads to telomere shortening (Luke et al. 2008). The telomere loss associated with inducible transcription of TERRA is additive with that due to the inactivation of telomerase (Maicher et al. 2012), suggesting that, in addition to telomerase inhibition, TERRA transcription can shorten telomeres.

Of note, TERRA harbors the sequence motif that can adapt a four-stranded (quadruplex) structure. Both DNA and RNA, including TERRA, can form quadruplex structures in solution, but the presence of quadruplex structures in vivo is enigmatic (Maizels and Gray 2013; Xu et al. 2010).

However, a light-switching pyrene probe confirmed the presence of TERRA quadruplex structures at telomeres in living human cells (Xu et al. 2010). Furthermore, in budding yeast, telomere instability due to the depletion of Cdc13, a component of the telomere-capping complex, can be alleviated by drugs that stabilize quadruplex structures (Smith et al. 2011). Since quadruplex structures are resistant to exonucleatic digestion, such structures might be adapted by TERRA to stabilize the very end of chromosomes (Maizels and Gray 2013), at least in the absence of Cdc13.

6.5 Coordinated Transcription of mRNA and lncRNA in Response to Environmental Stimuli

Co-transcriptional recruitment of chromatin regulators couples lncRNA transcription, and the centromere and telomere are transcribed to maintain chromosome integrity. Finally, we will see the functional importance of lncRNA transcription in response to environmental stimuli. Proper utilization of carbon sources is critical for the survival of the cells, and yeast exploits lncRNA transcription in order to respond adequately to available carbon sources in the media. Simple eukaryotes utilize coordinated transcription of mRNAs and lncRNAs for proper adaptation to their surrounding environments, suggesting that lncRNA transcription is a part of integrated mechanisms of eukaryotic gene regulation in response to environmental signals.

In fission yeast, glucose starvation induces the expression of Fbp1, an enzyme required for gluconeogenesis. In the presence of glucose, *fbp1* mRNA is not transcribed; however, there is constitutive transcription of lncRNA transcribed from upstream of *fbp1* ORF. When glucose is depleted from the media, lncRNAs with discrete lengths are sequentially transcribed before the full activation of *fbp1* mRNA (Fig. 6.8a) (Hirota et al. 2008). In addition to *fbp1* locus, there are some genes whose activation is followed by stepwise transcription of lncRNAs upon glucose starvation, suggesting the presence of a group of lncRNAs that are transcribed prior to the full activation of mRNA when a particular metabolic source is deficient in the media. We named such RNAs as *m*etabolic-stress-induced *lo*ng *n*oncoding RNAs (mlonRNAs), and identified glucose-deficient-dependent mlonRNAs in a subset of loci (Oda et al. 2015).

In budding yeast, *GAL1* and *GAL10* are located in the *GAL* cluster. The *GAL* cluster is "induced" in the absence of glucose and the presence of galactose, "non-induced" in the absence of glucose and the presence of other carbon sources other than glucose, and "repressed" in the presence of glucose. In the presence of glucose, lncRNA is generated from the 3′ end of *GAL10*, running through *GAL10* ORF in the antisense direction and *GAL1* ORF in the sense direction (Fig. 6.8b) (Houseley et al. 2008). Different carbon sources alter genome-wide expression pattern of lncRNAs as well as mRNAs (Xu et al. 2009). As such, when the cells are grown in the galactose media depleted with glucose, lncRNA transcription at the *GAL* cluster ceases and *GAL1* and *GAL10* mRNAs are induced. The coordinated transcriptional switch is partly explained by the regulation mentioned in the previous section. When the cells are grown in glucose medium, high levels of H3K36me3 are enriched over *GAL10* and *GAL1* coding regions, and transcription of lncRNA reduces acetylation of the regions, preventing the transcription of *GAL10* and *GAL1* mRNAs in the presence of glucose (Houseley et al. 2008).

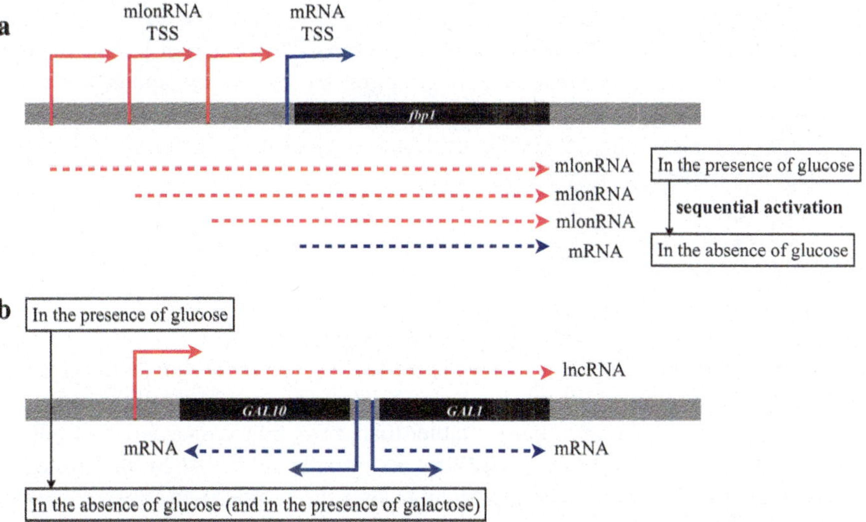

Fig. 6.8 Transcriptional switches from lncRNA to mRNA in response to an environmental stimulus. (**a**) In fission yeast, the longest mlonRNA is constitutively expressed in the presence of glucose. When glucose is depleted from the media, sequential activation of mlonRNAs of discrete lengths is followed by the full activation of *fbp1* mRNA, the products of which are necessary for gluconeogenesis. (**b**) In budding yeast, lncRNA transcribed from the *GAL* cluster represses expression of *GAL1* and *GAL10* mRNAs. In the absence of glucose and the presence of galactose, lncRNA transcription ceases and *GAL1* and *GAL10* mRNAs are expressed

An open question in this field is how the transcriptional switch is induced when the environment is altered. One obvious hypothesis is the recruitment of transcriptional activators. In the previous examples, transcription factors and chromatin modifiers were recruited in response to the environmental stimuli (Hirota et al. 2008; Houseley et al. 2008). Recent study has demonstrated that three-dimensional structures of the genome also regulate the interplay between lncRNA and mRNA transcription (Nadal-Ribelles et al. 2014). In response to high osmolarity, Hog1, a member of the stress-activated protein kinases, is activated in yeast to control the cell cycle and gene expression. Hog1 is recruited to the promoters of stress-induced lncRNAs to stimulate their expression. Among such lncRNAs, *CDC28* lncRNA is positively correlated with *CDC28* mRNA expression. *CDC28* lncRNA is transcribed from the 3′ end of *CDC28* ORF in an antisense direction, and both *CDC28* lncRNA expression and Hog1 recruitment are required to induce *CDC28* mRNA expression. Hog1 is recruited to promoters of both *CDC28* mRNA and lncRNA. However, in the absence of lncRNA, Hog1 is not recruited to the promoter of mRNA, but it is recruited to that of lncRNA, suggesting the presence of Hog1 transfer that is mediated by transcription of lncRNA (Nadal-Ribelles et al. 2014).

Gene loop formation depends on Ssu72, a component of the CPF (cleavage and polyadenylation factor) complex that plays an essential role in 3′-end formation of mRNA. Gene loops juxtapose promoters and terminators in a transcription-dependent manner (Ansari and Hampsey 2005; O'Sullivan et al. 2004; Singh and Hampsey 2007; Tan-Wong et al. 2012). Formation of gene looping via transcription of lncRNA recruits Hog1 at the promoter of *CDC28* mRNA, thereby inducing the expression of *CDC28* mRNA. As this example shows, three-dimensional structures of the genome play important roles in coupling lncRNA transcription to mRNA activation. In summary, lncRNA transcription has both positive and negative effects on mRNA transcription, and diverse utilization of lncRNA enables yeast genomes to adapt to a rapidly changing environment.

6.6 Perspectives

Although lncRNAs are broadly defined as noncoding transcripts longer than 200 bp, there are groups of lncRNAs that are functionally and mechanistically distinct. For some lncRNAs, transcription per se, rather than RNA transcripts, activates mRNA expression because transcribed lncRNAs are rapidly degraded by the exosome. These lncRNAs recruit histone modifiers to prevent cryptic transcription, thereby repressing background mRNA expression. Centromeric and telomeric transcripts affect chromatin organization by recruiting protein complexes. The functional importance of the centromeric and telomeric transcripts is evolutionally conserved, although the mechanisms might differ from one species to another. There is yet another group of lncRNAs whose transcription is coupled to the subsequent activation of mRNA in a particular environmental condition.

Throughout this review, we have focused on the roles of lncRNAs that might underlie the diverse biological complexity of eukaryotes. The presence of lncRNAs is ubiquitous among eukaryotes, but utilization of such transcripts is diverse from one species to another, and molecular mechanisms of lncRNA function described in this review are probably a small part of the possible repertoires of lncRNA usage in yeast genomes. Indeed, genome-wide studies of yeast genomes have revealed widespread transcription of lncRNAs, most of which are still intangible about their functions and mechanisms.

A recent study (Tuck and Tollervey 2013) demonstrated that more than one-third of SUTs are transported to the cytoplasm. Some lncRNAs containing ORF may also be transported to the cytoplasm, as exemplified by the cytoplasmic localization of *fbp1* lncRNA (Galipon et al. 2013). Such lncRNAs that contain ORF are difficult to identify from RNA-seq data, and it is possible that ORF-containing lncRNAs comprise a significant portion of cytoplasmic RNAs. Since such transcripts evaded the nuclear degradation machineries, they are, at most, as stable as mRNAs, and such stability may suggest the functional importance of noncoding transcripts in the cytoplasm.

Indeed, a recent proteomic study of human cells revealed that many "noncoding" RNAs are translated into short peptides (Wilhelm et al. 2014). The functions of these peptides in human cells, if any, are unknown, although pioneering work on *Drosophila* demonstrated the functional importance of short peptides in epithelial morphogenesis (Kondo et al. 2007). In yeast, it is possible that some lncRNAs are translated into short peptides as well. In fission yeast, *fbp1* lncRNA recruits ribosomes to upstream ORFs (Galipon et al. 2013), raising the possibility that uORFs are translated into short peptides. The functional importance of such short peptides might be revealed from yeast genetics, which possibly would yield insight into the functional importance of recently identified short peptides generated from "noncoding" RNAs in human cells.

The proportion of noncoding regions in genomes increases proportionally to the intuitive perception of biological complexity. In particular, transcription of lncRNAs is ubiquitous in all lineages of eukaryotes, but a large part of their functions and mechanisms is still concealed even in the simplest genomes of eukaryotes. Thus, understanding the molecular mechanisms of lncRNA functions from yeast to human will surely provide insight into the divergent biological complexity of eukaryotes.

Acknowledgments We thank Prof. R. Kurokawa for giving us the opportunity to contribute to this book. This work was supported by grants from the Japan Society for the Promotion of Science (nos. 23114003, 21241046, 26291018) to K.O. This work was also supported by a Grant-in-Aid for Scientific Research on Innovative Areas (no. 221S0002) and by the Platform for Dynamic Approaches to Living Systems from the Ministry of Education, Culture, Sports, Science and Technology, Japan.

References

Allshire RC, Karpen GH (2008) Epigenetic regulation of centromeric chromatin: old dogs, new tricks? Nat Rev Genet 9:923–937

Ansari A, Hampsey M (2005) A role for the CPF 3′-end processing machinery in RNAP II-dependent gene looping. Genes Dev 19:2969–2978

Azzalin CM et al (2007) Telomeric repeat-containing RNA and RNA surveillance factors at mammalian chromosome ends. Science 318:798–801

Blasco MA (2007) The epigenetic regulation of mammalian telomeres. Nat Rev Genet 8:299–309

Bühler M, Moazed D (2007) Transcription and RNAi in heterochromatic gene silencing. Nat Struct Mol Biol 14:1041–1048

Castel SE, Martienssen (2013) RNA interference in the nucleus: roles for small RNAs in transcription, epigenetics and beyond. Nat Rev Genet 14:100–112

Chen ES et al (2008) Cell cycle control of centromeric repeat transcription and heterochromatin assembly. Nature 451:734–737

Choi ES et al (2011) Identification of noncoding transcripts from within CENP-A chromatin at fission yeast centromeres. J Biol Chem 286:23600–23607

Claverie JM (2001) What if there are only 30,000 human genes? Science 291:1255–1257

Covitz PA, Mitchell AP (1993) Repression by the yeast meiotic inhibitor RME1. Genes Dev 7:1598–1608

Covitz PA et al (1991) The yeast *RME1* gene encodes a putative zinc finger protein that is directly repressed by a1-α2. Genes Dev 5:1982–1989

Davis CA, Ares M Jr (2006) Accumulation of unstable promoter-associated transcripts upon loss of the nuclear exosome subunit Rrp6 in *Saccharomyces cerevisiae*. Proc Natl Acad Sci U S A 103:3262–3267

Fischer TF et al (2009) Diverse roles of HP1 proteins in heterochromatin assembly and functions in fission yeast. Proc Natl Acad Sci U S A 106:8998–9003

Folco HD et al (2008) Heterochromatin and RNAi are required to establish CENP-A chromatin at centromeres. Science 319:94–97

Galipon J et al (2013) Stress-induced lncRNAs evade nuclear degradation and enter the translational machinery. Genes Cells 18:353–368

Gent JI, Dawe RK (2012) RNA as a structural and regulatory component of the centromere. Annu Rev Genet 46:443–453

Govind CK et al (2007) Gcn5 promotes acetylation, eviction, and methylation of nucleosomes in transcribed coding regions. Mol Cell 25:31–42

Govind CK et al (2010) Phosphorylated Pol II CTD recruits multiple HDACs, including Rpd3C(S), for methylation-dependent deacetylation of ORF nucleosomes. Mol Cell 39:234–246

Grewal SIS, Jia S (2007) Heterochromatin revisited. Nat Rev Genet 8:35–46

Hall IM et al (2002) Establishment and maintenance of a heterochromatin domain. Science 297:2232–2237

Hayashi T et al (2004) Mis16 and Mis18 are required for CENP-A loading and histone deacetylation at centromeres. Cell 118:715–729

Hayashi T et al (2014) *Schizosaccharomyces pombe* centromere protein Mis19 links Mis16 and Mis18 to recruit CENP-A through interacting with NMD factors and the SWI/SNF complex. Genes Cells 19:541–554

Hirota K et al (2008) Stepwise chromatin remodelling by a cascade of transcription initiation of non-coding RNAs. Nature 456:130–134

Honigberg SM, Purnapatre K (2003) Signal pathway integration in the switch from the mitotic cell cycle to meiosis in yeast. J Cell Sci 116:2137–2147

Houseley J et al (2008) A ncRNA modulates histone modification and mRNA induction in the yeast *GAL* gene cluster. Mol Cell 32:685–695

Jensen TH et al (2013) Dealing with pervasive transcription. Mol Cell 52:473–484

Kassir Y et al (1988) *IME1*, a positive regulator gene of meiosis in S. cerevisiae. Cell 52:853–862

Kim T, Buratowski S (2009) Dimethylation of H3K4 by Set1 recruits the Set3 histone deacetylase complex to 5′ transcribed regions. Cell 137:259–272

Kim T et al (2012) Set3 HDAC mediates effects of overlapping noncoding transcription on gene induction kinetics. Cell 150:1158–1169

Kirmizis A et al (2007) Arginine methylation at histone H3R2 controls deposition of H3K4 tri-methylation. Nature 449:928–932

Kondo T et al (2007) Small peptide regulators of actin-based cell morphogenesis encoded by a polycistronic mRNA. Nat Cell Biol 9:660–665

Kueng S et al (2013) SIR proteins and the assembly of silent chromatin in budding yeast. Annu Rev Genet 47:275–306

Luke B et al (2008) The Rat1p 5′ to 3′ exonuclease degrades telomeric repeat-containing RNA and promotes telomere elongation in *Saccharomyces cerevisiae*. Mol Cell 32:465–477

Maicher A et al (2012) Deregulated telomere transcription causes replication-dependent telomere shortening and promotes cellular senescence. Nucleic Acid Res 40:6649–6659

Maizels N, Gray LT (2013) The G4 genome. PLoS Genet 9:e1003468

Marguerat S et al (2012) Quantitative analysis of fission yeast transcriptomes and proteomes in proliferating and quiescent cells. Cell 151:671–683

Marina DB et al (2013) A conserve ncRNA-binding protein recruits silencing factors to heterochromatin through an RNAi-independent mechanism. Genes Dev 27:1851–1856

Mattick JS (2004) RNA regulation: a new genetics? Nature 5:316–323

Motamedi MR et al (2008) HP1 proteins form distinct complexes and mediate heterochromatic gene silencing by nonoverlapping mechanisms. Mol Cell 32:778–790

Murray SC et al (2012) A pre-initiation complex at the 3′-end of genes drives antisense transcription independent of divergent sense transcription. Nucleic Acid Res 40:2432–2444

Nadal-Ribelles M et al (2014) Control of Cdc28 CDK1 by a stress-induced lncRNA. Mol Cell 53:549–561

Neil H et al (2009) Widespread bidirectional promoters are the major source of cryptic transcripts in yeast. Nature 457:1038–1042

Oda A et al (2015) Dynamic transition of transcription and chromatin landscape during fission yeast adaptation to glucose starvation. Genes Cells. doi:10.1111/gtc.12229

O'Sullivan JM et al (2004) Gene loops juxtapose promoters and terminators in yeast. Nat Genet 36:1014–1018

Ohkuni K, Kitagawa K (2011) Endogenous transcription at the centromere facilitates centromere activity in budding yeast. Curr Biol 21:1695–1703

Ohkuni K, Kitagawa K (2012) Role of transcription at centromeres in budding yeast. Transcription 3:193–197

Pokholok DK et al (2005) Genome-wide map of nucleosome acetylation and methylation in yeast. Cell 122:517–527

Reyes-Turcu FE, Grewal SIS (2012) Different means, same end — heterochromatin formation by RNAi and RNAi-independent RNA processing factors in fission yeast. Curr Opin Genet Dev 22:156–163

Rhee HS, Pugh BF (2012) Genome-wide structure and organization of eukaryotic pre-initiation complexes. Nature 483:295–301

Rhind N et al (2011) Comparative functional genomics of the fission yeasts. Science 332:930–936

Rinn JL, Chang HY (2012) Genome regulation by long noncoding RNAs. Annu Rev Biochem 81:145–166

Singh BN, Hampsey M (2007) A transcription-independent role for TFIIB in gene looping. Mol Cell 27:806–816

Smekalova E, Baumann P (2013) TERRA —a calling card for telomerase. Mol Cell 51:703–704

Smith JS et al (2011) Rudimentary G-quadruplex-based telomere capping in *Saccharomyces cerevisiae*. Nat Struct Mol Biol 18:478–485

Sugiyama T et al (2007) SHREC, an effector complex for heterochromatic transcriptional silencing. Cell 128:491–504

Tan-Wong SM et al (2012) Gene loops enhance transcriptional directionality. Science 338:671–675

Thornton JL et al (2014) Context dependency of Set1/COMPASS-mediated histone H3 Lys4 trimethylation. Genes Dev 28:115–120

Tuck AC, Tollervey D (2013) A transcriptome-wide atlas of RNP composition reveals diverse classes of mRNAs and lncRNAs. Cell 154:996–1009

van Dijk et al (2011) XUTs are a class of Xrn1-sensitive antisense regulatory non-coding RNA in yeast. Nature 475:114–117

van Werven FJ, Amon A (2011) Regulation of entry into gametogenesis. Phil Trans R Soc B 366:3521–3531

van Werven FJ et al (2012) Transcription of two long noncoding RNAs mediates mating-type control of gametogenesis in budding yeast. Cell 150:1170–1181

Venkatesh S, Workman JL (2013) Non-coding transcription SETs up regulation. Cell Res 23:311–313

Venkatesh S et al (2012) Set2 methylation of histone H3 lysine 36 suppresses histone exchange on transcribed genes. Nature 489:452–455

Volpe TA et al (2002) Regulation of heterochromatic silencing and histone H3 lysine-9 methylation by RNAi. Science 297:1833–1837

Wilhelm BT et al (2008) Dynamic repertoire of a eukaryotic transcriptome surveyed at single-nucleotide resolution. Nature 453:1239–1243

Wilhelm M et al (2014) Mass-spectrometry-based draft of the human proteome. Nature 509:582–587

Wyers F et al (2005) Cryptic Pol II transcripts are degraded by a nuclear quality control pathway involving a new poly(A) polymerase. Cell 121:725–737

Xu Z et al (2009) Bidirectional promoters generate pervasive transcription in yeast. Nature 457:1033–1037

Xu Y et al (2010) Telomeric repeat-containing RNA structure in living cels. Proc Natl Acad Sci U S A 107:14579–14584

Yamada T et al (2005) The nucleation and maintenance of heterochromatin by a histone deacetylase in fission yeast. Mol Cell 20:173–185

Ye J et al (2014) Transcriptional outcome of telomere signalling. Nat Rev Genet 15:491–503

Zofall M et al (2012) RNA elimination machinery targeting meiotic mRNAs promotes facultative heterochromatin formation. Science 335:96–100

Chapter 7
Long Noncoding RNAs as Structural and Functional Components of Nuclear Bodies

Taro Mannen*, **Takeshi Chujo***, **and Tetsuro Hirose**

Abstract The mammalian cell nucleus harbors various membraneless suborganelles named nuclear bodies that are characterized by distinct sets of resident proteins. Nuclear bodies are thought to serve as sites for the biogenesis, assembly, and storage of specific proteins and RNAs. In the last decade, multiple nuclear bodies were found to contain long noncoding RNAs (lncRNAs), and the physiological and molecular functions of these lncRNAs have been elucidated. Numerous lncRNAs are induced in response to cellular stresses, presumably as a mechanism to cope with environmental changes. Some lncRNAs play architectural or structural roles to construct and sustain nuclear bodies; these lncRNAs exert their physiological functions by sequestering specific regulatory proteins in nuclear bodies. Other lncRNAs do not contribute to the integrity of the nuclear body structure but play significant roles in the transcriptional and post-transcriptional regulation of genes by modulating the function and localization of related regulatory proteins. In this review, we focus on the recently unveiled roles of lncRNAs that act as structural and/or functional components of nuclear bodies.

Keywords Nuclear body • Subnuclear structure • Long noncoding RNA • RNA-binding protein • Ribonucleoprotein complex • Architectural RNA • Molecular sponge • Stress response

7.1 Introduction

The cell nucleus contains genetic material organized into multiple chromosomes. Chromosomes do not diffuse randomly in the nucleus but occupy discrete territories during interphase (Cremer and Cremer 2010). The mammalian cell nucleus also harbors nuclear bodies, a range of spherical, subnuclear structures located within the interchromatin space (Fig. 7.1). Multiple nuclear bodies have been characterized to date, including nucleoli, Cajal bodies, promyelocytic leukemia bodies, nuclear

*Author contributed equally with all other contributors.

T. Mannen • T. Chujo • T. Hirose (✉)
Institute for Genetic Medicine, Hokkaido University, Sapporo 060-0815, Japan
e-mail: hirose@igm.hokudai.ac.jp

R. Kurokawa (ed.), *Long Noncoding RNAs*, DOI 10.1007/978-4-431-55576-6_7

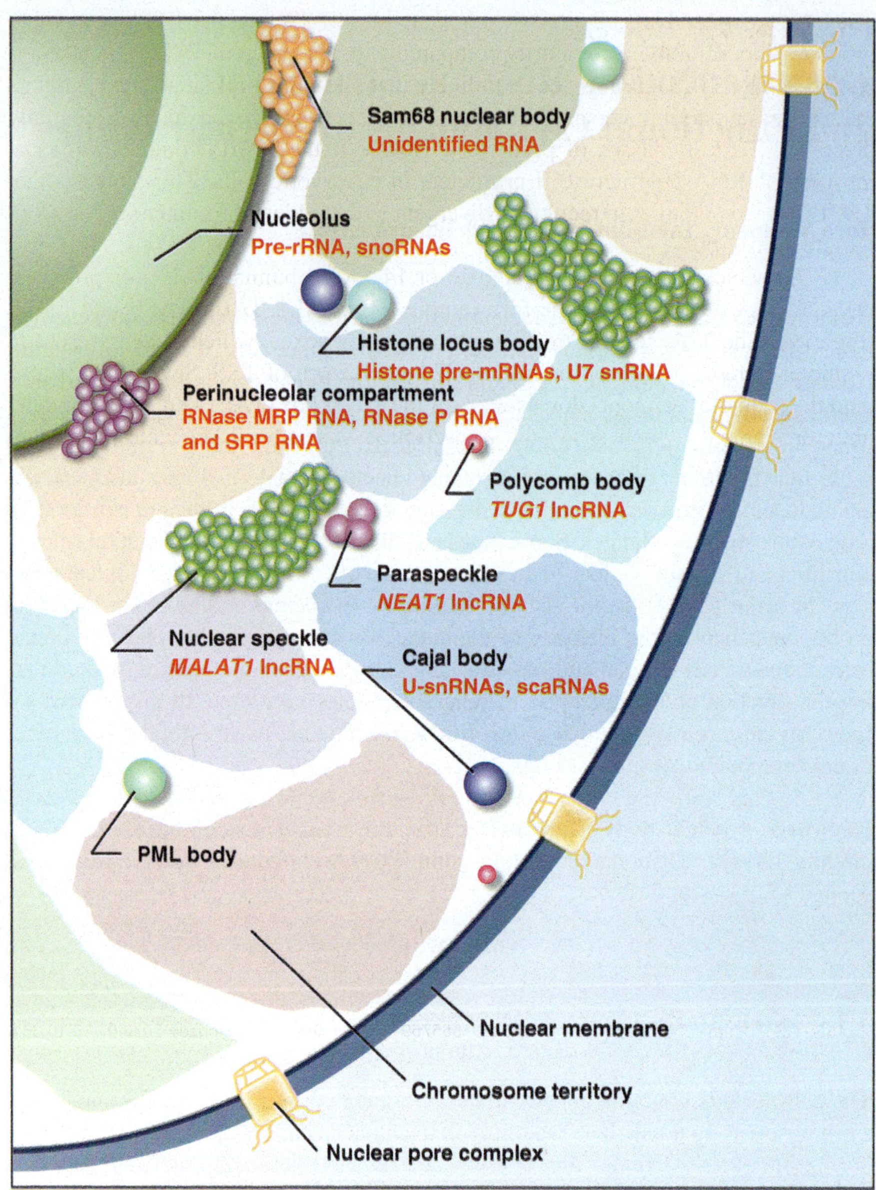

Fig. 7.1 Nuclear bodies in mammalian cells. A schematic illustration of mammalian nuclear structures, including chromosome territories and representative nuclear bodies. Many nuclear bodies harbor RNA components (shown in *red*). *MRP* mitochondrial RNA processing, *scaRNA* small Cajal body-specific RNA, *snoRNA* small nucleolar RNA, *snRNA* small nuclear RNA, *SRP* signal recognition particle

speckles, paraspeckles and polycomb bodies. These nuclear bodies contain a number of proteins and RNA factors that are essential for the complex regulation of various nuclear events, including gene expression (Table 7.1) (Dundr and Misteli 2010; Mao et al. 2011b). Unlike some cytoplasmic organelles, nuclear bodies are not compartmentalized by lipid membranes, and their structural integrity is maintained by protein–protein, protein–RNA, and/or protein–DNA interactions. Nuclear bodies are present at a steady state and respond dynamically to basic physiological processes, as well as various forms of stress and alterations in metabolic conditions or cellular signaling.

Following the completion of the Human Genome Project, extensive transcriptome analyses revealed that three quarters of the human genome is transcribed to produce not only mRNAs but also non-protein-coding transcripts such as long noncoding RNAs (lncRNAs), which are over 200 nucleotides in length (Djebali et al. 2012). Version 7 of the GENCODE gene set (December 2010 freeze, GRCh37) included 9,277 manually curated annotated lncRNAs (Derrien et al. 2012), and that number has continued to grow in subsequent versions (14,470 lncRNA genes in GENCODE version 20, April 2014 freeze, GRCh38). Although the molecular functions of these lncRNAs remain largely unknown, emerging evidence has indicated that they play diverse roles in the regulation of various cellular processes, both in the nucleus and the cytoplasm. In the nucleus, multiple lncRNAs act as regulators of epigenetic histone modifications, which are related to various physiological phenomena and diseases (Batista and Chang 2013; Geisler and Coller 2013). In addition, a subset of lncRNAs function as structural components of nuclear bodies. Here, we review the molecular and physiological roles of lncRNAs that are contained in nuclear bodies, with a particular focus on lncRNAs that serve as platforms for nuclear body formation.

7.2 LncRNAs as Structural Components of Nuclear Bodies

Over the last decade, several lncRNAs have been found to function as architectural RNAs for nuclear body formation. Among them, nuclear-enriched abundant transcript 1 (*NEAT1*), which functions as a scaffold for the construction of paraspeckles, has been studied extensively.

7.2.1 NEAT1 *lncRNA Functions as a Platform for Paraspeckle Formation*

Paraspeckles, nuclear bodies that are often located adjacent to nuclear speckles, were initially identified by a proteomic study aimed at characterizing human nucleoli (Fox et al. 2002). With an average diameter of approximately 0.36 μm, paraspeckles are found in most cultured mammalian cells (Souquere et al. 2010) and

Table 7.1 Representative components and functions of RNA-containing nuclear bodies

Body name	Defining components		(Putative) function
	Protein	RNA	
Paraspeckle	PSPC1	*NEAT1* lncRNA	Regulation of transcription by protein sequestration and retention of hyper-edited mRNAs
Nucleolus	FBL	Pre-rRNA, snoRNAs	Ribosome biogenesis
Nuclear speckle	SRSF2/SC35	*MALAT1* lncRNA	Storage and recycling of splicing factors
Nuclear stress body	HSF1	SatIII lncRNA	Regulation of transcription and splicing under stress
Cajal body	COIL	U-snRNAs, scaRNAs	Biogenesis and recycling of snRNPs
Perinucleolar compartment	PTBP1	MRP RNA, SRP RNA, RNase P RNA	Post-transcriptional regulation of a subset of pol III-transcribed RNAs
Sam68 nuclear body	KHDRBS1/Sam68	n.d.	Splicing regulation?
Polycomb body	BMI1	*TUG1* lncRNA	Transcriptional regulation?
PML body	PML	n.d.	Regulation of genome stability and DNA repair
Histone locus body	NPAT	Histone pre-mRNAs, U7 snRNA	Histone mRNA transcription and 3′ end processing
Nuclear gem	SMN	n.d.	Biogenesis and/or recycling of snRNPs?
Nucleolar DC	VHL, HSP70	IGS lncRNAs	Sequestration of proteins under stress to halt ribosome biogenesis

COIL coilin, *FBL* fibrillarin, *HSF1* heat shock transcription factor 1, *KHDRBS1* KH domain-containing, RNA-binding, signal transduction associated 1, *MRP* mitochondrial RNA processing, *n.d.* not determined, *NPAT* nuclear protein, ataxia-telangiectasia locus, *PML* promyelocytic leukemia, *PSPC1* paraspeckle component 1, *PTBP1* polypyrimidine tract binding protein 1, *scaRNA* small Cajal body-specific RNA, *SMN* survival of motor neuron, *SRP* signal recognition particle, *SRSF2* serine/arginine-rich splicing factor 2, *VHL* von Hippel-Lindau tumour suppressor, *E3* ubiquitin-protein ligase were confirmed to be equivalent to the interchromatin granule-associated zone, which is an electron-dense structure when viewed by electron microscopy (Souquere et al. 2010; Visa et al. 1993). Paraspeckles contain several well-characterized members of the Drosophila behavior human splicing (DBHS) family of proteins, including paraspeckle component 1 (PSPC1), non-POU domain-containing octamer binding protein (NONO; also known as $p54^{nrb}$), and splicing factor proline- and glutamine-rich (SFPQ; also known as PSF) (Fox et al. 2002). Inhibition of RNA polymerase II transcription by treatment of cells with actinomycin D or 5,6-dichloro-1-β-D-ribofuranosylbenzimidazole (DRB) results in the rapid disintegration of paraspeckles, suggesting that their formation is dependent on ongoing transcription (Fox et al. 2005; Sasaki et al. 2009; Sunwoo et al. 2009).

In 2009, several groups independently reported that the *NEAT1* lncRNA localizes specifically to and functions as an essential structural component of paraspeckles (Chen and Carmichael 2009; Clemson et al. 2009; Sasaki et al. 2009; Sunwoo et al. 2009). The two isoforms of *NEAT1*, namely *NEAT1_1* (3.7 kb) and *NEAT1_2* (23 kb), are expressed from a common promoter and differ at their 3′ ends, which are processed by canonical polyadenylation and RNase P cleavage, respectively (Sunwoo et al. 2009). In embryonic fibroblasts from *Neat1* knockout mice, transient expression of *Neat1_2*, but not *Neat1_1*, is essential for the *de novo* formation of paraspeckles (Naganuma et al. 2012). By contrast, overexpression of *NEAT1_1* increases the number of paraspeckles substantially, indicating that it plays a supplementary role in the formation of these nuclear bodies (Clemson et al. 2009; Naganuma et al. 2012).

In addition to DBHS family proteins, more than 40 other proteins localize to paraspeckles (Fong et al. 2013; Naganuma et al. 2012); most of these paraspeckle proteins (PSPs) have RNA-binding domains and some interact directly with *NEAT1*. RNA interference analyses revealed that seven PSPs, namely, SFPQ, NONO, RNA-binding protein 14 (RBM14), heterogeneous nuclear ribonucleoprotein K (HNRNPK), DAZ-associated protein 1 (DAZAP1), fused in sarcoma (FUS), and heterogeneous nuclear ribonucleoprotein H3 (HNRNPH3), are necessary for paraspeckle formation and maintenance (Naganuma et al. 2012). Later, Fong et al. (2013) reported that the following PSPs also contribute to paraspeckle structural integrity: E3 ubiquitin-protein ligase HECT domain-containing protein 3 (HECTD3); family with sequence similarity 53, member B (FAM53B); zinc finger protein 24 (ZNF24); X-linked inhibitor of apoptosis protein (XIAP); and ecto-NOX disulfide-thiol exchanger 1 (ENOX1). Alternate processing of the 3′ end of the *NEAT1* precursor RNA is controlled by HNRNPK; specifically, HNRNPK binds to the short pyrimidine stretch located between the canonical polyadenylation signal for *NEAT1_1* and the upstream cleavage factor Im (CFIm) binding cluster, where it displaces nucleoside diphosphate linked moiety X-type motif 21 (NUDT21), a PSP from the functional CFIm complex. This displacement suppresses the 3′ end processing of the *NEAT1_1* RNA, thereby promoting *NEAT1_2* transcription; consequently, knockdown of HNRNPK reduces and increases the levels of *NEAT1_2* and *NEAT1_1* transcripts, respectively. NONO and SFPQ are required for the accumulation of *NEAT1_2* specifically, suggesting their involvement in the stabilization of this isoform. Whereas the category 1A PSPs HNRNPK, NONO, and SFPQ are required for *NEAT1_2* biogenesis and accumulation, other PSPs contribute to paraspeckle formation in different ways. For example, knockdown of DAZAP1, FUS, or HNRNPH3 leads to the disintegration of paraspeckles without affecting the steady-state level of *NEAT1_2*, suggesting that these category 1B PSPs are required for nuclear body formation (Fig. 7.2).

The formation of paraspeckles is coupled with the transcription of *NEAT1* (Mao et al. 2011a); transcriptional inhibition quickly disrupts paraspeckle integrity, resulting in relocation of their associated proteins to the perinucleolar cap and diffuse expression of *NEAT1* in the nucleoplasm (Sasaki et al. 2009). Live cell imaging revealed that PSPs are recruited to the site of *NEAT1* transcription, and as the size of the newly formed paraspeckle increases, new paraspeckles bud off from the original structure, resulting in the formation of clusters of paraspeckles at the site of transcription. Fluorescent recovery after photobleaching analyses showed that *NEAT1* is exchanged at a much slower rate than the protein components of paraspeckles; these studies identified hierarchical roles of *NEAT1* lncRNA and protein components, and revealed that *NEAT1* is the architectural RNA involved in paraspeckle assembly.

7.2.2 The Biological Functions of NEAT1 *lncRNA*

Although paraspeckle biogenesis and the involvement of *NEAT1* in this process have been studied extensively, the physiological roles of both *NEAT1* and paraspeckles are just beginning to be elucidated. A detailed in situ hybridization analysis of *Neat1* expression in mouse tissues revealed that *Neat1_1* is expressed widely, whereas *Neat1_2* expression is restricted to specific subpopulations of cells (Hutchinson et al. 2007; Nakagawa et al. 2011). Surprisingly, *Neat1* knockout mice are viable, are fertile, and show no apparent phenotype. Furthermore, the integrity of tissues containing cells with high levels of *Neat1_2* is not affected by the absence of *Neat1*, suggesting that the expression of *Neat1* and formation of paraspeckles are not essential for mice living under laboratory conditions. However, the induction of *NEAT1_2* transcription and the formation of paraspeckles might have functional roles under specific conditions, for example, in response to specific stresses or stimuli. This proposal is supported by studies demonstrating that *NEAT1* expression and paraspeckle formation are induced in the central nervous system after viral infection, in patients with frontotemporal lobar degeneration, and during the early stage of amyotrophic lateral sclerosis (Nishimoto et al. 2013; Saha et al. 2006; Tollervey et al. 2011).

Paraspeckles regulate gene expression through the sequestration of specific proteins and hyper-edited mRNAs (Fig. 7.3). Under normal conditions, the expression of adenosine-to-inosine (A-to-I) hyper-edited mRNAs, such as the mouse CTN-RNA, is inhibited by their retention in nuclear paraspeckles. The mouse CTN-RNA, which is a long isoform of the mouse cationic amino acid transporter 2 (m*Cat2*)

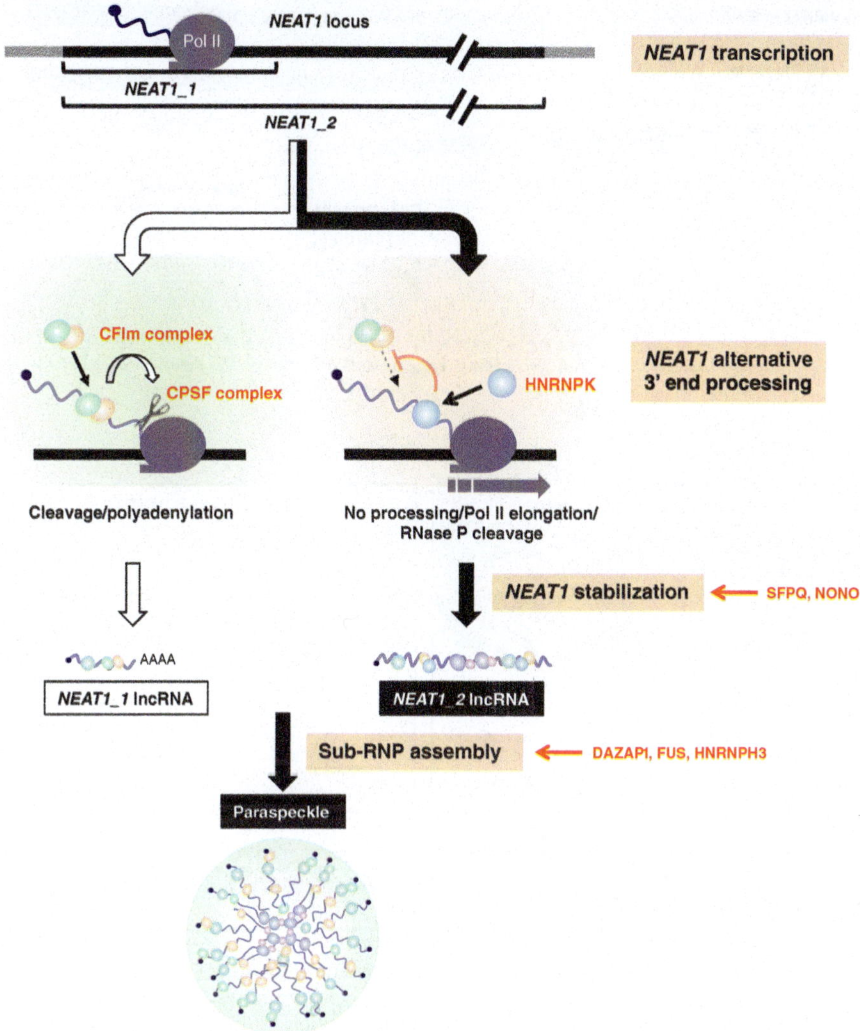

Fig. 7.2 Current model of paraspeckle formation. Paraspeckles are formed in conjunction with *NEAT1_2* biogenesis, *NEAT1_2* stabilization, and ribonucleoprotein assembly. *NEAT1_2* biogenesis includes transcription of the *NEAT1* gene by RNA polymerase II (Pol II) and alternative 3′ end processing. Ongoing *NEAT1* transcription is required for paraspeckle formation. Processing of the 3′ end of the *NEAT1_1* isoform is mediated by the CFIm complex (including NUDT21 and CPSF6), which binds to the *NEAT1* precursor and recruits the cleavage and polyadenylation specificity factor (CPSF) complex to promote cleavage and polyadenylation. For *NEAT1_2* synthesis, HNRNPK binds to the *NEAT1* precursor and displaces NUDT21 from the functional CFIm complex, thereby preventing binding of the complex to the *NEAT1* precursor. Cleavage of the 3′ end of *NEAT1_2* is performed by RNase P. PSPs such as SFPQ and NONO are essential for *NEAT1* stabilization. PSPs such as DAZAP1, FUS, and HNRNPH3 contribute to the assembly of *NEAT1* ribonucleoproteins to form paraspeckles

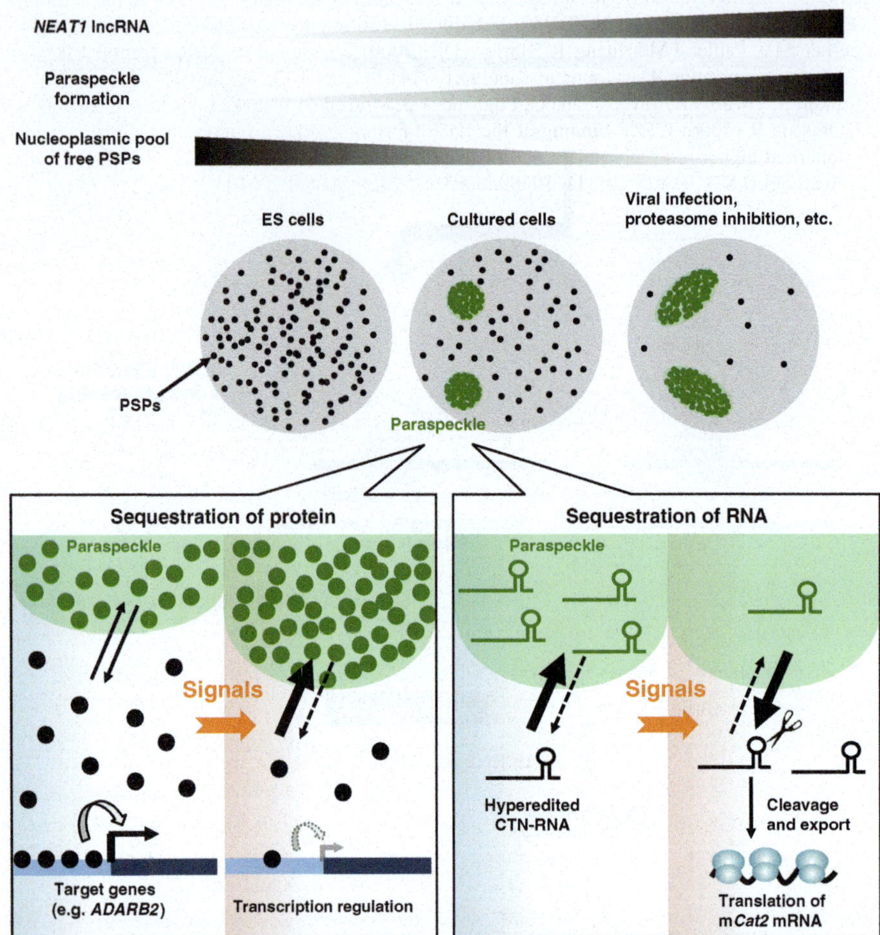

Fig. 7.3 Overview of paraspeckle functions. The expression level of *NEAT1* dictates the size and shape of paraspeckles. *NEAT1* expression is induced by cellular stresses, including viral infection and proteasome inhibition, resulting in the de novo formation or enlargement of paraspeckles. *NEAT1* sequesters PSPs such as SFPQ (shown as *small black dots*) in paraspeckles, resulting in a reduction in the level of freely available PSPs in the nucleoplasm. Consequently, the expression levels of SFPQ target genes, such as *ADARB2*, are downregulated at the transcriptional level. Paraspeckles also regulate gene expression through the retention of hyper-edited mRNAs, such as CTN-RNA, which comprises the mCAT2 protein-coding region and an extended 3′ UTR containing inverted repeats. Under normal conditions, PSPs such as NONO retain the CTN-RNA within paraspeckles; however, under certain stress conditions, the mCAT2 coding region is released by endonucleolytic cleavage of the CTN-RNA and subsequently exported from the nucleus and translated into mCAT2 protein

transcript (Prasanth et al. 2005), contains an extended 3′ untranslated region (UTR) harboring two short interspersed nuclear elements in an inverted orientation; these inverted repeats form an intra-molecular double-stranded RNA that is recognized by adenosine deaminase, which converts adenosine-to-inosine (A-to-I editing). In response to certain stress signals, the CTN-RNA is released from paraspeckles, cleaved at the 3′ UTR, and released into the cytoplasm, where it is translated into the mCAT2 protein. Although human CAT2 does not contain an inverted repeat sequence in its 3′ UTR, many other human mRNAs are reported to contain hyper-edited 3′ UTRs (Kim et al. 2004); in human embryonic stem cells, which do not express *NEAT1* and contain no typical paraspeckles, these mRNAs are exported efficiently to the cytoplasm. In differentiated cells, knockdown of *NEAT1* disrupts paraspeckles, and the retained mRNAs are exported to the cytoplasm, suggesting that hyper-edited mRNAs are retained in the nuclei of differentiated cells through paraspeckle function (Chen and Carmichael 2009). Notably, NONO has inosine-binding activity and is involved in the retention of hyper-edited mRNAs within paraspeckles (Chen and Carmichael 2009; Zhang and Carmichael 2001). These findings suggest that paraspeckles act as a reservoir of A-to-I hyper-edited mRNAs, which can be exported to the cytoplasm in response to stress to trigger their prompt translation.

Under certain stress conditions, *NEAT1* is upregulated markedly and modulates gene expression by sequestering specific transcriptional regulatory proteins in paraspeckles. Two groups performed microarray analyses of *NEAT1* knockdown cells and found that *NEAT1* regulates the transcription of several genes (Hirose et al. 2014; Imamura et al. 2014), including the gene encoding adenosine deaminase B2 (ADARB2). Under stress conditions caused by treatment of cells with the proteasome inhibitor MG132, paraspeckles become dramatically enlarged; this enlargement is caused mainly by the transcriptional upregulation of *NEAT1* and results in an accumulation of PSPs, such as SFPQ and NONO, in paraspeckles. Concomitant with a reduction in binding of SFPQ to the *ADARB2* promoter, the expression level of the *ADARB2* mRNA is reduced upon enhanced sequestration of SFPQ in paraspeckles. Moreover, *NEAT1* knockout fibroblasts are more sensitive to proteasome inhibition, leading to enhanced cell death under these conditions. These findings confirm that paraspeckles are stress-responsive nuclear bodies.

7.2.3 Intergenic Spacer (IGS) lncRNAs Are Required for the Formation of Nucleolar Detention Centers (DCs)

The nucleolus is the classical subnuclear domain involved in ribosome biogenesis. The nucleolus assembles approximately 400 tandem repeats of ribosomal DNA that comprises an enhancer, promoter, ribosomal RNA (rRNA) sequences, and large IGSs. The nucleolus functions not only in ribosome biogenesis but also in cellular stress responses. For example, exposure of cells to heat shock, hypoxia, acidosis, aspirin, serum starvation, or DNA damage induces the nucleolar capture and immobilization of various proteins with nucleolar detention peptide sequences (Mekhail et al. 2004; Stark and Dunlop 2005; Welch and Feramisco 1984). Consequently, a stress-induced subnucleolar structure named the DC is generated (Audas et al. 2012; Jacob et al. 2013). During this process, lncRNAs derived from IGSs play important architectural roles; the IGS lncRNAs are necessary for stress-induced DC formation and subsequent remodeling of the nucleolus and arrest of ribosome biogenesis. For example, acidosis induces the expression of the IGS_{28} lncRNA (approximately 400 nt in length) from a region located 28 kb downstream of the rRNA transcriptional start site in the IGS; this lncRNA captures and retains proteins such as von Hippel-Lindau, DNA methyltransferase 1 (DNMT1) and the DNA polymerase subunit POLD1 in the DC. Similarly, following heat shock stress, the HSP70 protein interacts with regions of the IGS_{16} and IGS_{22} lncRNAs, which are generated from regions located 16 and 22 kb downstream of the rRNA transcription start site, respectively. Inhibition of a specific IGS lncRNA does not affect the ability of other IGS lncRNAs to sequester their target proteins, suggesting that the functions of individual IGS lncRNAs are independent.

7.2.4 The Roles of Satellite III (satIII) lncRNAs in Nuclear Stress Body Formation

Heat shock stress not only targets specific proteins to the nucleolus but also induces the *de novo* formation of nuclear stress bodies (nSBs) (Jolly et al. 2002). Heat shock induces the transcriptional activation of tandem arrays of SatIII repeats located at the pericentromeric regions of chromosome 9q12 (Jolly et al. 2004; Rizzi et al. 2004). Upon induction, SatIII lncRNAs remain associated with the chromosome locus and function as a scaffold for the recruitment of various proteins to form nSBs, including transcription factors such as heat shock factor 1 (HSF1) and splicing factors such as serine/arginine-rich splicing factor 1 (SRSF1) and scaffold attachment factor B (SAFB) (Denegri et al. 2001), thereby potentially affecting global gene expression. Knockdown of SatIII lncRNA suppresses the recruitment of SRSF1 and SAFB to nSBs (Valgardsdottir et al. 2005), confirming that this lncRNA functions as an architectural RNA that captures specific protein factors in heat-stressed cells.

7.2.5 Histone mRNA Precursors in Histone Locus Bodies (HLBs)

HLBs, the sites of histone pre-mRNA transcription and processing, are another example of nuclear bodies formed on architectural RNA. Histone pre-mRNAs and the proteins required for processing of their 3′ ends are enriched in HLBs

(Bongiorno-Borbone et al. 2008; Nizami et al. 2010). In a previous study, artificial tethering of histone H2b pre-mRNAs tagged with bacteriophage-derived MS2 stem-loop structures to a specific engineered site in a HeLa cell chromosome resulted in the recruitment of HLB proteins, such as nuclear protein ataxia-telangiectasia locus (NPAT) and FADD-like IL-1β-converting enzyme-associated huge protein (FLASH), to the tethering site (Shevtsov and Dundr 2011), suggesting that the presence of histone pre-mRNA is sufficient to trigger HLB formation. This finding suggests that histone pre-mRNA serves as the architectural RNA for HLB formation.

7.3 LncRNAs as Functional Components of Nuclear Bodies

In some nuclear bodies, lncRNAs do not have architectural roles but interact with various proteins to regulate distinct steps of gene expression.

*7.3.1 Metastasis Associated Lung Adenocarcinoma Transcript 1 (*MALAT1*) lncRNA in Nuclear Speckles and Taurine Upregulated Gene 1 (*TUG1*) lncRNA in Polycomb Bodies*

Nuclear speckles are nuclear bodies enriched in splicing-related proteins and the *MALAT1* lncRNA. *MALAT1* was originally identified as one of several genes that are upregulated in metastatic non-small cell lung cancer (Ji et al. 2003). Although *MALAT1* is not essential for nuclear speckle formation or integrity in human and mouse cells (Nakagawa et al. 2012; Tripathi et al. 2010; Zhang et al. 2012), it exhibits a high degree of conservation from mammals to zebrafish (Hutchinson et al. 2007; Ulitsky et al. 2011), indicating that it may have important roles in biological processes. The *MALAT1* lncRNA interacts with and influences the phosphorylation level and distribution of serine/arginine proteins to nuclear speckles, and depletion of *MALAT1* alters the splicing pattern of a subset of endogenous pre-mRNAs in HeLa cells. Furthermore, the *MALAT1* lncRNA plays a role in gene activation by promoting the relocation of growth control genes between nuclear speckles and nuclear polycomb bodies via binding to unmethylated Polycomb 2 (Pc2) protein, a component of Polycomb repressive complex 1 (Yang et al. 2011). Pc2 interacts with distinct lncRNAs depending on its methylation status; specifically, unmethylated Pc2 interacts with the *MALAT1* lncRNA and localizes to nuclear speckles, where it promotes the expression of growth control genes, and methylated Pc2 interacts with the *TUG1* lncRNA, leading to the retention of both Pc2 and growth control genes in polycomb bodies, where their expression is repressed. The presence of the *MALAT1* and *TUG1* lncRNAs in nuclear bodies is necessary for the specific retention of Pc2 and growth control gene promoters bound by Pc2 in response to signaling pathways.

7.3.2 Gomafu *lncRNA Plays a Role in Splicing Factor 1 Retention*

Gomafu (also known as *MIAT*) is a nuclear-retained lncRNA that is specifically expressed in subsets of neurons (Blackshaw et al. 2004; Rapicavoli et al. 2010) and is located in a novel nuclear body (Sone et al. 2007). Splicing factor 1 (SF1), which recognizes splicing branch points, binds to tandem copies of UACUAAC repeats in the *Gomafu* lncRNA (Tsuiji et al. 2011). In vitro studies revealed that sequestration of SF1 by binding to *Gomafu* delays splicing, suggesting that this interaction may affect splicing regulation in neurons.

7.3.3 *Prader-Willi Syndrome (PWS) Region Small Nucleolar lncRNAs (sno-lncRNAs) Function as a Molecular Sink for Splicing Factor*

A recent study showed that an imprinted region in chromosome 15 (15q11–q13), which is specifically deleted in PWS, encodes a class of lncRNAs named sno-lncRNAs (Yin et al. 2012). Sno-lncRNAs are generated by exonucleolytic trimming of excised intronic RNA, resulting in the formation of lncRNAs that contain stable snoRNP sequences at both ends but lack 5′ cap structures and 3′ poly(A) tails. Unlike snoRNAs and small Cajal body-specific RNAs, which localize to nucleoli and Cajal bodies, respectively, PWS region sno-lncRNAs are accumulated near their sites of transcription. Knockdown of PWS region sno-lncRNAs has little effect on the expression of nearby genes, suggesting that these sno-lncRNAs do not affect gene expression *in cis*. On the other hand, PWS region sno-lncRNAs contain multiple binding sites for the splicing factor Fox2 and seem to create a subnuclear structure at sites where Fox2 is pooled. Indeed, altering the levels of PWS region sno-lncRNAs leads to a redistribution of Fox2 in the nucleus and changes mRNA splicing patterns; hence, these sno-lncRNAs appear to function as a molecular sink for Fox2 and participate in the regulation of splicing in specific subnuclear domains. In PWS patients, Fox splicing factors are distributed uniformly throughout the nucleus, resulting in altered patterns of splicing regulation during early embryonic development and adulthood. Consequently, the lack of PWS region sno-lncRNAs may be implicated in the abnormal development of PWS patients.

7.3.4 *Colon Cancer Associated Transcript 1 (*CCAT1*) lncRNA*

The *CCAT1* lncRNA, which was originally identified using a representational difference analysis and cDNA cloning (Nissan et al. 2012), is located in the region upstream of the *MYC* gene and is highly associated with all stages of colon cancer tumorigenesis, including premalignancy (Alaiyan et al. 2013; Nissan et al. 2012). Recently, two isoforms of *CCAT1* (*CCAT1-S* and the novel long isoform *CCAT1-L*) were identified (Xiang et al. 2014); these lncRNAs are expressed from a single promoter and differ at their 3′ ends. Notably, *CCAT1-S* is located in the cytoplasm, whereas *CCAT1-L* accumulates at its site of transcription in the nucleus and forms nuclear foci. Although knockdown of *CCAT1-L* reduces *MYC* transcription, transient overexpression of *CCAT1-L* does not affect the expression level of this gene. These findings suggest that *CCAT1-L* regulates *MYC* transcription by functioning *in cis* at the *MYC* enhancer. In addition, *CCAT1-L* interacts with the CTCF transcription factor and modulates the long-range chromatin interaction between the *MYC* enhancer and promoter.

7.3.5 The meiRNA *lncRNA Controls Meiosis Initiation in* Schizosaccharomyces pombe

MeiRNA is an *S. pombe* lncRNA that promotes switching of the cell cycle mode from mitosis to meiosis. In fission yeast, the selective removal of meiosis-specific transcripts prevents untimely entry into meiosis. During vegetative growth, the YTH family RNA-binding protein Mmi1 binds to the determinant of selective removal region (DSR) of meiotic mRNAs and promotes their degradation (Harigaya et al. 2006). Mmi1 also localizes to multiple nuclear foci and promotes meiotic mRNA degradation through recruitment of the exosome (Harigaya et al. 2006; Yamanaka et al. 2010). In addition, binding of Mmi1 to specific nascent mRNAs containing the DSR motif promotes the formation of facultative heterochromatin islands at meiotic loci (Zofall et al. 2012). During the meiotic prophase, Mmi1 is sequestered to a structure known as the Mei2 dot, where its function is inhibited. The Mei2 dot contains the RNA-binding protein Mei2 and two lncRNA isoforms, *meiRNA-S* and *meiRNA-L*, which are transcribed from the *sme2* gene and attached to the *sme2* locus (Shimada et al. 2003; Watanabe and Yamamoto 1994; Yamashita et al. 1998). These lncRNAs are targets of Mmi1, suggesting that they can compete with meiotic targets for Mmi1 binding. These findings indicate that the transcription of specific lncRNAs shuts down the selective elimination of meiosis-specific transcripts, thereby playing a pivotal role in switching developmental gene expression in *S. pombe*.

The formation of bivalent chromosomes during meiosis allows the exchange of genetic material, and the Mei2 dot plays a role in homologous chromosome recognition (Ding et al. 2012). Deletion of the *sme2* locus, at which the Mei2 dot is formed, leads to a loss of robust chromosome pairing, while transposition of the *sme2* locus to other chromosomal sites promotes pairing at these ectopic sites in a manner that is dependent on the expression of the *meiRNA* lncRNAs. Although both *meiRNA-S* and *meiRNA-L* are accumulated at the sites of transcription, only *meiRNA-L* is required for chromosome pairing. In addition, the expression of *meiRNA* from both chromosomes is required for robust pairing. These findings suggest that mei2 dot structures act as site-specific identifiers in chromosomes.

7.4 Conclusions and Perspectives

Since RNA is the initial output of gene expression, using RNAs as the anchors of subcellular structures allows cells to respond rapidly to environmental or developmental cues. This strategy is used not only by lncRNAs that recruit histone modification complexes to specific loci, but also by architectural RNAs that form nuclear bodies in response to cellular stresses (Fig. 7.4). *NEAT1*, which is involved in paraspeckle formation along with more than 40 proteins, is the most extensively studied architectural RNA. Currently, the mechanism by which *NEAT1* ribonucleoproteins assemble to form an intact nuclear body and the *NEAT1* sequences to which these proteins bind are unclear. Answering these questions would help to elucidate the mechanisms by which architectural satIII and IGS lncRNAs are able to form nuclear stress bodies and nucleolar DCs, respectively, and may aid the discovery of novel

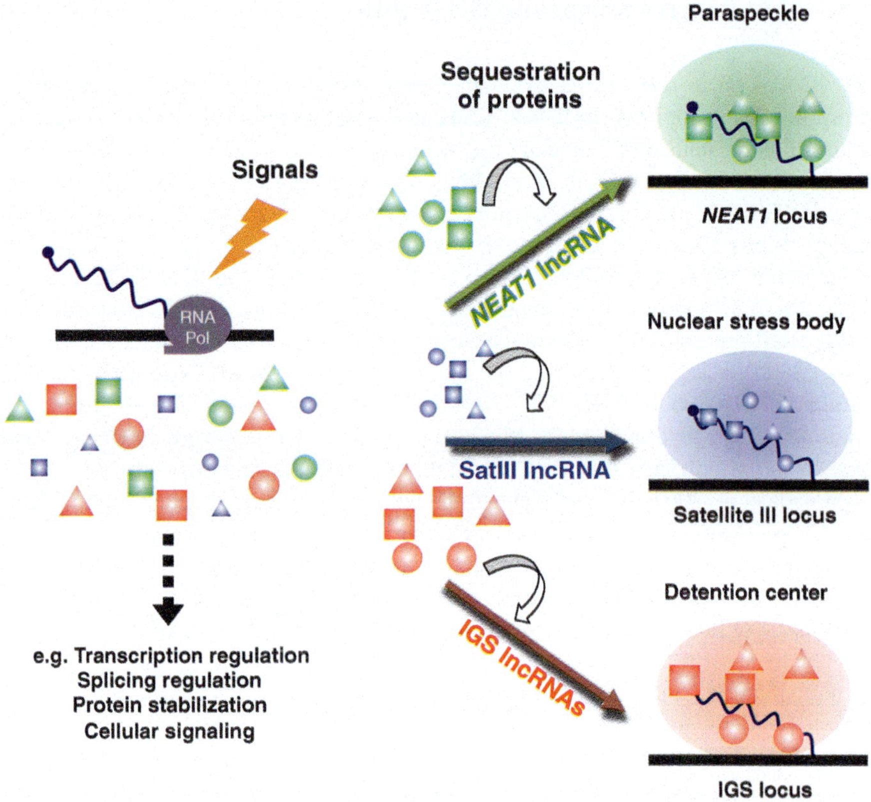

Fig. 7.4 Roles of architectural lncRNAs in nuclear body formation. Some lncRNAs serve as architectural RNAs by contributing to the biogenesis of specific nuclear bodies. In response to cellular stresses, architectural lncRNAs are upregulated and function as molecular sponges that sequester specific proteins in nuclear bodies such as paraspeckles, nuclear stress bodies and nucleolar DCs, thereby controlling transcriptional and splicing activities

architectural RNAs. Notably, *NEAT1*, SatIII, and IGS lncRNAs contain repetitive sequences that may function as RNA elements for efficient sequestration of multiple proteins on a single lncRNA molecule. A number of disease-associated nuclear foci are formed on transcripts from mutant genes that contain repetitive sequence expansions; these repeated RNAs are thought to cause hereditary neurological diseases (Wojciechowska and Krzyzosiak 2011), raising the "toxic RNA hypothesis". Nuclear RNA foci are formed on transcripts that are expressed at varying levels and harbor different types and lengths of repeated motifs, including CUG, CCUG, CGG, CAG, AUUCU, and UGGAA. Typically, these RNA foci contain specific RNA-binding proteins. Myotonic dystrophy type 1 (DM1), which is characterized by adult-onset muscular atrophy, is one example of a noncoding repeat expansion disorder; this disease is caused by an expansion of the CTG repeat in the 3′ UTR of the gene encoding dystrophia myotonica-protein kinase. The repeated RNA forms nuclear foci that sequester the muscleblind-like RNA splicing factor, resulting in the disturbance of gene expression. Additional investigations of endogenous lncRNA-containing nuclear bodies may contribute to current understanding of the manner in which disease-associated repetitive RNAs form nuclear foci.

Acknowledgments We thank all members of the Hirose laboratory for their encouragement and helpful discussions.

References

Alaiyan B, Ilyayev N, Stojadinovic A, Izadjoo M, Roistacher M, Pavlov V, Tzivin V, Halle D, Pan H, Trink B, Gure AO, Nissan A (2013) Differential expression of colon cancer associated transcript1 (CCAT1) along the colonic adenoma-carcinoma sequence. BMC Cancer 13:196. doi:10.1186/1471-2407-13-196

Audas TE, Jacob MD, Lee S (2012) Immobilization of proteins in the nucleolus by ribosomal intergenic spacer noncoding RNA. Mol Cell 45(2):147–157. doi:10.1016/j.molcel.2011.12.012

Batista PJ, Chang HY (2013) Long noncoding RNAs: cellular address codes in development and disease. Cell 152(6):1298–1307. doi:10.1016/j.cell.2013.02.012

Blackshaw S, Harpavat S, Trimarchi J, Cai L, Huang H, Kuo WP, Weber G, Lee K, Fraioli RE, Cho SH, Yung R, Asch E, Ohno-Machado L, Wong WH, Cepko CL (2004) Genomic analysis of mouse retinal development. PLoS Biol 2(9):E247. doi:10.1371/journal.pbio.0020247

Bongiorno-Borbone L, De Cola A, Vernole P, Finos L, Barcaroli D, Knight RA, Melino G, De Laurenzi V (2008) FLASH and NPAT positive but not Coilin positive Cajal Bodies correlate with cell ploidy. Cell Cycle 7(15):2357–2367

Chen LL, Carmichael GG (2009) Altered nuclear retention of mRNAs containing inverted repeats in human embryonic stem cells: functional role of a nuclear noncoding RNA. Mol Cell 35(4):467–478. doi:10.1016/j.molcel.2009.06.027

Clemson CM, Hutchinson JN, Sara SA, Ensminger AW, Fox AH, Chess A, Lawrence JB (2009) An architectural role for a nuclear noncoding RNA: NEAT1 RNA is essential for the structure of paraspeckles. Mol Cell 33(6):717–726. doi:10.1016/j.molcel.2009.01.026

Cremer TC, Cremer M (2010) Chromosome Territories. Cold Spring Harb Perspect Biol 2(3):a003889. doi:10.1101/cshperspect.a003889

Denegri M, Chiodi I, Corioni M, Cobianchi F, Riva S, Biamonti G (2001) Stress-induced nuclear bodies are sites of accumulation of pre-mRNA processing factors. Mol Biol Cell 12(11):3502–3514

Derrien T, Johnson R, Bussotti G, Tanzer A, Djebali S, Tilgner H, Guernec G, Martin D, Merkel A, Knowles DG, Lagarde J, Veeravalli L, Ruan X, Ruan Y, Lassmann T, Carninci P, Brown JB, Lipovich L, Gonzalez JM, Thomas M, Davis CA, Shiekhattar R, Gingeras TR, Hubbard TJ, Notredame C, Harrow J, Guigo R (2012) The GENCODE v7 catalog of human long noncoding RNAs: analysis of their gene structure, evolution, and expression. Genome Res 22(9):1775–1789. doi:10.1101/gr.132159.111

Ding DQ, Okamasa K, Yamane M, Tsutsumi C, Haraguchi T, Yamamoto M, Hiraoka Y (2012) Meiosis-specific noncoding RNA mediates robust pairing of homologous chromosomes in meiosis. Science 336(6082):732–736. doi:10.1126/science.1219518

Djebali S, Davis CA, Merkel A, Dobin A, Lassmann T, Mortazavi A, Tanzer A, Lagarde J, Lin W, Schlesinger F, Xue C, Marinov GK, Khatun J, Williams BA, Zaleski C, Rozowsky J, Roder M, Kokocinski F, Abdelhamid RF, Alioto T, Antoshechkin I, Baer MT, Bar NS, Batut P, Bell K, Bell I, Chakrabortty S, Chen X, Chrast J, Curado J, Derrien T, Drenkow J, Dumais E, Dumais J, Duttagupta R, Falconnet E, Fastuca M, Fejes-Toth K, Ferreira P, Foissac S, Fullwood MJ, Gao H, Gonzalez D, Gordon A, Gunawardena H, Howald C, Jha S, Johnson R, Kapranov P, King B, Kingswood C, Luo OJ, Park E, Persaud K, Preall JB, Ribeca P, Risk B, Robyr D, Sammeth M, Schaffer L, See LH, Shahab A, Skancke J, Suzuki AM, Takahashi H, Tilgner H, Trout D, Walters N, Wang H, Wrobel J, Yu Y, Ruan X, Hayashizaki Y, Harrow J, Gerstein M, Hubbard T, Reymond A, Antonarakis SE, Hannon G, Giddings MC, Ruan Y, Wold B, Carninci P, Guigo R, Gingeras TR (2012) Landscape of transcription in human cells. Nature 489(7414):101–108. doi:10.1038/nature11233

Dundr M, Misteli T (2010) Biogenesis of nuclear bodies. Cold Spring Harb Perspect Biol 2(12):a000711. doi:10.1101/cshperspect.a000711

Fong KW, Li Y, Wang W, Ma W, Li K, Qi RZ, Liu D, Songyang Z, Chen J (2013) Whole-genome screening identifies proteins localized to distinct nuclear bodies. J Cell Biol 203(1):149–164. doi:10.1083/jcb.201303145

Fox AH, Lam YW, Leung AK, Lyon CE, Andersen J, Mann M, Lamond AI (2002) Paraspeckles: a novel nuclear domain. Curr Biol: CB 12(1):13–25

Fox AH, Bond CS, Lamond AI (2005) P54nrb forms a heterodimer with PSP1 that localizes to paraspeckles in an RNA-dependent manner. Mol Biol Cell 16(11):5304–5315. doi:10.1091/mbc.E05-06-0587

Geisler S, Coller J (2013) RNA in unexpected places: long non-coding RNA functions in diverse cellular contexts. Nat Rev Mol Cell Biol 14(11):699–712. doi:10.1038/nrm3679

Harigaya Y, Tanaka H, Yamanaka S, Tanaka K, Watanabe Y, Tsutsumi C, Chikashige Y, Hiraoka Y, Yamashita A, Yamamoto M (2006) Selective elimination of messenger RNA prevents an incidence of untimely meiosis. Nature 442(7098):45–50. doi:10.1038/nature04881

Hirose T, Virnicchi G, Tanigawa A, Naganuma T, Li R, Kimura H, Yokoi T, Nakagawa S, Benard M, Fox AH, Pierron G (2014) NEAT1 long noncoding RNA regulates transcription via protein sequestration within subnuclear bodies. Mol Biol Cell 25(1):169–183. doi:10.1091/mbc.E13-09-0558

Hutchinson JN, Ensminger AW, Clemson CM, Lynch CR, Lawrence JB, Chess A (2007) A screen for nuclear transcripts identifies two linked noncoding RNAs associated with SC35 splicing domains. BMC Genomics 8:39. doi:10.1186/1471-2164-8-39

Imamura K, Imamachi N, Akizuki G, Kumakura M, Kawaguchi A, Nagata K, Kato A, Kawaguchi Y, Sato H, Yoneda M, Kai C, Yada T, Suzuki Y, Yamada T, Ozawa T, Kaneki K, Inoue T, Kobayashi M, Kodama T, Wada Y, Sekimizu K, Akimitsu N (2014) Long noncoding RNA NEAT1-dependent SFPQ relocation from promoter region to paraspeckle mediates IL8 expression upon immune stimuli. Mol Cell 53(3):393–406. doi:10.1016/j.molcel.2014.01.009

Jacob MD, Audas TE, Uniacke J, Trinkle-Mulcahy L, Lee S (2013) Environmental cues induce a long noncoding RNA-dependent remodeling of the nucleolus. Mol Biol Cell 24(18):2943–2953. doi:10.1091/mbc.E13-04-0223

Ji P, Diederichs S, Wang W, Boing S, Metzger R, Schneider PM, Tidow N, Brandt B, Buerger H, Bulk E, Thomas M, Berdel WE, Serve H, Muller-Tidow C (2003) MALAT-1, a novel noncoding

RNA, and thymosin beta4 predict metastasis and survival in early-stage non-small cell lung cancer. Oncogene 22(39):8031–8041. doi:10.1038/sj.onc.1206928

Jolly C, Konecny L, Grady DL, Kutskova YA, Cotto JJ, Morimoto RI, Vourc'h C (2002) In vivo binding of active heat shock transcription factor 1 to human chromosome 9 heterochromatin during stress. J Cell Biol 156(5):775–781. doi:10.1083/jcb.200109018

Jolly C, Metz A, Govin J, Vigneron M, Turner BM, Khochbin S, Vourc'h C (2004) Stress-induced transcription of satellite III repeats. J Cell Biol 164(1):25–33. doi:10.1083/jcb.200306104

Kim DD, Kim TT, Walsh T, Kobayashi Y, Matise TC, Buyske S, Gabriel A (2004) Widespread RNA editing of embedded alu elements in the human transcriptome. Genome Res 14(9):1719–1725. doi:10.1101/gr.2855504

Mao YS, Sunwoo H, Zhang B, Spector DL (2011a) Direct visualization of the co-transcriptional assembly of a nuclear body by noncoding RNAs. Nat Cell Biol 13(1):95–101. doi:10.1038/ncb2140

Mao YS, Zhang B, Spector DL (2011b) Biogenesis and function of nuclear bodies. Trends Genet 27(8):295–306. doi:10.1016/i.tig.2011.05.006

Mekhail K, Gunaratnam L, Bonicalzi ME, Lee S (2004) HIF activation by pH-dependent nucleolar sequestration of VHL. Nat Cell Biol 6(7):642–647. doi:10.1038/ncb1144

Naganuma T, Nakagawa S, Tanigawa A, Sasaki YF, Goshima N, Hirose T (2012) Alternative 3′-end processing of long noncoding RNA initiates construction of nuclear paraspeckles. EMBO J 31(20):4020–4034. doi:10.1038/emboj.2012.251

Nakagawa S, Naganuma T, Shioi G, Hirose T (2011) Paraspeckles are subpopulation-specific nuclear bodies that are not essential in mice. J Cell Biol 193(1):31–39. doi:10.1083/jcb.201011110

Nakagawa S, Ip JY, Shioi G, Tripathi V, Zong X, Hirose T, Prasanth KV (2012) Malat1 is not an essential component of nuclear speckles in mice. RNA 18(8):1487–1499. doi:10.1261/rna.033217.112

Nishimoto Y, Nakagawa S, Hirose T, Okano HJ, Takao M, Shibata S, Suyama S, Kuwako K, Imai T, Murayama S, Suzuki N, Okano H (2013) The long non-coding RNA nuclear-enriched abundant transcript 1_2 induces paraspeckle formation in the motor neuron during the early phase of amyotrophic lateral sclerosis. Mol Brain 6:31. doi:10.1186/1756-6606-6-31

Nissan A, Stojadinovic A, Mitrani-Rosenbaum S, Halle D, Grinbaum R, Roistacher M, Bochem A, Dayanc BE, Ritter G, Gomceli I, Bostanci EB, Akoglu M, Chen YT, Old LJ, Gure AO (2012) Colon cancer associated transcript-1: a novel RNA expressed in malignant and pre-malignant human tissues. Int J Cancer J Int du Cancer 130(7):1598–1606. doi:10.1002/ijc.26170

Nizami Z, Deryusheva S, Gall JG (2010) The Cajal body and histone locus body. Cold Spring Harb Perspect Biol 2(7):a000653. doi:10.1101/cshperspect.a000653

Prasanth KV, Prasanth SG, Xuan Z, Hearn S, Freier SM, Bennett CF, Zhang MQ, Spector DL (2005) Regulating gene expression through RNA nuclear retention. Cell 123(2):249–263. doi:10.1016/j.cell.2005.08.033

Rapicavoli NA, Poth EM, Blackshaw S (2010) The long noncoding RNA RNCR2 directs mouse retinal cell specification. BMC Dev Biol 10:49. doi:10.1186/1471-213X-10-49

Rizzi N, Denegri M, Chiodi I, Corioni M, Valgardsdottir R, Cobianchi F, Riva S, Biamonti G (2004) Transcriptional activation of a constitutive heterochromatic domain of the human genome in response to heat shock. Mol Biol Cell 15(2):543–551. doi:10.1091/mbc.E03-07-0487

Saha S, Murthy S, Rangarajan PN (2006) Identification and characterization of a virus-inducible non-coding RNA in mouse brain. J Gen Virol 87(Pt 7):1991–1995. doi:10.1099/vir.0.81768-0

Sasaki YT, Ideue T, Sano M, Mituyama T, Hirose T (2009) MENepsilon/beta noncoding RNAs are essential for structural integrity of nuclear paraspeckles. Proc Natl Acad Sci U S A 106(8):2525–2530. doi:10.1073/pnas.0807899106

Shevtsov SP, Dundr M (2011) Nucleation of nuclear bodies by RNA. Nat Cell Biol 13(2):167–173. doi:10.1038/ncb2157

Shimada T, Yamashita A, Yamamoto M (2003) The fission yeast meiotic regulator Mei2p forms a dot structure in the horse-tail nucleus in association with the sme2 locus on chromosome II. Mol Biol Cell 14(6):2461–2469. doi:10.1091/mbc.E02-11-0738

Sone M, Hayashi T, Tarui H, Agata K, Takeichi M, Nakagawa S (2007) The mRNA-like noncoding RNA Gomafu constitutes a novel nuclear domain in a subset of neurons. J Cell Sci 120(Pt 15):2498–2506. doi:10.1242/jcs.009357

Souquere S, Beauclair G, Harper F, Fox A, Pierron G (2010) Highly ordered spatial organization of the structural long noncoding NEAT1 RNAs within paraspeckle nuclear bodies. Mol Biol Cell 21(22):4020–4027. doi:10.1091/mbc.E10-08-0690

Stark LA, Dunlop MG (2005) Nucleolar sequestration of RelA (p65) regulates NF-kappaB-driven transcription and apoptosis. Mol Cell Biol 25(14):5985–6004. doi:10.1128/MCB.25.14.5985-6004.2005

Sunwoo H, Dinger ME, Wilusz JE, Amaral PP, Mattick JS, Spector DL (2009) MEN epsilon/beta nuclear-retained non-coding RNAs are up-regulated upon muscle differentiation and are essential components of paraspeckles. Genome Res 19(3):347–359. doi:10.1101/gr.087775.108

Tollervey JR, Curk T, Rogelj B, Briese M, Cereda M, Kayikci M, Konig J, Hortobagyi T, Nishimura AL, Zupunski V, Patani R, Chandran S, Rot G, Zupan B, Shaw CE, Ule J (2011) Characterizing the RNA targets and position-dependent splicing regulation by TDP-43. Nat Neurosci 14(4):452–458. doi:10.1038/nn.2778

Tripathi V, Ellis JD, Shen Z, Song DY, Pan Q, Watt AT, Freier SM, Bennett CF, Sharma A, Bubulya PA, Blencowe BJ, Prasanth SG, Prasanth KV (2010) The nuclear-retained noncoding RNA MALAT1 regulates alternative splicing by modulating SR splicing factor phosphorylation. Mol Cell 39(6):925–938. doi:10.1016/j.molcel.2010.08.011

Tsuiji H, Yoshimoto R, Hasegawa Y, Furuno M, Yoshida M, Nakagawa S (2011) Competition between a noncoding exon and introns: Gomafu contains tandem UACUAAC repeats and associates with splicing factor-1. Genes Cells: Devoted Mol Cell Mech 16(5):479–490. doi:10.1111/j.1365-2443.2011.01502.x

Ulitsky I, Shkumatava A, Jan CH, Sive H, Bartel DP (2011) Conserved function of lincRNAs in vertebrate embryonic development despite rapid sequence evolution. Cell 147(7):1537–1550. doi:10.1016/j.cell.2011.11.055

Valgardsdottir R, Chiodi I, Giordano M, Cobianchi F, Riva S, Biamonti G (2005) Structural and functional characterization of noncoding repetitive RNAs transcribed in stressed human cells. Mol Biol Cell 16(6):2597–2604. doi:10.1091/mbc.E04-12-1078

Visa N, Puvion-Dutilleul F, Bachellerie JP, Puvion E (1993) Intranuclear distribution of U1 and U2 snRNAs visualized by high resolution in situ hybridization: revelation of a novel compartment containing U1 but not U2 snRNA in HeLa cells. Eur J Cell Biol 60(2):308–321

Watanabe Y, Yamamoto M (1994) S. pombe mei2+ encodes an RNA-binding protein essential for premeiotic DNA synthesis and meiosis I, which cooperates with a novel RNA species meiRNA. Cell 78(3):487–498

Welch WJ, Feramisco JR (1984) Nuclear and nucleolar localization of the 72,000-dalton heat shock protein in heat-shocked mammalian cells. J Biol Chem 259(7):4501–4513

Wojciechowska M, Krzyzosiak WJ (2011) Cellular toxicity of expanded RNA repeats: focus on RNA foci. Hum Mol Genet 20(19):3811–3821. doi:10.1093/hmg/ddr299

Xiang JF, Yin QF, Chen T, Zhang Y, Zhang XO, Wu Z, Zhang S, Wang HB, Ge J, Lu X, Yang L, Chen LL (2014) Human colorectal cancer-specific CCAT1-L lncRNA regulates long-range chromatin interactions at the MYC locus. Cell Res 24(5):513–531. doi:10.1038/cr.2014.35

Yamanaka S, Yamashita A, Harigaya Y, Iwata R, Yamamoto M (2010) Importance of polyadenylation in the selective elimination of meiotic mRNAs in growing S. pombe cells. EMBO J 29(13):2173–2181. doi:10.1038/emboj.2010.108

Yamashita A, Watanabe Y, Nukina N, Yamamoto M (1998) RNA-assisted nuclear transport of the meiotic regulator Mei2p in fission yeast. Cell 95(1):115–123

Yang L, Lin C, Liu W, Zhang J, Ohgi KA, Grinstein JD, Dorrestein PC, Rosenfeld MG (2011) ncRNA- and Pc2 methylation-dependent gene relocation between nuclear structures mediates gene activation programs. Cell 147(4):773–788. doi:10.1016/j.cell.2011.08.054

Yin QF, Yang L, Zhang Y, Xiang JF, Wu YW, Carmichael GG, Chen LL (2012) Long noncoding RNAs with snoRNA ends. Mol Cell 48(2):219–230. doi:10.1016/j.molcel.2012.07.033

Zhang Z, Carmichael GG (2001) The fate of dsRNA in the nucleus: a p54(nrb)-containing complex mediates the nuclear retention of promiscuously A-to-I edited RNAs. Cell 106(4):465–475

Zhang B, Arun G, Mao YS, Lazar Z, Hung G, Bhattacharjee G, Xiao X, Booth CJ, Wu J, Zhang C, Spector DL (2012) The lncRNA Malat1 is dispensable for mouse development but its transcription plays a cis-regulatory role in the adult. Cell Rep 2(1):111–123. doi:10.1016/j.celrep.2012.06.003

Zofall M, Yamanaka S, Reyes-Turcu FE, Zhang K, Rubin C, Grewal SI (2012) RNA elimination machinery targeting meiotic mRNAs promotes facultative heterochromatin formation. Science 335(6064):96–100. doi:10.1126/science.1211651

Part IV
Biological Actions of lncRNAs

Chapter 8
Long Noncoding RNA in Epigenetic Gene Regulation

Yuko Hasegawa and Shinichi Nakagawa

Abstract Recent studies have revealed the functional significance of long noncoding RNA (lncRNA) in various biological processes including epigenetic gene regulations. Genomic imprinting is one of the epigenetic processes related to lncRNA in mammals, which controls parent-of-origin-specific gene expression essential for normal development. To date, over 100 genes have been recognized as imprinted genes, the majority of them form clusters on the genome. Each of these imprinting clusters contains DNA regulatory elements called imprinting control regions (ICRs), which are frequently located near the lncRNA genes. In some cases, genetically modified mice and human patients exhibiting imprinting disorder show aberrant expression of these lncRNAs, suggesting a close relationship between genomic imprinting and lncRNAs. DNA methylation and histone modifications are the principal molecular mechanisms of epigenetic gene regulation, and recent progress has uncovered diversities and complexities of the lncRNA actions in these processes. In this chapter, we summarize research on genomic imprinting by focusing on the role of lncRNAs to provide insight into lncRNA-mediated regulation of gene expression.

Keywords Long non-coding RNA • Genomic imprinting • Epigenetic gene regulation

8.1 Introduction

Early observations that the phenotypes of the progeny differ on the basis of whether they are derived from the father or the mother led to the discovery of imprinting in a variety of organisms, including insects, plants, and mammals (reviewed in Ferguson-Smith 2011).

In the imprinted genomic regions, genes are expressed from one of the two homologous chromosomes, resulting in mono-allelic gene expression. This parent-of-origin-specific gene regulation is accompanied by differential DNA methylation

Y. Hasegawa (✉) • S. Nakagawa
Nakagawa RNA Biology Laboratory, RIKEN, 2-1 Hirosawa, Wako, Saitama 351-0198, Japan
e-mail: yhasegawa@riken.jp; nakagawas@riken.jp

R. Kurokawa (ed.), *Long Noncoding RNAs*, DOI 10.1007/978-4-431-55576-6_8

in male and female gametes (Bourc'his et al. 2001; Howell et al. 2001; Hata et al. 2002; Kaneda et al. 2004) arising from differentially methylated regions (DMRs) (Sasaki et al. 1992; Stoger et al. 1993; Bartolomei et al. 1993; Ferguson-Smith et al. 1993). Approximately 150 imprinted genes have been identified to date (http://www.mousebook.org/mousebook-catalogs/imprinting-resource), and this number is likely to increase consequent to the development of the sequencing techniques that enable genome-wild and tissue-specific analyses of gene expression, distinguishing the maternally and paternally derived alleles using SNPs (single nucleotide polymorphisms). The majority of imprinted genes cluster on several chromosomal regions. These imprinting clusters contain at least one lncRNA gene together with several protein-coding genes (reviewed in Koerner et al. 2009). The imprinting clusters are well conserved between mouse and human. Several human diseases are caused by imprinting disorders, and mapping of the responsible genes has been done in patients with parental-origin effects (Hall 1990). Coupled with these efforts, studies using genetically modified mice have identified cis-acting elements called the imprinting control region (ICR; synonym: imprinted control element (ICE)) in each imprinting cluster (reviewed in Edwards and Ferguson-Smith 2007). Deletion of the ICR causes loss of the parental-specific expression pattern of the imprinted genes within the imprinting clusters. Therefore, the ICR is regarded as a regulatory element ensuring ordered genomic imprinting. The ICR contains DMR, wherein parent-specific DNA methylation alters the ICR properties and promotes asymmetric regulation of the homologous genes within the imprinting clusters. Three imprinting clusters (Igf2r, Kcnq1, and Gnas) acquire DNA methylation on their maternally inherited ICRs, whereas the paternal ICRs of two additional clusters (Igf2 and Dlk1) are methylated (Figs. 8.1, 8.2, 8.3, 8.4, 8.5, and 8.6). All of the ICRs within these imprinting clusters are located upstream of or in the promoters of the lncRNA genes; thus, whether these lncRNAs play functional roles in genomic imprinting has been a central question in this field. To date, the roles of

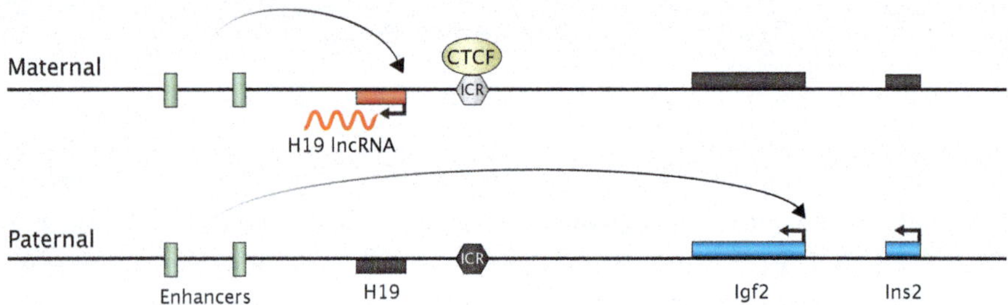

Fig. 8.1 Igf2 imprinting cluster. Enhancers located downstream of H19 interact with the Igf2 promoter exclusively when the ICR on the same chromosome is methylated. The unmethylated ICR on the maternal allele recruits CTCF to prevent the enhancer–promoter interaction, resulting in the repression of Igf2

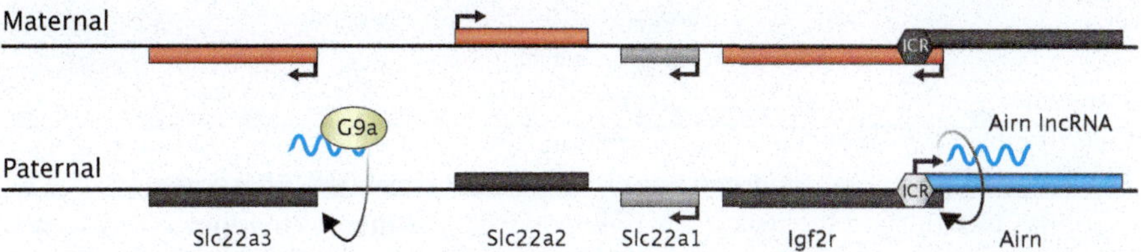

Fig. 8.2 Igf2r imprinting cluster. Slc22a3 and Slc22a2 are placenta-specific imprinted genes, whereas Igf2r is a ubiquitous imprinted gene. The ICR containing the Airn promoter is located within the intron region of Igf2r. Airn transcription intersects with the Igf2r promoter and represses transcription via transcription interference. Airn lncRNA interacts with G9a and recruits H3K9me3 to the Slc22a3 promoter. *Black arrows* indicate the transcriptional direction

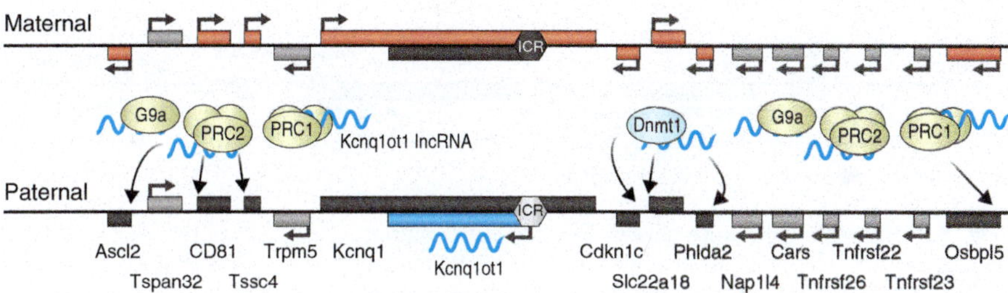

Fig. 8.3 Kcnq1 imprinting cluster. Ascl2, CD81, Tssc4, and Osbpl5 are placenta-specific imprinted genes. Kcnq1, Cdkn1c, Slc22a18 and Phlda2 are ubiquitous imprinted genes. Kcnq1ot1 lncRNA is expressed from the intron region of Kcnq1. In the placenta, Kcnq1ot1 represses placenta-specific imprinted genes via regulating histone modification. Ubiquitous imprinted genes require the DNA methyltransferase Dnmt1 for silencing, and Kcnq1ot1 regulates DNA methylation. *Black arrows* indicate the transcriptional direction

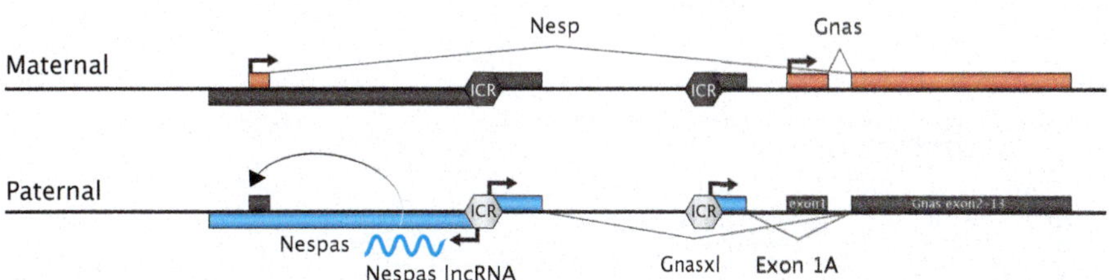

Fig. 8.4 Gnas imprinting cluster. Nesp, Gnasxl, Exon 1A and Gnas are variant transcripts of the Gnas protein-coding gene. Maternal-specific Gnas expression is limited in several tissues. The ICR located upstream of Exon 1A specifically regulates Gnas imprinted expression, whereas the ICR located upstream of Nespas regulates all imprinted genes. Nespas lncRNA is required for the repression of Nesp expression. *Black arrows* indicate the transcriptional direction

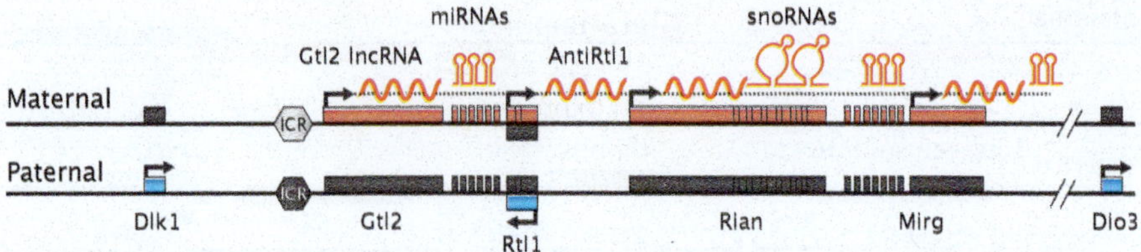

Fig. 8.5 Dlk1-Dio3 imprinting cluster. Anti-Rtl1 and Mirg contain miRNA genes. Rian encodes box C/D snoRNAs. Many additional miRNA genes are located in this imprinting cluster. *Black arrows* indicate the transcriptional direction. A *dashed line* means continuous transcription

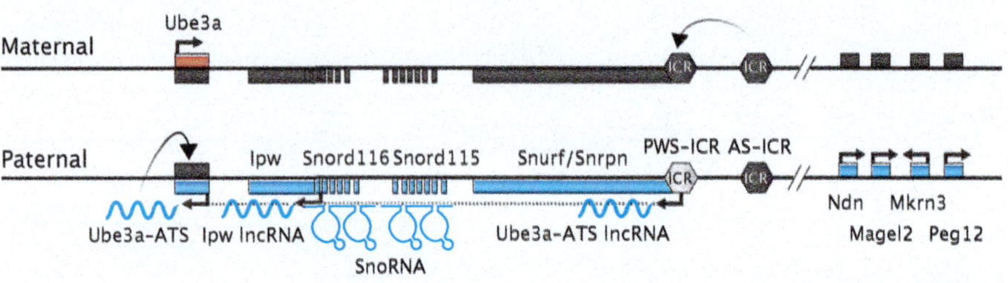

Fig. 8.6 PWS/AS imprinting cluster. The PWS/AS imprinting cluster has two ICRs. AS-ICR regulates PWS-ICR. Note: Both of the alleles of AS-ICR are methylated (Perk et al. 2002). Ube3a-ATS lncRNA is expressed from the nearby Snurf/Snrpn promoter and regulates Ube3a imprinted expression. Ube3a is a brain-specific imprinted gene. Snord115 and Snord116 encode snoRNAs. *Black arrows* indicate the transcriptional direction. A *dashed line* means continuous transcription

several lncRNAs in the regulation of imprinted genes have been assessed, and multiple types of regulatory mechanisms have been proposed. This review is a follow-up of what is known and how researchers have studied these lncRNAs involved in genomic imprinting, which represents one of the important functions of noncoding RNAs. The details of Xist and X-inactivation are discussed in Chap. 9.

8.2 Genomic Imprinting Cluster

8.2.1 Igf2 Imprinting Cluster

The most famous example of lncRNA-independent genomic imprinting regulation is the Igf2 imprinting cluster, which contains paternally expressed protein-coding genes and maternally expressed H19 lncRNA (Fig. 8.1). H19 is an approximately 2 kb transcript that is spliced and exported to the cytoplasm; however, the transcript does not contain conserved potential open reading frames and does not associate with ribosomes (Brannan et al. 1990). Mice lacking the maternal H19 gene and an additional 10 kb upstream flanking sequence exhibit loss of imprinted expression of the protein-coding genes in the Igf2 imprinting cluster (Leighton et al. 1995). However, the replacement of the H19 gene with a luciferase gene and the promoter deletion did not affect Igf2-imprinted expression (Jones et al. 1998; Schmidt et al. 1999); thus, the imprinting inhibition observed in the mice is thought to be caused by the loss of the H19 upstream element, which was subsequently identified as the ICR (Thorvaldsen et al. 1998). These results indicate that the transcript expressed from the H19 gene locus is dispensable for regulating the Igf2 imprinting cluster. In addition, the ICR in the Igf2 cluster exhibits insulator activity in a DNA methylation-dependent manner (Bell and Felsenfeld 2000; Hark et al. 2000; Kaffer et al. 2000; Szabo et al. 2000). These studies reported that the unmethylated maternal ICR in the Igf2 imprinting cluster interacts with the insulator protein CTCF (CCCTC binding factor) to prevent enhancers located downstream of H19 and Igf2r from binding to the Igf2 promoter and activating expression, thereby alternatively allowing the enhancers to interact with the H19 promoter to activate H19 lncRNA expression. DNA methylation on the paternally inherited ICR inhibits CTCF binding and is thought to promote the engagement of the enhancers with the Igf2 promoter to promote Igf2 expression (Fig. 8.1). These long-range chromosomal interactions were assessed using chromosome conformation capture (3C) (Murrell et al. 2004; Kurukuti et al. 2006). Thus, the Igf2 imprinting cluster has been recognized as an insulator model wherein genomic imprinting is independent of noncoding RNA. Additionally, these studies indicate that changes in the three-dimensional chromatin structure facilitate epigenetic gene regulation.

8.2.2 *Igf2r Imprinting Cluster*

Igf2r is one of the earliest identified imprinted genes in mice (Barlow et al. 1991). The Igf2r imprinting cluster on mouse chromosome 17 contains three maternally expressed protein-coding genes and one paternally expressed Airn (antisense Igf2r RNA) lncRNA (Fig. 8.2). Among the three protein-coding genes, only Igf2r is ubiquitously imprinted, and the other two genes exhibit imprinted expression specifically in the placenta (Zwart et al. 2001). Airn lncRNA is transcribed from the promoter located inside the intronic region of the Igf2r protein-coding gene in an antisense orientation overlapping the Igf2r promoter. Although Airn shares the common features of other polymerase II transcripts (5′ capped and polyadenylated), the majority of the transcripts are not spliced, resulting in production of huge transcripts at the size of approximately 100 kb (Seidl et al. 2006). Although the minor spliced variants are exported to the cytoplasm in a fashion similar to canonical mRNAs, a fraction of unspliced Airn remains in the nucleus and accumulates near its transcriptional site (Seidl et al. 2006; Nagano et al. 2008). The ICR in the Igf2r imprinting cluster contains the Airn promoter (Wutz et al. 1997). Airn promoter deletion or the truncation of the Airn transcript by inserting a polyadenylation cassette causes the loss of parent-specific expression of additional protein-coding genes that are located upstream or downstream of the Airn gene, demonstrating that

Airn lncRNA functions as a bidirectional silencer (Wutz et al. 1997; Zwart et al. 2001; Sleutels et al. 2002). The role of Airn lncRNA in the regulation of the Slc22a3 protein-coding gene was thoroughly investigated (Nagano et al. 2008). In the earlier phases of development, Airn lncRNA interacts with the Slc22a3 promoter in the placenta, but this interaction is significantly reduced during later stages of development when Slc22a3 mono-allelic expression is no longer observed. Importantly, this promoter-lncRNA interaction was not observed in the other two imprinted genes, Slc22a2 and Igf2r. The Airn distribution pattern across the Slc22a3 promoter highly correlates with H3K9me3 enrichment, wherein histone modification is catalyzed by G9a histone methyltransferase. Airn interacts with G9a, and G9a-deficient mice exhibit aberrant Slc22a3 imprinted expression. This finding implies that Airn silences the paternally inherited Slc22a3 allele possibly via interaction with G9a. Airn nuclear localization may be required for G9a recruitment to specific sites within the Igf2r imprinting cluster. In contrast, G9a depletion does not affect Igf2r imprinted gene expression (Nagano et al. 2008), despite the fact that all of the imprinted genes within the Igf2r cluster require Airn for their mono-allelic expression (Wutz et al. 1997; Zwart et al. 2001; Sleutels et al. 2002). Thus, Airn is likely to silence Igf2r by different mechanisms that are independent of G9a. Because Airn overlaps with the Igf2r promoter but not with the promoter of Slc22a3 and Slc22a2, transcriptional interference (Shearwin et al. 2005) has been hypothesized as an alternative gene silencing mechanism. This possibility was assessed by truncation of Airn transcription by inserting polyadenylation cassette in various locations, or exchanges of Airn promoter location. Interestingly, mono-allelic expression of Igf2r was not observed when the transcription of Airn did not overlap with the Igf2r promoter, suggesting that transcription interference with the Igf2r promoter, but not Airn RNA itself, is required for the gene silencing (Latos et al. 2012). Transcriptional interference likely occurs in the absence of repressive chromatin markers because the repressed Igf2r promoter retains active chromatin features, including high sensitivity to DNase I treatment and H3K4me3 enrichment (Latos et al. 2012). Thus, Airn exploits a mechanism for gene repression; transcriptional interference for Igf2r and G9a histone methyltransferase-dependent mechanisms for Slc22a3 (Fig. 8.2).

8.2.3 *Kcnq1 Imprinting Cluster*

The Kcnq1 imprinting cluster is located on the distal end of mouse chromosome 7, and its homologous region is located on human chromosome 11p15.5. This region contains one paternally expressed lncRNA, referred to as Kcnq1ot1 (Kcnq1 opposite strand/antisense transcript 1), and several maternally expressing protein-coding genes (Fig. 8.3). The ICR in this imprinting cluster contains a CpG island near the Kcnq1ot1 promoter, and deletion of this island inhibits parent-specific expression of multiple imprinted genes (Fitzpatrick et al. 2002; Yatsuki et al. 2002). Kcnq1ot1 lncRNA is an approximately 90 kb unspliced noncoding RNA that is transcribed from the intronic region of the Kcnq1 protein-coding gene. This noncoding RNA was identified by several groups studying Beckwith-Wiedemann syndrome (BWS), which causes prenatal overgrowth and a predisposition to cancer (Lee et al. 1999; Mitsuya et al. 1999; Smilinich et al. 1999). BWS patients commonly exhibit loss of methylation at the ICR and bi-allelic expression of Kcnq1ot1 lncRNA (Lee et al. 1999; Smilinich et al. 1999; Engel et al. 2000). The association of Kcnq1ot1 lncRNA with the regulation of the imprinted genes was examined via promoter deletion and the induction of premature transcription termination (Fitzpatrick et al. 2002; Mancini-Dinardo et al. 2006). These results provide strong evidence supporting the notion that Kcnq1ot1 is required for the regulation of the Kcnq1 imprinting cluster. Imprinted genes in the Kcnq1 imprinting cluster are categorized into two groups. Genes that are located in the distal region of the Kcnq1 imprinting cluster exhibit parent-specific expression exclusively in the placenta, whereas their counterparts that are near the Kcnq1ot1 gene exhibit parent-specific expression in both the placenta and embryo, and are referred to as ubiquitously imprinted genes. All of the ubiquitously imprinted genes require DNA methyltransferase 1 (Dnmt1) to repress the paternal allele (Lewis et al. 2004), and deletion of the 1 kb region of Kcnq1ot1 lncRNA disrupts parent-specific DNA methylation (Mohammad et al. 2010). On the other hands, placenta-specific imprinted genes lack parent-of-origin-specific DNA methylation, and the imprinted expression of these genes is not lost in mouse lacking Dnmt1 (Lewis et al. 2004). These facts suggest that the placenta-specific imprinted gene expression is controlled by distinct mechanisms that control ubiquitously imprinted genes. Trimethylation at Lys27 on histone H3 (H3K27me3) and dimethylation at Lys9 on histone H3 (H3K9me2) are plausible components of the DNA methylation-independent silencing mechanism, given that they are recognized as repressive markers. ChIP (chromatin immunoprecipitation) analysis revealed that silenced alleles of the placenta-specific imprinted genes preferentially acquire these repressive histone modifications, compared with the active alleles (Umlauf et al. 2004). Moreover, high-resolution analysis revealed placenta-specific enrichment of histone methylation in several regions (Pandey et al. 2008). This finding implies that tissue-specific histone modifications across the Kcnq1 imprinting cluster are likely associated with imprinted gene regulation. Supporting these observations, mice strains lacking Ezh2 (embryonic ectoderm development), Rnf2 (ring finger protein 2), and G9a histone methyltransferase exhibit loss of placenta-specific imprinted gene expression (Terranova et al. 2008; Wagschal et al. 2008). Interestingly, DNA-FISH experiments indicate that paternally inherited repressed Kcnq1 imprinting cluster exhibits higher contraction compared with the active cluster on the maternally inherited chromosome, and this genomic contraction is disrupted in the absence of Ezh2 or Rnf2 (Terranova et al. 2008). This finding may suggest a connection between three-dimensional chromatin structure alterations and histone modifying factors. Parent-specific and tissue-specific histone modifications are disrupted by Kcnq1ot1 promoter deletion (Lewis et al. 2004; Pandey et al. 2008), suggesting a plausible role of Kcnq1ot1 in their regulation. Although no evidence currently suggests the direct binding of Kcnq1ot1 lncRNA to histone methyltransferase, RNA immunoprecipitation assays indicate that Kcnq1ot1 interacts with G9a,

Ezh2, and Suz12 specifically in the placenta (Pandey et al. 2008). As Kcnq1ot1 accumulates near its transcriptional site (Mohammad et al. 2008; Terranova et al. 2008), it can associate with these enzymes to establish a placenta-specific histone modification pattern across the Kcnq1 imprinting cluster. One puzzling aspect of this model is that some chromosomal regions within this cluster do not lose H3K27me3 enrichment without Kcnq1ot1 expression (Pandey et al. 2008). These sites may acquire H3K27me3 through Kcnq1ot1-independent mechanisms. The precise mechanisms of Kcnq1ot1 lncRNA-mediated gene regulation that accompanies histone modification and DNA methylation are likely to be diverse and complex. The Cdkn1c protein-coding gene represents such a complicated modulation, as described by numerous studies. The protein factors Dnmt1, Eed, Ezh2, and Rnf2 have been suggested to control Cdkn1c imprinted expression independently or additively, and their regulatory hierarchy remains unclear (Mager et al. 2003; Lewis et al. 2004; Terranova et al. 2008). Moreover, Cdkn1c imprinted expression depends on Kcnq1ot1 lncRNA; however, this dependency is not applicable to several tissues (Shin et al. 2008). The reason why the imprinted gene regulation in the Kcnq1 imprinting cluster is comprised of various mechanisms remains unclear. Mammalian organisms may possess these regulatory machineries to accommodate proper imprinting of divergent chromatin structures observed in the specific cells types. For better understanding of the primary role of Kcnq1ot1 lncRNA, it may be important to determine which genomic regions interact with Kcnq1ot1 and to uncover the site of Kcnq1ot1 action in the genome. This location can be identified using recently developed techniques, such as ChIRP (chromatin isolation by RNA purification), CHART (capture hybridization analysis of RNA targets) and RAP (RNA antisense purification) technology, which analyze genomic locations associated with lncRNAs (Chu et al. 2011; Simon et al. 2011; Engreitz et al. 2013).

8.2.4 Gnas Imprinting Cluster

The mouse Gnas imprinting cluster on chromosome 2 is a structurally complex domain that contains a paternally expressing noncoding RNA named Nespas (neuroendocrine secretory protein antisense antisense) and the Gnas protein-coding gene (Fig. 8.4). Gnas uses distinct promoters and the first exons to generate multiple transcripts referred to as Gnas, Exon 1A, Gnasxl, and Nesp (reviewed in Weinstein et al. 2004). The canonical Gnas transcript exhibits bi-allelic expression but is predominantly expressed from the maternally inherited allele in some tissues including the renal proximal tubule (Yu et al. 1998). Exon 1A and Gnasxl are expressed exclusively from the paternal allele, whereas Nesp is expressed from the maternally inherited chromosome (Hayward et al. 1998; Peters et al. 1999; Liu et al. 2000). Among these transcripts, Exon 1A does not contain an in-frame ATG codon and has been recognized as a noncoding RNA (Liu et al. 2000). Thus, the Gnas imprinting cluster contains two noncoding RNAs, Nespas and Exon 1A. Exon 1A and Nespas localization remains unknown; thus, it remains unclear whether these ncRNAs exhibit properties similar to Kcnq1ot1 and Airn lncRNA. The Gnas imprinting cluster contains two ICRs; one is near Exon 1A, and the other is near Nespas (Liu et al. 2000; Coombes et al. 2003). The ICR upstream of Exon 1A is required for Exon 1A expression and specifically regulates imprinted gene expression of canonical Gnas through unknown mechanisms (Williamson et al. 2004). Initially, it was assumed that Exon 1A competes with the Gnas promoter and that Exon 1A is exclusively expressed in the tissues where Gnas exhibits mono-allelic expression. However, Exon 1A is ubiquitously expressed (Liu et al. 2000), so its expression per se is unlikely to control Gnas expression. In contrast to the specialized function of the ICR upstream of Exon 1A in canonical Gnas regulation, deletion of the ICR containing the Nespas promoter not only abolishes the Nespas transcript but also causes aberrant expression of all other imprinted genes within the Gnas imprinting cluster (Williamson et al. 2006). Although it remains unclear whether Nespas lncRNA regulates all of the imprinted genes in this cluster, Nespas truncation via polyadenylation cassette insertion disrupts the establishment of DNA methylation in the Nesp promoter, resulting in the activation of the repressed allele (Williamson et al. 2011). Nesp derepression is possibly caused by the unusual enrichment of the active marker H3K4me3 (Williamson et al. 2011). Because methylation of histone H3 at lysine 4 prevents Dnmt3A from interacting with the chromatin (Ooi et al. 2007), the author proposed that Nespas lncRNA is primarily required to remove H3K4 methylation from the Nesp promoter by recruiting histone demethylases, before the acquisition of DNA methylation. Interestingly, ectopic expression of Nespas on the maternally inherited chromosome, where Nesp is originally active, represses Nesp in cis, despite the fact that the promoter remains unmethylated (Williamson et al. 2011).

8.2.5 Dlk1-Dio3 Imprinting Cluster

The Dlk1-Dio3 imprinting cluster contains multiple maternally expressed noncoding RNA genes, including Gtl2 (gene-trap locus 2; also known as Meg3, maternally expressed gene 3), AntiRtl1 (antisense transcript overlapping Rtl1 gene), Rian (RNA imprinted and accumulated in nucleus) containing the C/D-box snoRNA cluster, and Mirg (microRNA containing gene) (Fig. 8.5). All of these genes are transcribed in the same orientation and have intergenic transcripts, suggesting that lncRNAs in the Dlk1-Dio3 imprinting cluster are processed from one long polycistronic RNA that is transcribed from the Gtl2 promoter (Seitz et al. 2004). Gtl2 was the first identified lncRNA among the noncoding RNAs located in this imprinting cluster by analyzing mice that give rise to a parental-origin-dependent phenotype (Schuster-Gossler et al. 1998). The ICR in this imprinting cluster is located upstream of Gtl2 gene, and all of the imprinted genes are affected by ICR deletion in embryonic tissues (Lin et al. 2003, 2007). In contrast, only part of gene expression is changed in the placenta upon the deletion of the ICR (Lin et al. 2007), suggesting tissue-specific mechanism of action for the ICR in the Dlk1-Dio3 imprinting cluster. The functional role of Gtl2 lncRNA remains unclear. Gtl2 lncRNA localizes to the nucleus (Schuster-Gossler et al. 1998), indicating that it regulates gene expression similarly to other imprinting-associated lncRNAs. Two genetically modified mice strains containing deletions of the Gtl2 allele were created, but their phenotypes are inconsistent (Takahashi et al. 2009; Zhou et al. 2010). Although Gtl2 expression is abolished in both of these strains, only one strain exhibits a severe phenotype, including the reduction of other maternally expressed noncoding RNAs, activation of silenced genes on the paternal allele, and perinatal death. The author hypothesizes that the difference in the amount of residual noncoding RNAs downstream of the Gtl2 gene could cause the phenotypic variation (Zhou et al. 2010). However, the detailed function of other noncoding RNAs, including miRNAs and snoRNAs, encoded in the Dlk1-Dio3 imprinting cluster remains unclear; thus, further analysis is required to understand the functional importance of these noncoding RNAs. Interestingly, miRNAs generated from antiRtl1 are involved in a trans-silencing of Rtl1 through a siRNA-mediated pathway (Davis et al. 2005).

8.2.6 PWS/AS Imprinting Cluster

The PWS/AS imprinting cluster that maps to mouse chromosome 7 and human 15q11-q13 is implicated in two distinct neurobehavioral disorders, PWS (Prader-Willi syndrome) and AS (Angelman syndrome). This imprinting cluster contains Ube3a-ATS (Ube3a antisense RNA transcript) lncRNA, IPW (Imprinted in Prader-Willi syndrome) lncRNA, and the snoRNA clusters Snord115 and Snord116 (Fig. 8.6). PWS is caused by a deficiency in paternal gene expression, whereas AS results from a deficiency in maternal gene expression. The causative gene for the PWS phenotype has not been described yet, and the condition is thought to arise from the additive effect of multiple genes. In contrast, the major phenotypic defect of AS can be explained by the loss of maternal-specific expression of the Ube3a protein-coding gene (Mabb et al. 2011). Ube3a mono-allelic expression is detected specifically in the brain, and it exhibits bi-allelic expression in other tissues (Rougeulle et al. 1997; Vu and Hoffman 1997). Paternal Ube3a silencing is thought to be associated with a paternally expressed large antisense transcript called Ube3a-ATS because this lncRNA is also expressed specifically in the brain where Ube3a exhibits mono-allelic expression (Rougeulle et al. 1998; Chamberlain and Brannan 2001; Yamasaki et al. 2003). Ube3a-ATS is transcribed from a nearby ICR located upstream of Snurf/Snrpn protein-coding gene in the antisense direction of Ube3a, overlapping IPW and the snoRNA cluster (Runte et al. 2001). Given that all of these noncoding RNAs exhibit neuron-specific expression and are transcribed exclusively from the paternal allele, Ube3a-ATS may serve as a host transcript for them. Decreased Ube3a-ATS expression or early termination of Ube3a-ATS transcription inhibits paternal Ube3a silencing (Meng et al. 2012). A topoisomerase inhibitor reactivates Ube3a expression potentially through the reduction of Ube3a-ATS (Huang et al. 2012). Although the mechanism by which Ube3a-ATS represses Ube3a expression is unknown, Ube3a-ATS lncRNA is localized to the nucleus (Powell et al. 2013), and host lncRNAs of snoRNAs encoded in Snord115 and

Snord116 also accumulate near their transcription sites, possibly indicating a cis-regulatory function (Vitali et al. 2010). Although the role of these 'RNA clouds' remains obscure, current studies stress the trans-acting function of snoRNAs. These snoRNAs exhibit the structural hallmarks of box C/D snoRNAs but do not contain complementary sequences that could direct a pseudouridylation or methylation of rRNA or snRNAs (Cavaille et al. 2000; de los Santos et al. 2000; Filipowicz 2000). Instead, snoRNAs encoded by Snord115 exhibit sequence complementarity to the alternatively spliced exon of the serotonin receptor 5-HT_{2C}R that is likely to promote this exon inclusion (Kishore and Stamm 2006). 5-HT_{2C}R mRNA undergoes RNA editing, which affects the inclusion efficiency of this exon. However, these snoRNAs might induce change in the splicing pattern through editing-independent mechanisms because there is no evidence indicating that these snoRNAs cause the editing alternation (Kishore and Stamm 2006). Likewise, lncRNAs encoded by the other SNORD116 cluster are also associated with splicing regulation. In human ES cells, a new class of lncRNAs whose ends correspond to the positions of the snoRNAs (sno-lncRNAs) expressed from the human SNORD116 cluster were discovered by massively parallel sequencing analysis (Yin et al. 2012). Given that sno-lncRNAs localize near their gene loci, it was initially expected that they regulate imprinted gene expression within the PWS/AS cluster; however, no influences were observed upon knockdown of sno-lncRNAs with ASO (antisense oligonucleotide). Further analysis demonstrated that sno-lncRNAs interact with the alternative splicing regulator Fox2 and alter the Fox-regulated gene splicing possibly through sequestering Fox2 and subsequently inhibiting its activity. These findings indicate trans-acting functions of lncRNAs encoded in the PWS/AS imprinting cluster. Surprisingly, a recent study using PWS human fibroblast-derived induced pluripotent stem cells (iPSCs) indicated that human IPW lncRNA regulates expression of genes located in distinct imprinting clusters located in a different chromosome, the DLK1-DIO3 imprinting cluster (Stelzer et al. 2014). In the PWS-iPSCs, expression of the lncRNAs encoded in the PWS/AS imprinting cluster is lost, whereas maternally expressed noncoding RNAs in the DLK1-DIO3 imprinting cluster are overexpressed. Given that parent-specific DNA methylation and mono-allelic expression of imprinted genes in the DLK1-DIO3 imprinting cluster are retained, this excessive expression is hypothesized to be caused by enhancing the transcription of the already active allele. This aberrant expression was suppressed by exogenously expressed IPW lncRNA. Additionally, IPW lncRNA interacts with G9a histone methyltransferase and may act as a trans-acting factor controlling H3K9me3 modification in the DLK1-DIO3 imprinting cluster (Stelzer et al. 2014).

8.3 Conclusions and Future Directions

The importance of lncRNAs in various biological processes has been clearly demonstrated; however, the molecular basis for their actions remains an open question. The lncRNAs associated with genomic imprinting loci are relatively well studied and are useful as the model cases for lncRNA-mediated gene expression regulation. Of note, the lncRNAs encoded in the imprinting cluster frequently exhibit common features, including nuclear localization and retention near their transcriptional sites, regardless of whether they are cis-acting or trans-acting factors or whether their transcripts or transcription is required. The reduced splicing efficiency of lncRNAs compared with canonical mRNAs may explain this property, because the efficient export of mRNA to the cytoplasm requires splicing (Reed and Hurt 2002). As described in this review, lncRNAs associated with the imprinted loci interact with multiple regulatory factors, including DNA methyltransferase and histone modifiers. Given that these epigenetic regulators are closely associated with each other and therefore it is difficult to deduce a cause-and-effect relationship, it is important to distinguish the primary role of lncRNAs. The identification of genomic regions interacting with lncRNAs would provide valuable information regarding the site of lncRNA action. Elucidation of the tissue-specific function of lncRNAs and their regulatory mechanisms is also important to understand why lncRNAs employ multiple factors for epigenetic gene regulation. These approaches will undoubtedly provide deeper insight into gene expression regulation mediated by lncRNAs.

References

Barlow DP, Stoger R, Herrmann BG, Saito K, Schweifer N (1991) The mouse insulin-like growth factor type-2 receptor is imprinted and closely linked to the Tme locus. Nature 349(6304):84–87. doi:10.1038/349084a0

Bartolomei MS, Webber AL, Brunkow ME, Tilghman SM (1993) Epigenetic mechanisms underlying the imprinting of the mouse H19 gene. Genes Dev 7(9):1663–1673

Bell AC, Felsenfeld G (2000) Methylation of a CTCF-dependent boundary controls imprinted expression of the Igf2 gene. Nature 405(6785):482–485. doi:10.1038/35013100

Bourc'his D, Xu GL, Lin CS, Bollman B, Bestor TH (2001) Dnmt3L and the establishment of maternal genomic imprints. Science 294(5551):2536–2539. doi:10.1126/science.1065848

Brannan CI, Dees EC, Ingram RS, Tilghman SM (1990) The product of the H19 gene may function as an RNA. Mol Cell Biol 10(1):28–36

Cavaille J, Buiting K, Kiefmann M, Lalande M, Brannan CI, Horsthemke B, Bachellerie JP, Brosius J, Huttenhofer A (2000) Identification of brain-specific and imprinted small nucleolar RNA genes exhibiting an unusual genomic organization. Proc Natl Acad Sci U S A 97(26):14311–14316. doi:10.1073/pnas.250426397

Chamberlain SJ, Brannan CI (2001) The Prader-Willi syndrome imprinting center activates the paternally expressed murine Ube3a antisense transcript but represses paternal Ube3a. Genomics 73(3):316–322. doi:10.1006/geno.2001.6543

Chu C, Qu K, Zhong FL, Artandi SE, Chang HY (2011) Genomic maps of long noncoding RNA occupancy reveal principles of RNA-chromatin interactions. Mol Cell 44(4):667–678. doi:10.1016/j.molcel.2011.08.027

Coombes C, Arnaud P, Gordon E, Dean W, Coar EA, Williamson CM, Feil R, Peters J, Kelsey G (2003) Epigenetic properties and identification of an imprint mark in the Nesp-Gnasxl domain of the mouse Gnas imprinted locus. Mol Cell Biol 23(16):5475–5488

Davis E, Caiment F, Tordoir X, Cavaille J, Ferguson-Smith A, Cockett N, Georges M, Charlier C (2005) RNAi-mediated allelic trans-interaction at the imprinted Rtl1/Peg11 locus. Curr Biol 15(8):743–749. doi:10.1016/j.cub.2005.02.060

de los Santos T, Schweizer J, Rees CA, Francke U (2000) Small evolutionarily conserved RNA, resembling C/D box small nucleolar RNA, is transcribed from PWCR1, a novel imprinted gene in the Prader-Willi deletion region, which is highly expressed in brain. Am J Hum Genet 67(5):1067–1082. doi:10.1086/303106

Edwards CA, Ferguson-Smith AC (2007) Mechanisms regulating imprinted genes in clusters. Curr Opin Cell Biol 19(3):281–289. doi:10.1016/j.ceb.2007.04.013

Engel JR, Smallwood A, Harper A, Higgins MJ, Oshimura M, Reik W, Schofield PN, Maher ER (2000) Epigenotype-phenotype correlations in Beckwith-Wiedemann syndrome. J Med Genet 37(12):921–926

Engreitz JM, Pandya-Jones A, McDonel P, Shishkin A, Sirokman K, Surka C, Kadri S, Xing J, Goren A, Lander ES, Plath K, Guttman M (2013) The Xist lncRNA exploits three-dimensional genome architecture to spread across the X chromosome. Science 341(6147):1237973. doi:10.1126/science.1237973

Ferguson-Smith AC (2011) Genomic imprinting: the emergence of an epigenetic paradigm. Nat Rev Genet 12(8):565–575. doi:10.1038/nrg3032

Ferguson-Smith AC, Sasaki H, Cattanach BM, Surani MA (1993) Parental-origin-specific epigenetic modification of the mouse H19 gene. Nature 362(6422):751–755. doi:10.1038/362751a0

Filipowicz W (2000) Imprinted expression of small nucleolar RNAs in brain: time for RNomics. Proc Natl Acad Sci U S A 97(26):14035–14037. doi:10.1073/pnas.97.26.14035

Fitzpatrick GV, Soloway PD, Higgins MJ (2002) Regional loss of imprinting and growth deficiency in mice with a targeted deletion of KvDMR1. Nat Genet 32(3):426–431. doi:10.1038/ng988

Hall JG (1990) Genomic imprinting: review and relevance to human diseases. Am J Hum Genet 46(5):857–873

Hark AT, Schoenherr CJ, Katz DJ, Ingram RS, Levorse JM, Tilghman SM (2000) CTCF mediates methylation-sensitive enhancer-blocking activity at the H19/Igf2 locus. Nature 405(6785):486–489. doi:10.1038/35013106

Hata K, Okano M, Lei H, Li E (2002) Dnmt3L cooperates with the Dnmt3 family of de novo DNA methyltransferases to establish maternal imprints in mice. Development 129(8):1983–1993

Hayward BE, Moran V, Strain L, Bonthron DT (1998) Bidirectional imprinting of a single gene: GNAS1 encodes maternally, paternally, and biallelically derived proteins. Proc Natl Acad Sci U S A 95(26):15475–15480

Howell CY, Bestor TH, Ding F, Latham KE, Mertineit C, Trasler JM, Chaillet JR (2001) Genomic imprinting disrupted by a maternal effect mutation in the Dnmt1 gene. Cell 104(6):829–838

Huang HS, Allen JA, Mabb AM, King IF, Miriyala J, Taylor-Blake B, Sciaky N, Dutton JW Jr, Lee HM, Chen X, Jin J, Bridges AS, Zylka MJ, Roth BL, Philpot BD (2012) Topoisomerase inhibitors unsilence the dormant allele of Ube3a in neurons. Nature 481(7380):185–189. doi:10.1038/nature10726

Jones BK, Levorse JM, Tilghman SM (1998) Igf2 imprinting does not require its own DNA methylation or H19 RNA. Genes Dev 12(14):2200–2207. doi:10.1101/gad.12.14.2200

Kaffer CR, Srivastava M, Park KY, Ives E, Hsieh S, Batlle J, Grinberg A, Huang SP, Pfeifer K (2000) A transcriptional insulator at the imprinted H19/Igf2 locus. Genes Dev 14(15):1908–1919

Kaneda M, Okano M, Hata K, Sado T, Tsujimoto N, Li E, Sasaki H (2004) Essential role for de novo DNA methyltransferase Dnmt3a in paternal and maternal imprinting. Nature 429(6994):900–903. doi:10.1038/nature02633

Kishore S, Stamm S (2006) The snoRNA HBII-52 regulates alternative splicing of the serotonin receptor 2C. Science 311(5758):230–232. doi:10.1126/science.1118265

Koerner MV, Pauler FM, Huang R, Barlow DP (2009) The function of non-coding RNAs in genomic imprinting. Development 136(11):1771–1783. doi:10.1242/dev.030403

Kurukuti S, Tiwari VK, Tavoosidana G, Pugacheva E, Murrell A, Zhao Z, Lobanenkov V, Reik W, Ohlsson R (2006) CTCF binding at the H19 imprinting control region mediates maternally inherited higher-order chromatin conformation to restrict enhancer access to Igf2. Proc Natl Acad Sci U S A 103(28):10684–10689. doi:10.1073/pnas.0600326103

Latos PA, Pauler FM, Koerner MV, Senergin HB, Hudson QJ, Stocsits RR, Allhoff W, Stricker SH, Klement RM, Warczok KE, Aumayr K, Pasierbek P, Barlow DP (2012) Airn transcriptional overlap, but not its lncRNA products, induces imprinted Igf2r silencing. Science 338(6113):1469–1472. doi:10.1126/science.1228110

Lee MP, DeBaun MR, Mitsuya K, Galonek HL, Brandenburg S, Oshimura M, Feinberg AP (1999) Loss of imprinting of a paternally expressed transcript, with antisense orientation to KVLQT1, occurs frequently in Beckwith-Wiedemann syndrome and is independent of insulin-like growth factor II imprinting. Proc Natl Acad Sci U S A 96(9):5203–5208

Leighton PA, Ingram RS, Eggenschwiler J, Efstratiadis A, Tilghman SM (1995) Disruption of imprinting caused by deletion of the H19 gene region in mice. Nature 375(6526):34–39. doi:10.1038/375034a0

Lewis A, Mitsuya K, Umlauf D, Smith P, Dean W, Walter J, Higgins M, Feil R, Reik W (2004) Imprinting on distal chromosome 7 in the placenta involves repressive histone methylation independent of DNA methylation. Nat Genet 36(12):1291–1295. doi:10.1038/ng1468

Lin SP, Youngson N, Takada S, Seitz H, Reik W, Paulsen M, Cavaille J, Ferguson-Smith AC (2003) Asymmetric regulation of imprinting on the maternal and paternal chromosomes at the Dlk1-Gtl2 imprinted cluster on mouse chromosome 12. Nat Genet 35(1):97–102. doi:10.1038/ng1233

Lin SP, Coan P, da Rocha ST, Seitz H, Cavaille J, Teng PW, Takada S, Ferguson-Smith AC (2007) Differential regulation of imprinting in the murine embryo and placenta by the Dlk1-Dio3 imprinting control region. Development 134(2):417–426. doi:10.1242/dev.02726

Liu J, Yu S, Litman D, Chen W, Weinstein LS (2000) Identification of a methylation imprint mark within the mouse Gnas locus. Mol Cell Biol 20(16):5808–5817

Mabb AM, Judson MC, Zylka MJ, Philpot BD (2011) Angelman syndrome: insights into genomic imprinting and neurodevelopmental phenotypes. Trends Neurosci 34(6):293–303. doi:10.1016/j.tins.2011.04.001

Mager J, Montgomery ND, de Villena FP, Magnuson T (2003) Genome imprinting regulated by the mouse Polycomb group protein Eed. Nat Genet 33(4):502–507. doi:10.1038/ng1125

Mancini-Dinardo D, Steele SJ, Levorse JM, Ingram RS, Tilghman SM (2006) Elongation of the Kcnq1ot1 transcript is required for genomic imprinting of neighboring genes. Genes Dev 20(10):1268–1282. doi:10.1101/gad.1416906

Meng L, Person RE, Beaudet al (2012) Ube3a-ATS is an atypical RNA polymerase II transcript that represses the paternal expression of Ube3a. Hum Mol Genet 21(13):3001–3012. doi:10.1093/hmg/dds130

Mitsuya K, Meguro M, Lee MP, Katoh M, Schulz TC, Kugoh H, Yoshida MA, Niikawa N, Feinberg AP, Oshimura M (1999) LIT1, an imprinted antisense RNA in the human KvLQT1 locus identified by screening for differentially expressed transcripts using monochromosomal hybrids. Hum Mol Genet 8(7):1209–1217

Mohammad F, Pandey RR, Nagano T, Chakalova L, Mondal T, Fraser P, Kanduri C (2008) Kcnq1ot1/Lit1 noncoding RNA mediates transcriptional silencing by targeting to the perinucleolar region. Mol Cell Biol 28(11):3713–3728. doi:10.1128/MCB.02263-07

Mohammad F, Mondal T, Guseva N, Pandey GK, Kanduri C (2010) Kcnq1ot1 noncoding RNA mediates transcriptional gene silencing by interacting with Dnmt1. Development 137(15):2493–2499. doi:10.1242/dev.048181

Murrell A, Heeson S, Reik W (2004) Interaction between differentially methylated regions partitions the imprinted genes Igf2 and H19 into parent-specific chromatin loops. Nat Genet 36(8):889–893. doi:10.1038/ng1402

Nagano T, Mitchell JA, Sanz LA, Pauler FM, Ferguson-Smith AC, Feil R, Fraser P (2008) The Air noncoding RNA epigenetically silences transcription by targeting G9a to chromatin. Science 322(5908):1717–1720. doi:10.1126/science.1163802

Ooi SK, Qiu C, Bernstein E, Li K, Jia D, Yang Z, Erdjument-Bromage H, Tempst P, Lin SP, Allis CD, Cheng X, Bestor TH (2007) DNMT3L connects unmethylated lysine 4 of histone H3 to de novo methylation of DNA. Nature 448(7154):714–717. doi:10.1038/nature05987

Pandey RR, Mondal T, Mohammad F, Enroth S, Redrup L, Komorowski J, Nagano T, Mancini-Dinardo D, Kanduri C (2008) Kcnq1ot1 antisense noncoding RNA mediates lineage-specific transcriptional silencing through chromatin-level regulation. Mol Cell 32(2):232–246. doi:10.1016/j.molcel.2008.08.022

Perk J, Makedonski K, Lande L, Cedar H, Razin A, Shemer R (2002) The imprinting mechanism of the Prader-Willi/Angelman regional control center. EMBO J 21(21):5807–5814

Peters J, Wroe SF, Wells CA, Miller HJ, Bodle D, Beechey CV, Williamson CM, Kelsey G (1999) A cluster of oppositely imprinted transcripts at the Gnas locus in the distal imprinting region of mouse chromosome 2. Proc Natl Acad Sci U S A 96(7):3830–3835

Powell WT, Coulson RL, Gonzales ML, Crary FK, Wong SS, Adams S, Ach RA, Tsang P, Yamada NA, Yasui DH, Chedin F, LaSalle JM (2013) R-loop formation at Snord116 mediates topotecan inhibition of Ube3a-antisense and allele-specific chromatin decondensation. Proc Natl Acad Sci U S A 110(34):13938–13943. doi:10.1073/pnas.1305426110

Reed R, Hurt E (2002) A conserved mRNA export machinery coupled to pre-mRNA splicing. Cell 108(4):523–531

Rougeulle C, Glatt H, Lalande M (1997) The Angelman syndrome candidate gene, UBE3A/E6-AP, is imprinted in brain. Nat Genet 17(1):14–15. doi:10.1038/ng0997-14

Rougeulle C, Cardoso C, Fontes M, Colleaux L, Lalande M (1998) An imprinted antisense RNA overlaps UBE3A and a second maternally expressed transcript. Nat Genet 19(1):15–16. doi:10.1038/ng0598-15

Runte M, Huttenhofer A, Gross S, Kiefmann M, Horsthemke B, Buiting K (2001) The IC-SNURF-SNRPN transcript serves as a host for multiple small nucleolar RNA species and as an antisense RNA for UBE3A. Hum Mol Genet 10(23):2687–2700

Sasaki H, Jones PA, Chaillet JR, Ferguson-Smith AC, Barton SC, Reik W, Surani MA (1992) Parental imprinting: potentially active chromatin of the repressed maternal allele of the mouse insulin-like growth factor II (Igf2) gene. Genes Dev 6(10):1843–1856

Schmidt JV, Levorse JM, Tilghman SM (1999) Enhancer competition between H19 and Igf2 does not mediate their imprinting. Proc Natl Acad Sci U S A 96(17):9733–9738

Schuster-Gossler K, Bilinski P, Sado T, Ferguson-Smith A, Gossler A (1998) The mouse Gtl2 gene is differentially expressed during embryonic development, encodes multiple alternatively spliced transcripts, and may act as an RNA. Dev Dyn 212(2):214–228. doi:10.1002/(SICI)1097-0177(199806)212:2<214::AID-AJA6>3.0.CO;2-K

Seidl CI, Stricker SH, Barlow DP (2006) The imprinted Air ncRNA is an atypical RNAPII transcript that evades splicing and escapes nuclear export. EMBO J 25(15):3565–3575. doi:10.1038/sj.emboj.7601245

Seitz H, Royo H, Bortolin ML, Lin SP, Ferguson-Smith AC, Cavaille J (2004) A large imprinted microRNA gene cluster at the mouse Dlk1-Gtl2 domain. Genome Res 14(9):1741–1748. doi:10.1101/gr.2743304

Shearwin KE, Callen BP, Egan JB (2005) Transcriptional interference – a crash course. Trends Genet 21(6):339–345. doi:10.1016/j.tig.2005.04.009

Shin JY, Fitzpatrick GV, Higgins MJ (2008) Two distinct mechanisms of silencing by the KvDMR1 imprinting control region. EMBO J 27(1):168–178. doi:10.1038/sj.emboj.7601960

Simon MD, Wang CI, Kharchenko PV, West JA, Chapman BA, Alekseyenko AA, Borowsky ML, Kuroda MI, Kingston RE (2011) The genomic binding sites of a noncoding RNA. Proc Natl Acad Sci U S A 108(51):20497–20502. doi:10.1073/pnas.1113536108

Sleutels F, Zwart R, Barlow DP (2002) The non-coding Air RNA is required for silencing autosomal imprinted genes. Nature 415(6873):810–813. doi:10.1038/415810a

Smilinich NJ, Day CD, Fitzpatrick GV, Caldwell GM, Lossie AC, Cooper PR, Smallwood AC, Joyce JA, Schofield PN, Reik W, Nicholls RD, Weksberg R, Driscoll DJ, Maher ER, Shows TB, Higgins MJ (1999) A maternally methylated CpG island in KvLQT1 is associated with an antisense paternal transcript and loss of imprinting in Beckwith-Wiedemann syndrome. Proc Natl Acad Sci U S A 96(14):8064–8069

Stelzer Y, Sagi I, Yanuka O, Eiges R, Benvenisty N (2014) The noncoding RNA IPW regulates the imprinted DLK1-DIO3 locus in an induced pluripotent stem cell model of Prader-Willi syndrome. Nat Genet 46(6):551–557. doi:10.1038/ng.2968

Stoger R, Kubicka P, Liu CG, Kafri T, Razin A, Cedar H, Barlow DP (1993) Maternal-specific methylation of the imprinted mouse Igf2r locus identifies the expressed locus as carrying the imprinting signal. Cell 73(1):61–71

Szabo P, Tang SH, Rentsendorj A, Pfeifer GP, Mann JR (2000) Maternal-specific footprints at putative CTCF sites in the H19 imprinting control region give evidence for insulator function. Curr Biol 10(10):607–610

Takahashi N, Okamoto A, Kobayashi R, Shirai M, Obata Y, Ogawa H, Sotomaru Y, Kono T (2009) Deletion of Gtl2, imprinted non-coding RNA, with its differentially methylated region induces lethal parent-origin-dependent defects in mice. Hum Mol Genet 18(10):1879–1888. doi:10.1093/hmg/ddp108

Terranova R, Yokobayashi S, Stadler MB, Otte AP, van Lohuizen M, Orkin SH, Peters AH (2008) Polycomb group proteins Ezh2 and Rnf2 direct genomic contraction and imprinted repression in early mouse embryos. Dev Cell 15(5):668–679. doi:10.1016/j.devcel.2008.08.015

Thorvaldsen JL, Duran KL, Bartolomei MS (1998) Deletion of the H19 differentially methylated domain results in loss of imprinted expression of H19 and Igf2. Genes Dev 12(23):3693–3702. doi:10.1101/gad.12.23.3693

Umlauf D, Goto Y, Cao R, Cerqueira F, Wagschal A, Zhang Y, Feil R (2004) Imprinting along the Kcnq1 domain on mouse chromosome 7 involves repressive histone methylation and recruitment of Polycomb group complexes. Nat Genet 36(12):1296–1300. doi:10.1038/ng1467

Vitali P, Royo H, Marty V, Bortolin-Cavaille ML, Cavaille J (2010) Long nuclear-retained non-coding RNAs and allele-specific higher-order chromatin organization at imprinted snoRNA gene arrays. J Cell Sci 123(Pt 1):70–83. doi:10.1242/jcs.054957

Vu TH, Hoffman AR (1997) Imprinting of the Angelman syndrome gene, UBE3A, is restricted to brain. Nat Genet 17(1):12–13. doi:10.1038/ng0997-12

Wagschal A, Sutherland HG, Woodfine K, Henckel A, Chebli K, Schulz R, Oakey RJ, Bickmore WA, Feil R (2008) G9a histone methyltransferase contributes to imprinting in the mouse placenta. Mol Cell Biol 28(3):1104–1113. doi:10.1128/MCB. 01111-07

Weinstein LS, Liu J, Sakamoto A, Xie T, Chen M (2004) Minireview: GNAS: normal and abnormal functions. Endocrinology 145(12):5459–5464. doi:10.1210/en.2004-0865

Williamson CM, Ball ST, Nottingham WT, Skinner JA, Plagge A, Turner MD, Powles N, Hough T, Papworth D, Fraser WD, Maconochie M, Peters J (2004) A cis-acting control region is required exclusively for the tissue-specific imprinting of Gnas. Nat Genet 36(8):894–899. doi:10.1038/ng1398

Williamson CM, Turner MD, Ball ST, Nottingham WT, Glenister P, Fray M, Tymowska-Lalanne Z, Plagge A, Powles-Glover N, Kelsey G, Maconochie M, Peters J (2006) Identification of an imprinting control region affecting the expression of all transcripts in the Gnas cluster. Nat Genet 38(3):350–355. doi:10.1038/ng1731

Williamson CM, Ball ST, Dawson C, Mehta S, Beechey CV, Fray M, Teboul L, Dear TN, Kelsey G, Peters J (2011) Uncoupling antisense-mediated silencing and DNA methylation in the imprinted Gnas cluster. PLoS Genet 7(3):e1001347. doi:10.1371/journal.pgen.1001347

Wutz A, Smrzka OW, Schweifer N, Schellander K, Wagner EF, Barlow DP (1997) Imprinted expression of the Igf2r gene depends on an intronic CpG island. Nature 389(6652):745–749. doi:10.1038/39631

Yamasaki K, Joh K, Ohta T, Masuzaki H, Ishimaru T, Mukai T, Niikawa N, Ogawa M, Wagstaff J, Kishino T (2003) Neurons but not glial cells show reciprocal imprinting of sense and antisense transcripts of Ube3a. Hum Mol Genet 12(8):837–847

Yatsuki H, Joh K, Higashimoto K, Soejima H, Arai Y, Wang Y, Hatada I, Obata Y, Morisaki H, Zhang Z, Nakagawachi T, Satoh Y, Mukai T (2002) Domain regulation of imprinting cluster in Kip2/Lit1 subdomain on mouse chromosome 7F4/F5: large-scale DNA methylation analysis reveals that DMR-Lit1 is a putative imprinting control region. Genome Res 12(12):1860–1870. doi:10.1101/gr.110702

Yin QF, Yang L, Zhang Y, Xiang JF, Wu YW, Carmichael GG, Chen LL (2012) Long noncoding RNAs with snoRNA ends. Mol Cell 48(2):219–230. doi:10.1016/j.molcel.2012.07.033

Yu S, Yu D, Lee E, Eckhaus M, Lee R, Corria Z, Accili D, Westphal H, Weinstein LS (1998) Variable and tissue-specific hormone resistance in heterotrimeric Gs protein alpha-subunit (Gsalpha) knockout mice is due to tissue-specific imprinting of the gsalpha gene. Proc Natl Acad Sci U S A 95(15):8715–8720

Zhou Y, Cheunsuchon P, Nakayama Y, Lawlor MW, Zhong Y, Rice KA, Zhang L, Zhang X, Gordon FE, Lidov HG, Bronson RT, Klibanski A (2010) Activation of paternally expressed genes and perinatal death caused by deletion of the Gtl2 gene. Development 137(16):2643–2652. doi:10.1242/dev.045724

Zwart R, Sleutels F, Wutz A, Schinkel AH, Barlow DP (2001) Bidirectional action of the Igf2r imprint control element on upstream and downstream imprinted genes. Genes Dev 15(18):2361–2366. doi:10.1101/gad.206201

Chapter 9
Mechanisms of Long Noncoding Xist RNA-Mediated Chromosome-Wide Gene Silencing in X-Chromosome Inactivation

Norishige Yamada and Yuya Ogawa

Abstract Chromatin modifications contribute to spatio-temporal gene expression during development in higher eukaryotes. The mechanism for how chromatin-modifying enzymes are recruited to their specific target loci remains largely unknown. Recent findings using deep sequencing analysis revealed that various chromatin-modifying enzymes interact with a variety of long noncoding RNAs (lncRNAs), suggesting the potential role of lncRNA in targeting chromatin-modifying enzymes to their target loci. X-chromosome inactivation (X-inactivation) is a great model of lncRNA-mediated epigenetic gene regulation. In X- inactivation, Xist RNA is exclusively expressed from the inactive X-chromosome (Xi), spreads over the entire Xi *in cis*, and recruits various chromatin modifiers to the Xi, leading to a unique epigenetic landscape over the entire Xi. Recent developments have unveiled key steps involved in the process of Xist RNA-mediated chromosome-wide silencing and particularly the Xist RNA domains and protein factors essential for Xist RNA localization on the Xi. In this chapter, we describe recent novel findings in X-inactivation with a special focus on how Xist RNA localizes on the Xi *in cis* and recruits various chromatin-modifying enzymes, and we discuss the potential mechanism of lncRNA-mediated targeting of chromatin-modifying enzymes to their target loci.

Keywords Non-coding RNA • Chromatin modifying enzymes • Xist • X-chromosome inactivation

---

N. Yamada • Y. Ogawa, Ph.D. (✉)
Division of Reproductive Sciences, Cincinnati Children's Hospital Medical Center, Department of Pediatrics, University of Cincinnati College of Medicine, 3333 Burnet Avenue, Cincinnati, OH 45229, USA
e-mail: yuya.ogawa@cchmc.org

R. Kurokawa (ed.), *Long Noncoding RNAs*, DOI 10.1007/978-4-431-55576-6_9

9.1 Introduction

The central dogma of molecular biology was described by Francis Crick as "the detailed residue-to-residue transfer of sequential information" (Crick 1970) in which DNA is transcribed into RNA, and RNA is translated into protein. Furthermore, information is restricted to a one-way flow and cannot be transcribed or transferred backward along the line. In this view, the major role of RNA is as an intermediary transmitter of genetic information within DNA to protein; the only exception is made for certain types of functional RNA, such as ribosomal RNA and transfer RNA. However, recent transcriptome studies have revealed that more than two-thirds of the mammalian genome is capable of being transcribed during some stage of development (Carninci et al. 2005; Djebali et al. 2012). Since less than 2 % of the mammalian genome carries protein-coding potential (Lander et al. 2001; Waterston et al. 2002), a number of transcripts derived from the mammalian genome are classified as non-protein-coding (noncoding) transcripts. A functionally versatile class of molecules called long noncoding RNA (lncRNA) has emerged in recent years as a key regulator of gene expression at different levels of transcription (Batista and Chang 2013). The discovery of thousands of lncRNAs expressed from all over the mammalian genome has opened up a new field of RNA biology. LncRNAs are generally classified as RNA molecules with more than 200 nucleotide-length ncRNAs and are distinct from typical small ncRNAs such as siRNA, miRNA, piRNA, and snoRNA (Ghildiyal and Zamore 2009; Kim et al. 2009). Although the function of the majority of these intensively expressed lncRNAs remains largely unclear, a number of lncRNAs play important roles in many vital cellular processes (Wilusz et al. 2009; Ulitsky and Bartel 2013) such as X-inactivation and imprinting (Ogawa and Lee 2002; Lee and Bartolomei 2013), cell cycle regulation (Kitagawa et al. 2013), stem cell pluripotency (Ng and Stanton 2013), cellular differentiation and organ morphogenesis (Fatica and Bozzoni 2014), regulation of metabolism (Kornfeld and Bruning 2014), and immune response (Heward and Lindsay 2014); lncRNAs have also been found to contribute to disease conditions, such as cancer (Cheetham et al. 2013).

A growing amount of evidence in recent years has indicated that various chromatin-modifying enzymes bind to a number of lncRNAs (Khalil et al. 2009; Zhao et al. 2010; Guttman et al. 2011; Guil et al. 2012), suggesting that lncRNAs might play an important role in transcriptional regulation by guiding chromatin-modifying enzymes to specific target gene loci. Gene regulation is a very tightly concerted and controlled process in eukaryotic organisms, and a repertoire of enzymes act on both DNA and histones to change the epigenetic landscape (Bernstein et al. 2007). The homeotic gene clusters, HOX gene clusters, are one of the most extensively investigated genetic loci in terms of lncRNA-mediated transcriptional regulation (Dasen 2013). The HOX antisense intergenic RNA (*HOTAIR*) gene in the *HOXC* locus on chromosome 12 expresses a 2.2 kilobase (kb) lncRNA, and its transcripts interact with a variety of factors for transcriptional repression: polycomb repressive complex 2 (PRC2, a lysine methyltransferase complex acting at the histone H3K27, establishing a lysine tri-methylation [H3K27me3]), lysine-specific demethylase 1 (LSD1/KDM1, a demethylase aiding a mono- and dimethyl modification at both H3K4 and H3K9), and REST/CoREST repressor complex through its 5′ and 3′ domains. Interestingly, this RNA-protein (RNP) complex affects gene silencing in the *HOXD* cluster on chromosome 2 and a subset of genes on the other chromosomes *in trans* (Rinn et al. 2007; Gupta et al. 2010; Tsai et al. 2010). It was found that HOTAIR lncRNA preferentially occupies a GA-rich DNA motif by using chromatin isolation by RNA purification sequencing (ChIRP-Seq) analysis, which can effectively retrieve specific lncRNAs bound to proteins and DNA sequences, and helps to map the general occupancy of lncRNA on the chromatin (Chu et al. 2011). These results raise the possibility that the RNP complex might assemble at the *HOXD* gene cluster through a DNA–RNA interaction or a DNA-binding factor interacting with a GA-rich motif. In contrast to the *trans* action of HOTAIR RNA to represses *HOXD* cluster genes, the HOXA transcript at the distal tip (HOTTIP) lncRNA, which is transcribed from the 5′ tip of the *HOXA* locus, binds with WD repeat-containing protein 5 (WDR5), a component of mixed-lineage leukemia (MLL) histone H3K4 methyltransferase complex, and directs the MLL complex to its proximal targeted *HOXA* genes *in cis* (Wang and Chang 2011). Although it is unknown how HOTTIP RNA activates its proximal genes *in cis*, it is suggested that chromosomal looping brings the HOTTIP RNA and MLL complex close to its target HOXA genes, where it accelerates H3K4me3 modification and transcriptional activation. Another potential mechanism that has been proposed is how DNA-binding factors act to anchor lncRNA-chromatin modifiers to target genes.

In addition, many functional lncRNAs are also known to bind with various protein factors such as chromatin-modifying enzymes, transcription factors, and nuclear scaffold proteins in X-inactivation and imprinting (Nagano et al. 2008; Pandey et al. 2008; Zhao et al. 2008), cell cycle regulation (Hung et al. 2011), the maintenance of pluripotency in embryonic stem cells (Guttman et al. 2011), tumor suppression (Huarte et al. 2010), immune response (Carpenter et al. 2013; Imamura et al. 2014; Li et al. 2014), and cellular differentiation and development (Klattenhoff et al. 2013; Wang and Chang 2011; Ng et al. 2012, 2013). This evidence suggests that lncRNAs could function as molecular scaffolds for protein factors and guide chromatin-modifying enzymes and transcriptional factors to their target gene loci (Guttman and Rinn 2012). However, while the recruitment of RNP complexes can take place in any of the aforementioned paths, it still remains evasive what determines their specific interaction with target loci to be activated or repressed. The unique nucleotide sequence or secondary structures of lncRNAs might allow for their interaction with DNA, proteins, and even RNA. LncRNAs are suggested to play an intermediary role between chromatin modifiers and transcription factors, DNA, RNA, and other proteins. The recent burst in lncRNA biology has produced an ever-increasing list of roles and functions of lncRNA, requiring additional work to better understand their role. X-inactive-specific transcript (Xist) RNA is one such lncRNA and is essential for the initiation of X-inactivation *in cis* (Sado and Brockdorff 2013). In this review, we describe the recent findings of lncRNA-mediated gene regulation in X-inactivation and discuss potential mechanisms for how lncRNA guides RNP complex to the target loci and induces transcriptional regulation.

9.2 X-Inactivation: Paradigm of lncRNA-Regulated Gene Regulation

Xist lncRNA-induced transcriptional gene silencing of X-linked genes in X-inactivation is a great model of lncRNA-mediated transcriptional regulation. X-inactivation is a dosage compensation mechanism for female mammals to balance the expression level of X-linked genes between XX females and XY males, whereby one of the two X-chromosomes in females is transcriptionally inactivated (Lyon 1961). Maintenance of an appropriate X-linked gene dosage is critical for cellular development and viability; therefore, abnormal X-inactivation causes severe developmental defects and diseases such as cancer (Payer and Lee 2008; Agrelo and Wutz 2010; Chaligne and Heard 2014). In mice, imprinted X-inactivation occurs at an early embryonic stage whereby the paternal X-chromosome is inactivated (Huynh and Lee 2003; Okamoto et al. 2004). At the blastocyst stage, where imprinted X-inactivation is maintained in the trophectoderm and primitive endoderm, which contribute to extra-embryonic tissues such as the placenta, the imprinted X-inactivation is erased in the epiblast lineage of the inner cell mass, followed by random X-inactivation of either paternal or maternal X-chromosome in the epiblast (Mak et al. 2004; Okamoto et al. 2004; Rastan 1982; Takagi et al. 1982). Embryonic lethality and aberrant development of extra-embryonic tissues are caused by X-inactivation failure, which suggests that proper X-inactivation is fundamental for normal mammalian development (Marahrens et al. 1997). In both imprinted and random X-inactivation, *Xist* has a pivotal role in X-inactivation (Penny et al. 1996; Marahrens et al. 1997). The *Xist* has been mapped within the X-inactivation center (*Xic*), a genetic locus required for X-inactivation, and identified on the basis of its characteristic inactive X-specific expression pattern (Borsani et al. 1991; Brockdorff et al. 1991, 1992; Brown et al. 1991, 1992). X-inactivation is initiated by Xist lncRNA that is highly expressed from the future Xi, coats the Xi, and recruits multiple chromatin-modifying enzymes such as PRC2 for H3K27me3 modification onto the Xi to repress X-linked genes during early development (Borsani et al. 1991; Brown et al. 1991; Brockdorff et al. 1991; Clemson et al. 1996; Plath et al. 2003; Silva et al. 2003; Kohlmaier et al. 2004). Thereafter, gene silencing is established by sequential epigenetic modifications and maintained through multiple rounds of cell division (Chow and Heard 2009; Wutz 2011).

Early studies examining the effect of X-chromosome truncation and translocation on X-inactivation helped map the *Xic* locus on X-chromosomes (Rastan 1983; Rastan and Robertson 1985). The *Xic* locus has been defined and mapped within a 1–2 Mb region on the X-chromosome (Cooper et al. 1993); further studies revealed that a region of about 300–500 kb containing *Xist* in yeast artificial chromosome

(YAC) transgenes is capable of inducing X-inactivation, indicating that the *Xic* resides within this region (Lee et al. 1996; Heard et al. 1999). To date, the minimal functional region of the *Xic* has been narrowed down to an area that is less than 80 kb and contains *Xist* (Lee et al. 1999b). The *Xic* harbors a number of lncRNAs (*Xist*, *Tsix*, *Jpx*, *Ftx*, *Xite*, *RepA*, and *Tsx*) and consists of a complex interplay of each lncRNA that regulates *Xist* expression at the onset of X-inactivation (Fig. 9.1) (Froberg et al. 2013; Maclary et al. 2013). As mentioned above, *Xist* is a central player in initiating X-inactivation: it contributes to chromosome-wide silencing of X-linked genes by recruiting multiple chromatin modifications throughout the entire Xi. Additionally, the shorter transcript RepA RNA (approximately 1.6 kb) is transcribed through *Xist* repeat A. It is suggested that RepA RNA interacts with PRC2 and is involved in *Xist* upregulation and H3K27me3 deposition around the *Xist* promoter to induce a chromosome-wide repressive epigenetic landscape during the onset of X-inactivation (Zhao et al. 2008).

To induce mono-allelic *Xist* upregulation from the future Xi at the onset of X-inactivation, its antisense noncoding gene, *Tsix*, plays a critical role as an

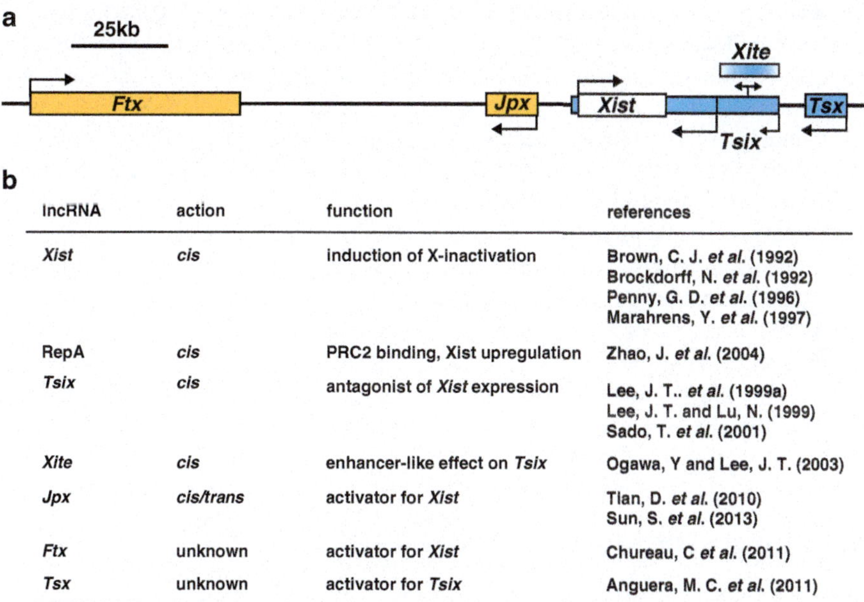

lncRNA	action	function	references
Xist	*cis*	induction of X-inactivation	Brown, C. J. *et al.* (1992) Brockdorff, N. *et al.* (1992) Penny, G. D. *et al.* (1996) Marahrens, Y. *et al.* (1997)
RepA	*cis*	PRC2 binding, Xist upregulation	Zhao, J. *et al.* (2004)
Tsix	*cis*	antagonist of *Xist* expression	Lee, J. T.. *et al.* (1999a) Lee, J. T. and Lu, N. (1999) Sado, T. *et al.* (2001)
Xite	*cis*	enhancer-like effect on *Tsix*	Ogawa, Y and Lee, J. T. (2003)
Jpx	*cis/trans*	activator for *Xist*	Tian, D. *et al.* (2010) Sun, S. *et al.* (2013)
Ftx	unknown	activator for *Xist*	Chureau, C *et al.* (2011)
Tsx	unknown	activator for *Tsix*	Anguera, M. C. *et al.* (2011)

Fig. 9.1 LncRNAs residing within the *Xic*. (**a**) A map of noncoding genes within the *Xic*. *Orange boxes* indicate genes that activate Xist expression. *Blue boxes* indicate genes and a locus that act to repress *Xist*. *Arrows* show transcription start sites and the direction of the transcription. *Tsix* has two alternative transcription start sites, a major and a minor transcription start site, which are shown by *large* and *small arrows*, respectively. The *Xite* region resides between the major and minor *Tsix* transcription start sites, and the *Xite* core region for activating *Tsix* is associated with bidirectional transcripts. (**b**) Table representing a list of noncoding genes in the *Xic* and their function

antagonist for *Xist* (Lee et al. 1999a; Lee and Lu 1999; Lee 2000; Sado et al. 2001). Although Tsix RNA and low levels of Xist RNA are expressed from both X-chromosomes before random X-inactivation, once differentiation occurs the levels of *Tsix* and *Xist* alter drastically. On the future Xi, the expression of *Tsix* ceases while the expression of *Xist* is upregulated; meanwhile, on the future active X-chromosome (Xa), mono-allelic *Tsix* expression represses upregulation of *Xist* expression until *Xist* is extinguished from the Xa. Interestingly, *Tsix* expression also terminates on the Xa thereafter. Although many models have been proposed to explain how *Tsix* represses *Xist* (Avner and Heard 2001; Ogawa and Lee 2002), it is known that *Tsix* regulates epigenetic modification of the *Xist* promoter (Navarro et al. 2005; Ohhata et al. 2008; Sado et al. 2005; Sun et al. 2006). The noncoding *Tsix* gene itself is also regulated by an lncRNA-associated enhancer-like locus, *Xite* (Ogawa and Lee 2003; Stavropoulos et al. 2005). *Xite* promotes the asymmetric persistence of *Tsix* expression on the future Xa at the onset of X-inactivation *in cis*, which in turn blocks the *Xist* upregulation on the future Xa and regulates the choice decision of random X-inactivation. Another long noncoding *Tsx* gene in *Xic* is also implicated in *Tsix* regulation (Anguera et al. 2011).

Unlike the negative regulatory function of *Tsix* for *Xist* expression, noncoding *Jpx* and *Ftx* genes located upstream of the *Xist* act as positive regulators for *Xist* expression (Tian et al. 2010; Chureau et al. 2011), although *Ftx* has been found to be dispensable for imprinted X-inactivation (Soma et al. 2014). Interestingly, expression of both *Jpx* and *Ftx* is upregulated at the onset of X-inactivation (*Ftx* upregulation is female-specific), and both are able to escape X-inactivation. Therefore, *Jpx* and *Ftx* are thought to induce and maintain high-level *Xist* expression during X-inactivation. Although *Jpx* deletion indicates *cis*-preferential effect on *Xist* expression similar to the *cis*-restricted action of *Tsix* on *Xist* repression, Jpx RNA can promote *Xist* expression *in trans* (Tian et al. 2010). In the current model, Jpx RNA extricates CTCF, which represses *Xist* induction before X-inactivation, from the *Xist* promoter on the future Xi, resulting in mono-allelic *Xist* upregulation on the future Xi at the initiation of X-inactivation (Sun et al. 2013). The complex crosstalk among multiple lncRNAs within the *Xic* during the initiation of X-inactivation successfully establishes mono-allelic *Xist* expression from the Xi.

9.3 Epigenetic Modifications on the Inactive X-Chromosome

During X-inactivation, the Xi is intensively decorated by various kinds of repressive epigenetic modifications in mammals (Fig. 9.2) (Chow and Heard 2009; Wutz 2011). A cascade of events occurs prior to X-linked gene silencing, starting with H3K4 demethylation, H3K9 hypoacetylation, and depletion of RNA polymerase II. These events trigger the recruitment of the polycomb group (PcG) proteins, followed by a series of methylations at H3K27 and H3K9, ubiquitination of H2A, and accumulation of the histone variant macroH2A. DNA methylation of the Xi is also important for X-inactivation maintenance. The synergistic effect of this multi-layer epigenetic modification contributes to stable maintenance of chromosome-wide

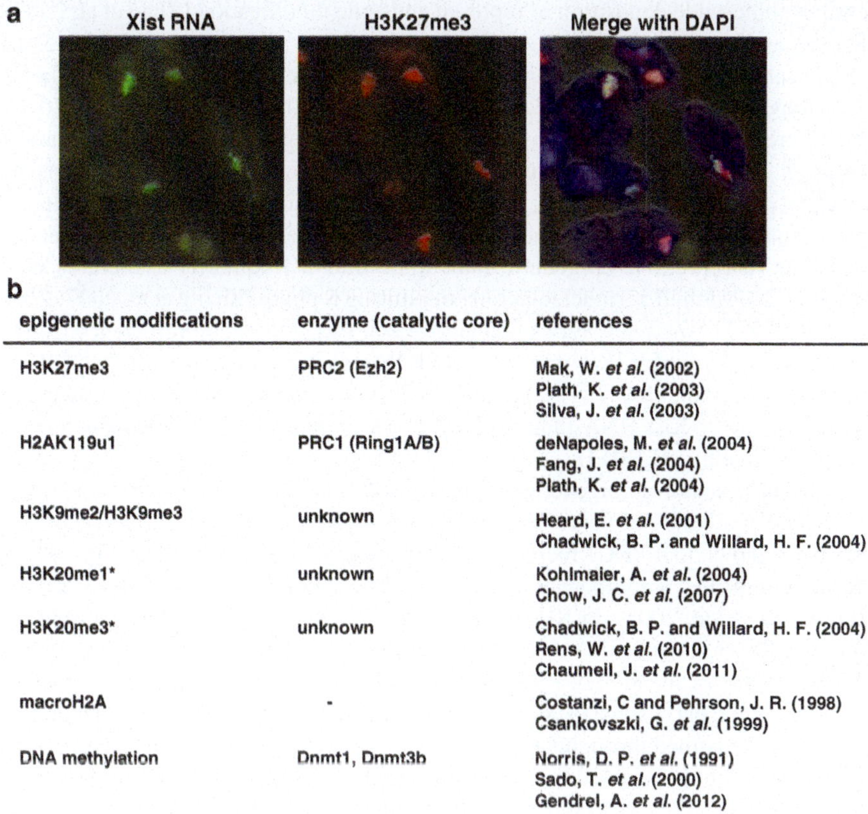

epigenetic modifications	enzyme (catalytic core)	references
H3K27me3	PRC2 (Ezh2)	Mak, W. *et al.* (2002) Plath, K. *et al.* (2003) Silva, J. *et al.* (2003)
H2AK119u1	PRC1 (Ring1A/B)	deNapoles, M. *et al.* (2004) Fang, J. *et al.* (2004) Plath, K. *et al.* (2004)
H3K9me2/H3K9me3	unknown	Heard, E. *et al.* (2001) Chadwick, B. P. and Willard, H. F. (2004)
H3K20me1*	unknown	Kohlmaier, A. *et al.* (2004) Chow, J. C. *et al.* (2007)
H3K20me3*	unknown	Chadwick, B. P. and Willard, H. F. (2004) Rens, W. *et al.* (2010) Chaumeil, J. *et al.* (2011)
macroH2A	-	Costanzi, C and Pehrson, J. R. (1998) Csankovszki, G. *et al.* (1999)
DNA methylation	Dnmt1, Dnmt3b	Norris, D. P. *et al.* (1991) Sado, T. *et al.* (2000) Gendrel, A. *et al.* (2012)

Fig. 9.2 Repressive epigenetic modifications on the Xi in mammals. (**a**) Representative images of immuno-FISH to detect Xist RNA and H3K27me3 during X-chromosome inactivation. Xist RNA clouds (*green*) are co-localized with H3K27me3 modification (*red*) on the Xi. DNA is counter-stained by DAPI (*blue*). (**b**) Table representing a list of repressive epigenetic modifications accumulating on the Xi during X-chromosome inactivation in mammals. *While H3K20me1 modification is observed on the Xi in mice and human, H3K20me3 accumulates on the Xi in mammals, except for mice

silencing on the Xi through multiple rounds of cell division (Csankovszki et al. 2001). PcG proteins, which are crucial for transcriptional gene regulation during development in higher organisms (Simon and Kingston 2009), have a crucial role for the deposition of two representative epigenetic hallmarks on the Xi: H3K27me3 and monoubiquitylation of H2AK119 (H2AK119u1) modification. H3K27me3 deposited by PRC2 is the most extensively studied epigenetic modification on the Xi (Mak et al. 2002; Plath et al. 2003; Silva et al. 2003). PRC2 comprises four unique core-protein components for H3K27 methylation: Eed, RbAp46/48, Suz12, and the catalytic subunit Ezh2. PRC2 transiently localizes on the Xi at an early stage of X-inactivation and deposits an H3K27me3 modification onto the Xi. Xist RNA and RepA RNA directly bind to Ezh2 and recruit PRC2 to the Xi for the chromosome-wide deposition of repressive histone modification (Zhao et al. 2008). Recent findings indicate that Ezh2 is phosphorylated at threonine 345 (T345) by CDK1 and CDK2 in a cell cycle-dependent manner, and its phosphorylation enhances the interaction of Ezh2 with HOTAIR RNA (Kaneko et al. 2010). Therefore, the phosphorylation of Ezh2 at T345 might regulate the interaction between Xist RNA and Ezh2 during X-inactivation. In contrast to the transient localization of PRC2 on the Xi, H3K27me3 modification resides on the Xi during X-inactivation. The Xi might need to interact with perinucleolar compartment to maintain its repressive epigenetic state since the Xi frequently associates with the PRC2-enriched perinucleolar territory during S phase (Zhang et al. 2007; Zhao et al. 2008). Alternatively, very small amounts of PRC2 might be sufficient to maintain a stable level of H3K27me3 on the Xi. Recent works indicate that two PRC2-cofactors, polycomblike 2 (Pcl2) and jumonji AT-rich interactive domain 2 (Jarid2), are essential for Xist RNA-induced PRC2 recruitment to the Xi (Casanova et al. 2011; da Rocha et al. 2014).

The representative epigenetic hallmark, H2AK119u1, is deposited by catalytic RING1A/B subunits of polycomb repressive complex 1 (PRC1). In *Drosophila*, it has been shown that PRC1 recruitment to its target loci is mediated by the binding of polycomb (PC, a component of PRC1) to PRC2-dependent H3K27me3 modification (Cao et al. 2002). In X-inactivation in mice, Ring1A/B in PRC1 is required for H2AK119u1 deposition on the Xi (de Napoles et al. 2004; Fang et al. 2004). Similar to the localization of PRC2 and H3K27me3, PRC1 localization on the Xi is transient at the early stage of X-inactivation, whereas H2AK119u1 modification on the Xi is maintained throughout subsequent cell divisions (de Napoles et al. 2004). However, it should be noted that it is indicated that the transient localization of both Ring1B and H2AK119u1 on the Xi occurs only at the early phase of X-inactivation (Fang et al. 2004). Furthermore, in contrast to *Drosophila*, in mice PRC1 recruitment and H2AK119u1 deposition to the Xi occurs without functional PRC2 or H3K27me3 modification, although localization of some PCR1 components is PRC2 dependent (Hernandez-Munoz et al. 2005; Schoeftner et al. 2006; Tavares et al. 2012).

H3K9me2 and H3K9me3 are also representative repressive epigenetic modifications on the Xi in mammals (Heard et al. 2001; Chadwick and Willard 2004; Rens et al. 2010; Chaumeil et al. 2011) although the enzymes that deposit these repressive modifications on the Xi have not been identified yet. Interestingly, H3K9me3 distribution is distinct from the landscape of H3K27me3 modification on the Xi in placental mammals (Chadwick and Willard 2004; Shevchenko et al. 2009; Nozawa et al. 2013). H3K27me3 and H2AK119u1 modifications are deposited by PRC2 and PRC1, respectively, and occupy the same territory on the Xi as Xist RNA and histone variant macroH2A. Similarly, H3K9me3 and HP1 reside in the same compartment on the Xi, although in different regions where colocalization of H3K27me3/H2AK119u1/Xist RNA takes place.

DNA hypermethylation on the Xi is acquired at a later phase of X-inactivation than deposition of repressive histone modifications (Norris et al. 1991). DNA methylation on the Xi is not essential for the initiation of X-inactivation but is required for stable maintenance of random X-inactivation (Sado et al. 2000, 2004).

Interestingly, it has also been observed that Xist RNA, DNA methylation, and histone hypoacetylation synergistically contribute to the maintenance of the Xi (Csankovszki et al. 2001). For the proper acquisition of hypermethylated CpG islands on the Xi, the structural maintenance of chromosomes hinge domain containing 1 (Smchd1) has a crucial function (Blewitt et al. 2008). Smchd1 localizes on the Xi but is not required for the initiation of X-inactivation, Xist upregulation, or recruitment of PRC2. Instead, Smchd1 is essential for the induction of hypermethylated CpG islands on the Xi and stable long-term gene silencing of a subset of X-linked genes. Indeed, there are Smchd1-dependent and -independent pathways to induce chromosome-wide DNA hypermethylation of CpG islands (Gendrel et al. 2012, 2013). The de novo DNA methyltransferase Dnmt3b induces chromosome-wide DNA hypermethylation on the Xi in both Smchd1-dependent and -independent pathways. Interestingly, a recent report demonstrates that SMCHD1, together with HP1-binding protein 1 (HBiX1), contributes to the compaction of the Xi and to the synchronous late replication timing of the Xi in human X-inactivation (Nozawa et al. 2013), suggesting the multi-functional role of Smchd1 in transcriptional regulation and higher-order chromosome organization.

Just like other epigenetic modifications on the Xi, monomethylation of histone H4K20 (H4K20me1) is enriched on the Xi and shares the same territory with H3K27me3/H2AK119u1/Xist RNA in both mice and humans, although the role of H4K20me1 in X-inactivation remains unknown and the responsible enzyme for H4K20me1 modification has not yet been identified (Kohlmaier et al. 2004; Schoeftner et al. 2006; Chow et al. 2007). In mammals, with the exception of mice, H4K20me3 instead of H4K20me1 is also known to accumulate on the Xi and occupy the same compartment as H3K9me3 (Chadwick and Willard 2004; Rens et al. 2010; Chaumeil et al. 2011). The accumulation of histone macroH2A on the Xi is also a well-known epigenetic modification induced in a Xist RNA-dependent manner and results in the formation of a subnuclear structure referred to as a macrochromatin body (Costanzi and Pehrson 1998; Pullirsch et al. 2010). The enrichment of macroH2A on the Xi is thought to contribute to long-term gene silencing (Csankovszki et al. 1999). The large non-histone part of the protein has been reported to have putative RNA binding properties and may interact with Xist RNA (Gilbert et al. 2000; Pehrson and Fuji 1998). The highly concerted and interconnected histone hallmarks brought about by various chromatin modifiers provide a special ambient atmosphere for the stable silencing of X-linked genes on the Xi via both structural modification and signaling crosstalk.

9.4 Functional Domains of Xist RNA Required for Localization of Xist RNA on the Xi

While a number of repressive epigenetic modifications are induced to the Xi in an Xist RNA-dependent manner, the only chromatin-modifying enzymes that have been identified are PRC1 and PRC2. Two components of PRC2, Ezh2 and Suz12, are known to bind to Xist RNA (Zhao et al. 2008; Kaneko et al. 2010; Maenner et al. 2010), although PRC2 binding is not essential for the accumulation of Xist RNA on the Xi during random X-inactivation (Schoeftner et al. 2006; Tavares et al. 2012). The localization of Xist RNA on the Xi *in cis* is a critical step for chromosome-wide recruitment of chromatin-modifying enzymes and deposition of repressive epigenetic modifications on the Xi. Therefore, uncovering the factors involved is important to better understand how X-inactivation goes wrong to develop disease states, such as cancer.

The *Xist* gene in mice has seven exons and expresses 17-kb-long lncRNA exclusively from the Xi (Brockdorff et al. 1992; Hong et al. 1999). Xist RNA contains six *Xist*-specific repeat sequences (repeats A to F) that are well conserved among eutherian mammals (Yen et al. 2007), suggesting the importance of these sequences for *Xist* function. Therefore, many studies have been performed to dissect the function of Xist RNA, particularly *Xist* repeat sequences. One notable study sought to survey the domain compartments of Xist RNA required for X-inactivation by testing doxycycline-inducible *Xist* transgenes carrying systematic deletions of *Xist* in mouse male ES cells to determine whether mutant Xist RNA is capable of inducing X-inactivation (Wutz et al. 2002). This unique approach successfully identified the crucial function of conserved repeat A in inducting gene silencing but found it to be unnecessary for Xist RNA localization. Independent experiments indicate that repeat A is an important region for interaction between Xist RNA and PRC2, although PRC2 localization does not completely disappear from the induced Xist RNA clouds that lack a repeat A region (Zhao et al. 2008; Maenner et al. 2010; Plath et al. 2003; Kohlmaier et al. 2004).

The doxycycline-inducible *Xist* transgene assay also revealed that all repeat sequences except for repeat A are dispensable for both gene silencing and Xist RNA localization (Wutz et al. 2002). However, further large deletions or deletion combinations in Xist RNA affect Xist RNA localization and X-linked gene silencing, suggesting that redundant functional domains are involved in Xist RNA localization on the Xi. In contrast to the transgene assay, two independent reports demonstrated that the repeat C sequence of Xist RNA within *Xist* exon 1 is involved in localization of Xist RNA on the Xi (Beletskii et al. 2001; Sarma et al. 2010). Transfection of antisense peptide nucleic acid (PNA) or locked nucleic acids (LNAs) targeted against repeat C of Xist RNA, but not against repeats B, D, E, or F, caused loss of Xist RNA clouds and PRC2 deposition, and compromised X-linked gene silencing. Interestingly, LNAs against human repeat C do not influence human XIST RNA localization (Sarma et al. 2010). This might be attributed to differences in the number of repeat C between mouse Xist RNA (14 times) and human XIST RNA (once). It is unclear why interference of repeat C in Xist RNA by PNA or LNAs does affect Xist RNA localization, but the deletion of repeat C in the transgene assay is not evident. It might be possible that doxycycline-inducible Xist RNA from the transgene is more abundant than that from the endogenous *Xist* locus, and the abundant Xist RNA might compensate for the defect of mutant Xist RNA in RNA localization. Contrary to the transgene assay in mice, a similar transgene assay using inducible *XIST* in humans showed that XIST RNA lacking the 3′-end of exon 1 to the end of

XIST lost its focal localization on the Xi and was observed as dispersed signals by XIST RNA fluorescence in situ hybridization (FISH). This suggests that the functional domain (or redundant elements) for XIST RNA localization resides in the second half of XIST RNA in humans (Chow et al. 2007). Our recent unpublished data using endogenous *Xist* targeting also indicates that exon 7 of Xist RNA is essential for stable Xist RNA localization in mice.

9.5 hnRNP U as Anchor Points for Xist RNA on the Xi

While the localization of Xist RNA on the Xi has been explored in terms of functional RNA domains, the protein factors required for Xist RNA localization on the Xi have not yet been intensely investigated. As a candidate for protein factors to connect Xist RNA with the Xi, nuclear scaffold heterogeneous nuclear ribonucleoprotein U (hnRNP U, also referred to as SP120 or SAF-A) has been reported to localize on the Xi, although its function in X-inactivation was unknown (Helbig and Fackelmayer 2003). A nuclear matrix protein hnRNP U has a very unique molecular structure: at its N-terminal is a DNA-binding SAF domain, while its C-terminal contains a RGG RNA binding domain (Fackelmayer et al. 1994; Kipp et al. 2000; Helbig and Fackelmayer 2003). Since the SAF domain preferentially binds to AT-rich chromosomal domains, termed SARs (scaffold attachment regions) or MARs (matrix attachment regions), hnRNP U has been proposed to anchor Xist RNA to the MAR/SAR region of the Xi. Recently, hnRNP U has been identified as a key protein factor required for Xist RNA localization on the Xi, based on the screening using an siRNA library against various RNA binding proteins (Hasegawa et al. 2010). In UV-crosslinking RNA immunoprecipitation (RIP) analysis, hnRNP U has been shown to directly interact with Xist RNA through its RNA binding RGG domain. In addition, lacking either the SAF box or RGG domain resulted in delocalization of Xist RNA and loss of H3K27me3 on the Xi, suggesting the crucial function of hnRNP U for Xist RNA recruitment to the Xi. Interestingly, Xist RNA depletion leads to delocalization of hnRNP U from the Xi (Pullirsch et al. 2010). These findings suggest that the formation of an RNP complex with Xist RNA and hnRNP U is essential for the recruitment of Xist RNA to the Xi. Crosslinking RIP analysis has shown that the RGG domain of hnRNP U preferentially binds to the exon 1 and 7 in Xist RNA (Fig. 9.3b) (Hasegawa et al. 2010).

Our recent work (unpublished data) showed that hnRNP U binds to the entire exon 1 region of Xist RNA (preferentially to the second half of exon 1) and broadly across exon 7 excluding repeat E. Consistent with the idea that hnRNP U anchors Xist RNA to the Xi through its interaction with exons 1 and 7, Xist RNA lacking exon 7 results in compromised Xist RNA localization and X-linked gene silencing during X-inactivation. This suggests that exon 7 is essential for the stable localization of Xist RNA through its interaction with hnRNP U, in addition to the presence of exon 1. Together with Xist RNA and hnRNP U in X-inactivation, several functional lncRNAs have also been known to interact with hnRNP proteins (Huarte et al. 2010; Carpenter et al. 2013; Li et al. 2014; Hacisuleyman et al. 2014), implying the conserved role of hnRNP proteins for lncRNA function. These proteins might also play a critical role in bridging various lncRNAs associated with different effector proteins and their genomic targets.

9.6 Mechanism of Xist RNA-Induced Gene Silencing *in Cis*

One of the mysteries of X-inactivation is how Xist RNA functions and spreads along the entire X-chromosome '*in cis*'. To date, a number of lncRNAs have been found to function in either *cis*- or *trans*-action (Guttman and Rinn 2012; Lee 2012). Recently, Yin Yang 1 (YY1), a transcription factor for both repressive and active transcriptional regulation, has been identified as a factor required for Xist RNA localization and spreading *in cis* (Jeon and Lee 2011). The role of repeat F proximal neighboring region in tethering Xist RNA onto the Xi was suspected upon discovering the very unique ability of repeat F proximal region to squelch Xist RNAs from the Xi and accumulate them at the *Xist* transgene locus in female mice embryonic fibroblast cells. Three YY1 binding sites near the repeat F region were shown to be essential for this squelching activity, and the resulting loss of transgene squelching with YY1 knockdown suggests that YY1 plays a critical role in anchoring Xist RNA to the *Xist* locus. Furthermore, since YY1 was shown to interact with Xist RNA through the repeat C region in vitro and in vivo, it is proposed that YY1 binding sites near repeat F act as a nucleation center to anchor Xist RNA at the *Xist* gene on the Xi *in cis* (Fig. 9.3c). These results are consistent with the role of repeat C for the localization of Xist RNA on the Xi (Beletskii et al. 2001; Sarma et al. 2010). Since YY1 only binds to the YY1 binding sites near repeat F on the Xi (Jeon and Lee 2011), allele-specific modifications of the YY1 binding sites, such as DNA methylation, may be involved (Makhlouf et al. 2014). Furthermore, the role played by YY1 as a transcriptional activator of *Xist* in both human and mouse has been proposed because the activation of *Xist* upregulation was hampered by abolishing YY1 during X-inactivation (Makhlouf et al. 2014). However, a previous report indicated that YY1 does not affect *Xist* expression but instead impacts Xist RNA localization (Jeon and Lee 2011). Although the mechanism for how Xist RNA is anchored to the *Xist* gene through YY1 binding is highly suggestive of its contributing role to the chromosome-wide spreading of Xist RNA on the Xi *in cis*, a deeper investigation should be carried out to elucidate the mechanism behind YY1, Xist RNA, and *Xist* gene cooperation in X-inactivation.

9.7 Mechanism of Chromosome-Wide Spreading of Gene Silencing Across the Inactive X-Chromosome

Ectopic expression of *Xist* is sufficient to induce a wide range of gene silencing *in cis*, although the spread of gene silencing is not expanded extensively in the autosomal background as compared with the X-chromosome (Wutz and Jaenisch 2000). One

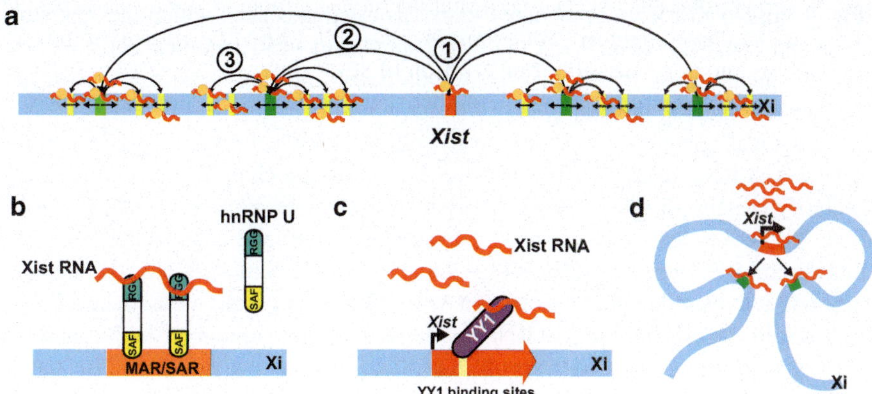

Fig. 9.3 Model of the molecular process for Xist RNA localization and spreading across the Xi *in cis*. (**a**) Three-step model for X-chromosome inactivation spreading on the Xi: Step 1: Xist RNA binds to the *Xist* gene as a nuclear center for Xist RNA spreading (see **c** for detail); Step 2: Xist RNA binds to the strong canonical binding sites (*green bars*) (see **d**); Step 3: Further propagation of Xist RNA to a number of non-canonical binding sites (*yellow bars*) and the entire X-chromosome. *Blue* and *red bars* indicate the Xi and *Xist*, respectively. *Red wavy lines* indicate Xist RNA; *orange circles* indicate chromatin-modifying enzymes. (**b**) Role of hnRNP U as a bridge between Xist RNA and the MAR/SAR region in the Xi. (**c**) YY1-dependent Xist RNA localization to the *Xist* gene as a nucleation center for X-inactivation *in cis*. (**d**) Spatial proximity between *Xist* and a number of hubs of Xist RNA binding sites for quick Xist RNA spreading across the entire Xi

distinct difference between Xist RNA- and other lncRNA-mediated gene regulation is that Xist RNA induces chromosome-wide gene silencing along the entire X-chromosome. Several models for Xist RNA propagation across the Xi have been proposed. One simple model for Xist RNA spreading is that Xist RNA associated with chromatin modification enzymes gradually spreads outward from the endogenous *Xist* locus throughout the entire Xi. The other popular concept is the 'way-station model' where Xist RNAs first bind to a number of way-stations on the X-chromosome and then spread outward from each anchor point along the adjacent nucleosome to the entire X-chromosome (Gartler and Riggs 1983).

Recent works have attempted to clarify the process of Xist RNA-mediated chromosome-wide gene silencing, specifically how Xist RNA and associated chromatin-modifying enzymes localize and spread across the entire Xi (Engreitz et al. 2013; Simon et al. 2013; Pinter et al. 2012). Chromatin immunoprecipitation (ChIP) of PRC2/Ezh2 and H3K27me3 during X-inactivation revealed that PRC2 is initially localized at –150 strong canonical sites that frequently overlap with CpG islands along the X-chromosome (Pinter et al. 2012). As X-inactivation progresses, PRC2 occupies further –4,000 non-canonical sites, mostly at intergenic regions, and finally propagates across the Xi. These PRC2 binding sites are reminiscent of the Xist RNA propagation hubs in the way-station model. Based on these findings, a three-step model for X-chromosome spreading has been proposed (Pinter et al. 2012)

(Fig. 9.3a): (1) Xist RNA is recruited to the "nucleation center" within the *Xist* gene (Jeon and Lee 2011) (Fig. 9.3c); (2) PRC2 spreads to –150 canonical sites; and (3) further spreading of PRC2 to –4,000 non-canonical sites occurs to propagate X-inactivation along the entire Xi.

Additional recent works based on hybridization of antisense oligonucleotides against Xist RNA to capture the Xist RNA-nucleosome complex have contributed to clarify certain aspects of X-inactivation spreading (Engreitz et al. 2013; Simon et al. 2013). Termed RNA antisense purification (RAP) and capture hybridization analysis of RNA targets (CHART) (Chu et al. 2011; Simon et al. 2011), these methods uncovered a genomic region with which Xist RNA is associated. The high-resolution maps of the Xist RNA-associated genomic region on the Xi obtained by deep sequencing revealed the profiles of Xist RNA binding sites across the entire Xi during X-inactivation. These works showed that Xist RNA was preferentially localized to a limited number of gene-rich sites across the entire X-chromosome at an early stage of X-inactivation and later propagated to the entire X-chromosome. H3K27me3 levels also showed a strong correlation with Xist RNA enrichment across the entire Xi. In addition, Xist RNA predominantly localizes at gene-dense regions but excludes escape genes that are expressed from the Xi in differentiated female cells. Interestingly, Hi-C analysis, also known as genome-wide chromosome confirmation capture (3C) (Lieberman-Aiden et al. 2009), revealed that Xist RNA recruitment sites at the initiation of X-inactivation reside in close spatial proximity to the endogenous *Xist* locus, suggesting that Xist RNA propagates quickly across the entire X-chromosome, using the spatial proximity between the *Xist* locus and hubs for Xist RNA binding (Fig. 9.3d). Furthermore, RAP experiments using inducible mutant Xist RNA lacking repeat A exhibited inefficient spreading of the repeat A-mutant Xist RNA across the Xi compared with wildtype Xist RNA. This is consistent with the observation of small Xist RNA clouds and H3K27me3 foci on the Xi in the cells expressing mutant Xist RNA lacking repeat A by FISH and immunofluorescence, which indicates that Xist RNA lacking the repeat A region results in defective Xist RNA spreading (Kohlmaier et al. 2004). These results raise the possibility that the repeat A motif may be crucial for Xist RNA to access and localize across the entire X-chromosome. On the basis of the abovementioned findings, a new model has been proposed: Xist RNA localizes to chromatin and spreads across the entire Xi by three-dimensionally scanning modifying chromatin and chromosome structure, thus expanding to newly accessible locations and crossing the entire Xi (Engreitz et al. 2013).

9.8 Conclusions

Over the past two decades since the discovery of Xist, intensive molecular analysis has revealed a number of findings in the mechanism of X-inactivation. A growing amount of evidence in the lncRNA field supports the theory that lncRNAs regulate gene expression through interaction with chromatin-modifying enzymes; however, little is understood about how these lncRNAs guide associated chromatin-modifying enzymes to specific genomic sites. Transcriptome analysis of lncRNAs that bind to Ezh2 has indicated no consensus on the identity of the RNA motif for Ezh2 binding, suggesting that the secondary structure of lncRNA might be crucial for the recognition of lncRNAs by Ezh2 (Khalil et al. 2009; Zhao et al. 2010). Furthermore, the CHART and RAP approaches for Xist RNA found no characteristic genomic sequence for Xist RNA binding sites, although Xist RNAs are recruited to a limited number of sites at the initiation of X-inactivation (Engreitz et al. 2013; Simon et al. 2013). Further investigation using different approaches will shed light on the molecular mechanism of lncRNA-mediated targeting of chromatin modifiers to its target loci. X-inactivation has greatly progressed our understanding of the molecular function of lncRNA-mediated gene regulation. Perspectives and clues obtained from the studies of lncRNA in other fields will produce a synergistic effect toward understanding the molecular mechanism of lncRNA-mediated gene regulation in X-inactivation.

Acknowledgments We would like to thank all Ogawa lab members for their helpful comments and Serenity Curtis for editing the manuscript. N.Y. is supported by a Postdoctoral Fellowship for Research Abroad of the Japan Society for the Promotion of Science (JSPS). This work was supported by grants from the NIH (no. RO1-GM102184) and the March of Dimes Research Foundation (#6-FY12-337) to Y.O.

References

Agrelo R, Wutz A (2010) X inactivation and disease. Semin Cell Dev Biol 21:194–200

Anguera MC, Ma W, Clift D et al (2011) Tsx produces a long noncoding RNA and has general functions in the germline, stem cells, and brain. PLoS Genet 7:e1002248

Avner P, Heard E (2001) X-chromosome inactivation: counting, choice and initiation. Nat Rev Genet 2:59–67

Batista PJ, Chang HY (2013) Long noncoding RNAs: cellular address codes in development and disease. Cell 152:1298–1307

Beletskii A, Hong YK, Pehrson J et al (2001) PNA interference mapping demonstrates functional domains in the noncoding RNA Xist. Proc Natl Acad Sci U S A 98:9215–9220

Bernstein BE, Meissner A, Lander ES (2007) The mammalian epigenome. Cell 128:669–681

Blewitt ME, Gendrel AV, Pang Z et al (2008) SmcHD1, containing a structural-maintenance-of-chromosomes hinge domain, has a critical role in X inactivation. Nat Genet 40:663–669

Borsani G, Tonlorenzi R, Simmler MC et al (1991) Characterization of a murine gene expressed from the inactive X chromosome. Nature 351:325–329

Brockdorff N, Ashworth A, Kay GF et al (1991) Conservation of position and exclusive expression of mouse Xist from the inactive X chromosome. Nature 351:329–331

Brockdorff N, Ashworth A, Kay GF et al (1992) The product of the mouse Xist gene is a 15 kb inactive X-specific transcript containing no conserved ORF and located in the nucleus. Cell 71:515–526

Brown CJ, Ballabio A, Rupert JL et al (1991) A gene from the region of the human X inactivation centre is expressed exclusively from the inactive X chromosome. Nature 349:38–44

Brown CJ, Hendrich BD, Rupert JL et al (1992) The human XIST gene: analysis of a 17 kb inactive X-specific RNA that contains conserved repeats and is highly localized within the nucleus. Cell 71:527–542

Cao R, Wang L, Wang H et al (2002) Role of histone H3 lysine 27 methylation in polycomb-group silencing. Science 298:1039–1043

Carninci P, Kasukawa T, Katayama S et al (2005) The transcriptional landscape of the mammalian genome. Science 309:1559–1563

Carpenter S, Aiello D, Atianand MK et al (2013) A long noncoding RNA mediates both activation and repression of immune response genes. Science 341:789–792

Casanova M, Preissner T, Cerase A et al (2011) Polycomblike 2 facilitates the recruitment of PRC2 polycomb group complexes to the inactive X chromosome and to target loci in embryonic stem cells. Development 138:1471–1482

Chadwick BP, Willard HF (2004) Multiple spatially distinct types of facultative heterochromatin on the human inactive X chromosome. Proc Natl Acad Sci U S A 101:17450–17455

Chaligne R, Heard E (2014) X-chromosome inactivation in development and cancer. FEBS Lett 588:2514–2522

Chaumeil J, Waters PD, Koina E et al (2011) Evolution from XIST-independent to XIST-controlled X-chromosome inactivation: epigenetic modifications in distantly related mammals. PLoS One 6:e19040

Cheetham SW, Gruhl F, Mattick JS et al (2013) Long noncoding RNAs and the genetics of cancer. Br J Cancer 108:2419–2425

Chow J, Heard E (2009) X inactivation and the complexities of silencing a sex chromosome. Curr Opin Cell Biol 21:359–366

Chow JC, Hall LL, Baldry SE et al (2007) Inducible XIST-dependent X-chromosome inactivation in human somatic cells is reversible. Proc Natl Acad Sci U S A 104:10104–10109

Chu C, Qu K, Zhong FL et al (2011) Genomic maps of long noncoding RNA occupancy reveal principles of RNA-chromatin interactions. Mol Cell 44:667–678

Chureau C, Chantalat S, Romito A et al (2011) Ftx is a non-coding RNA which affects Xist expression and chromatin structure within the X-inactivation center region. Hum Mol Genet 20:705–718

Clemson CM, McNeil JA, Willard HF et al (1996) XIST RNA paints the inactive X chromosome at interphase: evidence for a novel RNA involved in nuclear/chromosome structure. J Cell Biol 132:259–275

Cooper P, Keer JT, McCabe VM et al (1993) Physical mapping of 2000 kb of the mouse X chromosome in the vicinity of the Xist locus. Genomics 15:570–575

Costanzi C, Pehrson JR (1998) Histone macroH2A1 is concentrated in the inactive X chromosome of female mammals. Nature 393:599–601

Crick F (1970) Central dogma of molecular biology. Nature 227:561–563

Csankovszki G, Panning B, Bates B et al (1999) Conditional deletion of Xist disrupts histone macroH2A localization but not maintenance of X inactivation. Nat Genet 22:323–324

Csankovszki G, Nagy A, Jaenisch R (2001) Synergism of Xist RNA, DNA methylation, and histone hypoacetylation in maintaining X chromosome inactivation. J Cell Biol 153:773–784

da Rocha ST, Boeva V, Escamilla-Del-Arenal M et al (2014) Jarid2 is implicated in the initial Xist-induced targeting of PRC2 to the inactive X chromosome. Mol Cell 53:301–316

Dasen JS (2013) Long noncoding RNAs in development: solidifying the Lncs to Hox gene regulation. Cell Rep 5:1–2

de Napoles M, Mermoud JE, Wakao R et al (2004) Polycomb group proteins Ring1A/B link ubiquitylation of histone H2A to heritable gene silencing and X inactivation. Dev Cell 7:663–676

Djebali S, Davis CA, Merkel A et al (2012) Landscape of transcription in human cells. Nature 489:101–108

Engreitz JM, Pandya-Jones A, McDonel P et al (2013) The Xist lncRNA exploits three-dimensional genome architecture to spread across the X chromosome. Science 341:1237973

Fackelmayer FO, Dahm K, Renz A et al (1994) Nucleic-acid-binding properties of hnRNP-U/SAF-A, a nuclear-matrix protein which binds DNA and RNA in vivo and in vitro. Eur J Biochem 221:749–757

Fang J, Chen T, Chadwick B et al (2004) Ring1b-mediated H2A ubiquitination associates with inactive X chromosomes and is involved in initiation of X inactivation. J Biol Chem 279:52812–52815

Fatica A, Bozzoni I (2014) Long non-coding RNAs: new players in cell differentiation and development. Nat Rev Genet 15:7–21

Froberg JE, Yang L, Lee JT (2013) Guided by RNAs: X-inactivation as a model for lncRNA function. J Mol Biol 425:3698–3706

Gartler SM, Riggs AD (1983) Mammalian X-chromosome inactivation. Annu Rev Genet 17:155–190

Gendrel AV, Apedaile A, Coker H et al (2012) Smchd1-dependent and -independent pathways determine developmental dynamics of CpG island methylation on the inactive X chromosome. Dev Cell 23:265–279

Gendrel AV, Tang YA, Suzuki M et al (2013) Epigenetic functions of smchd1 repress gene clusters on the inactive X chromosome and on autosomes. Mol Cell Biol 33:3150–3165

Ghildiyal M, Zamore PD (2009) Small silencing RNAs: an expanding universe. Nat Rev Genet 10:94–108

Gilbert SL, Pehrson JR, Sharp PA (2000) XIST RNA associates with specific regions of the inactive X chromatin. J Biol Chem 275:36491–36494

Guil S, Soler M, Portela A et al (2012) Intronic RNAs mediate EZH2 regulation of epigenetic targets. Nat Struct Mol Biol 19:664–670

Gupta RA, Shah N, Wang KC et al (2010) Long non-coding RNA HOTAIR reprograms chromatin state to promote cancer metastasis. Nature 464:1071–1076

Guttman M, Rinn JL (2012) Modular regulatory principles of large non-coding RNAs. Nature 482:339–346

Guttman M, Donaghey J, Carey BW et al (2011) lincRNAs act in the circuitry controlling pluripotency and differentiation. Nature 477:295–300

Hacisuleyman E, Goff LA, Trapnell C et al (2014) Topological organization of multichromosomal regions by the long intergenic noncoding RNA Firre. Nat Struct Mol Biol 21:198–206

Hasegawa Y, Brockdorff N, Kawano S et al (2010) The matrix protein hnRNP U is required for chromosomal localization of Xist RNA. Dev Cell 19:469–476

Heard E, Mongelard F, Arnaud D et al (1999) Xist yeast artificial chromosome transgenes function as X-inactivation centers only in multicopy arrays and not as single copies. Mol Cell Biol 19:3156–3166

Heard E, Rougeulle C, Arnaud D et al (2001) Methylation of histone H3 at Lys-9 is an early mark on the X chromosome during X inactivation. Cell 107:727–738

Helbig R, Fackelmayer FO (2003) Scaffold attachment factor A (SAF-A) is concentrated in inactive X chromosome territories through its RGG domain. Chromosoma 112:173–182

Hernandez-Munoz I, Lund AH, van der Stoop P et al (2005) Stable X chromosome inactivation involves the PRC1 polycomb complex and requires histone MACROH2A1 and the CULLIN3/SPOP ubiquitin E3 ligase. Proc Natl Acad Sci U S A 102:7635–7640

Heward JA, Lindsay MA (2014) Long non-coding RNAs in the regulation of the immune response. Trends Immunol 35:408–419

Hong YK, Ontiveros SD, Chen C et al (1999) A new structure for the murine Xist gene and its relationship to chromosome choice/counting during X-chromosome inactivation. Proc Natl Acad Sci U S A 96:6829–6834

Huarte M, Guttman M, Feldser D et al (2010) A large intergenic noncoding RNA induced by p53 mediates global gene repression in the p53 response. Cell 142:409–419

Hung T, Wang Y, Lin MF et al (2011) Extensive and coordinated transcription of noncoding RNAs within cell-cycle promoters. Nat Genet 43:621–629

Huynh KD, Lee JT (2003) Inheritance of a pre-inactivated paternal X chromosome in early mouse embryos. Nature 426:857–862

Imamura K, Imamachi N, Akizuki G et al (2014) Long noncoding RNA NEAT1-dependent SFPQ relocation from promoter region to paraspeckle mediates IL8 expression upon immune stimuli. Mol Cell 53:393–406

Jeon Y, Lee JT (2011) YY1 tethers Xist RNA to the inactive X nucleation center. Cell 146:119–133

Kaneko S, Li G, Son J et al (2010) Phosphorylation of the PRC2 component Ezh2 is cell cycle-regulated and up-regulates its binding to ncRNA. Genes Dev 24:2615–2620

Khalil AM, Guttman M, Huarte M et al (2009) Many human large intergenic noncoding RNAs associate with chromatin-modifying complexes and affect gene expression. Proc Natl Acad Sci U S A 106:11667–11672

Kim VN, Han J, Siomi MC (2009) Biogenesis of small RNAs in animals. Nat Rev Mol Cell Biol 10:126–139

Kipp M, Gohring F, Ostendorp T et al (2000) SAF-Box, a conserved protein domain that specifically recognizes scaffold attachment region DNA. Mol Cell Biol 20:7480–7489

Kitagawa M, Kitagawa K, Kotake Y et al (2013) Cell cycle regulation by long non-coding RNAs. Cell Mol Life Sci 70:4785–4794

Klattenhoff CA, Scheuermann JC, Surface LE et al (2013) Braveheart, a long noncoding RNA required for cardiovascular lineage commitment. Cell 152:570–583

Kohlmaier A, Savarese F, Lachner M et al (2004) A chromosomal memory triggered by Xist regulates histone methylation in X inactivation. PLoS Biol 2:E171

Kornfeld JW, Bruning JC (2014) Regulation of metabolism by long, non-coding RNAs. Front Genet 5:57

Lander ES, Linton LM, Birren B et al (2001) Initial sequencing and analysis of the human genome. Nature 409:860–921

Lee JT (2000) Disruption of imprinted X inactivation by parent-of-origin effects at Tsix. Cell 103:17–27

Lee JT (2012) Epigenetic regulation by long noncoding RNAs. Science 338:1435–1439

Lee JT, Bartolomei MS (2013) X-inactivation, imprinting, and long noncoding RNAs in health and disease. Cell 152:1308–1323

Lee JT, Lu N (1999) Targeted mutagenesis of Tsix leads to nonrandom X inactivation. Cell 99:47–57

Lee JT, Strauss WM, Dausman JA et al (1996) A 450 kb transgene displays properties of the mammalian X-inactivation center. Cell 86:83–94

Lee JT, Davidow LS, Warshawsky D (1999a) Tsix, a gene antisense to Xist at the X-inactivation centre. Nat Genet 21:400–404

Lee JT, Lu N, Han Y (1999b) Genetic analysis of the mouse X inactivation center defines an 80-kb multifunction domain. Proc Natl Acad Sci U S A 96:3836–3841

Li Z, Chao TC, Chang KY et al (2014) The long noncoding RNA THRIL regulates TNFalpha expression through its interaction with hnRNPL. Proc Natl Acad Sci U S A 111:1002–1007

Lieberman-Aiden E, van Berkum NL, Williams L et al (2009) Comprehensive mapping of long-range interactions reveals folding principles of the human genome. Science 326:289–293

Lyon MF (1961) Gene action in the X-chromosome of the mouse (Mus musculus L.). Nature 190:372–373

Maclary E, Hinten M, Harris C et al (2013) Long noncoding RNAs in the X-inactivation center. Chromosome Res 21:601–614

Maenner S, Blaud M, Fouillen L et al (2010) 2-D structure of the A region of Xist RNA and its implication for PRC2 association. PLoS Biol 8:e1000276

Mak W, Baxter J, Silva J et al (2002) Mitotically stable association of polycomb group proteins eed and enx1 with the inactive x chromosome in trophoblast stem cells. Curr Biol 12:1016–1020

Mak W, Nesterova TB, de Napoles M et al (2004) Reactivation of the paternal X chromosome in early mouse embryos. Science 303:666–669

Makhlouf M, Ouimette JF, Oldfield A et al (2014) A prominent and conserved role for YY1 in Xist transcriptional activation. Nat Commun 5:4878

Marahrens Y, Panning B, Dausman J et al (1997) Xist-deficient mice are defective in dosage compensation but not spermatogenesis. Genes Dev 11:156–166

Nagano T, Mitchell JA, Sanz LA et al (2008) The air noncoding RNA epigenetically silences transcription by targeting G9a to chromatin. Science 322:1717–1720

Navarro P, Pichard S, Ciaudo C et al (2005) Tsix transcription across the Xist gene alters chromatin conformation without affecting Xist transcription: implications for X-chromosome inactivation. Genes Dev 19:1474–1484

Ng SY, Stanton LW (2013) Long non-coding RNAs in stem cell pluripotency. Wiley Interdiscip Rev RNA 4:121–128

Ng SY, Johnson R, Stanton LW (2012) Human long non-coding RNAs promote pluripotency and neuronal differentiation by association with chromatin modifiers and transcription factors. EMBO J 31:522–533

Ng SY, Bogu GK, Soh BS et al (2013) The long noncoding RNA RMST interacts with SOX2 to regulate neurogenesis. Mol Cell 51:349–359

Norris DP, Brockdorff N, Rastan S (1991) Methylation status of CpG-rich islands on active and inactive mouse X chromosomes. Mamm Genome 1:78–83

Nozawa RS, Nagao K, Igami KT et al (2013) Human inactive X chromosome is compacted through a PRC2-independent SMCHD1-HBiX1 pathway. Nat Struct Mol Biol 20:566–573

Ogawa Y, Lee JT (2002) Antisense regulation in X inactivation and autosomal imprinting. Cytogenet Genome Res 99:59–65

Ogawa Y, Lee JT (2003) Xite, X-inactivation intergenic transcription elements that regulate the probability of choice. Mol Cell 11:731–743

Ohhata T, Hoki Y, Sasaki H et al (2008) Crucial role of antisense transcription across the Xist promoter in Tsix-mediated Xist chromatin modification. Development 135:227–235

Okamoto I, Otte AP, Allis CD et al (2004) Epigenetic dynamics of imprinted X inactivation during early mouse development. Science 303:644–649

Pandey RR, Mondal T, Mohammad F et al (2008) Kcnq1ot1 antisense noncoding RNA mediates lineage-specific transcriptional silencing through chromatin-level regulation. Mol Cell 32:232–246

Payer B, Lee JT (2008) X chromosome dosage compensation: how mammals keep the balance. Annu Rev Genet 42:733–772

Pehrson JR, Fuji RN (1998) Evolutionary conservation of histone macroH2A subtypes and domains. Nucleic Acids Res 26:2837–2842

Penny GD, Kay GF, Sheardown SA et al (1996) Requirement for Xist in X chromosome inactivation. Nature 379:131–137

Pinter SF, Sadreyev RI, Yildirim E et al (2012) Spreading of X chromosome inactivation via a hierarchy of defined polycomb stations. Genome Res 22:1864–1876

Plath K, Fang J, Mlynarczyk-Evans SK et al (2003) Role of histone H3 lysine 27 methylation in X inactivation. Science 300:131–135

Plath K, Talbot D, Hamer KM et al (2004) Developmentally regulated alterations in Polycomb repressive complex 1 proteins on the inactive X chromosome. J Cell Biol 167:1025–1035

Pullirsch D, Hartel R, Kishimoto H et al (2010) The Trithorax group protein Ash2l and Saf-A are recruited to the inactive X chromosome at the onset of stable X inactivation. Development 137:935–943

Rastan S (1982) Timing of X-chromosome inactivation in postimplantation mouse embryos. J Embryol Exp Morphol 71:11–24

Rastan S (1983) Non-random X-chromosome inactivation in mouse X-autosome translocation embryos–location of the inactivation centre. J Embryol Exp Morphol 78:1–22

Rastan S, Robertson EJ (1985) X-chromosome deletions in embryo-derived (EK) cell lines associated with lack of X-chromosome inactivation. J Embryol Exp Morphol 90:379–388

Rens W, Wallduck MS, Lovell FL et al (2010) Epigenetic modifications on X chromosomes in marsupial and monotreme mammals and implications for evolution of dosage compensation. Proc Natl Acad Sci U S A 107:17657–17662

Rinn JL, Kertesz M, Wang JK et al (2007) Functional demarcation of active and silent chromatin domains in human HOX loci by noncoding RNAs. Cell 129:1311–1323

Sado T, Brockdorff N (2013) Advances in understanding chromosome silencing by the long noncoding RNA Xist. Philos Trans R Soc Lond B Biol Sci 368:20110325

Sado T, Fenner MH, Tan SS et al (2000) X inactivation in the mouse embryo deficient for Dnmt1: distinct effect of hypomethylation on imprinted and random X inactivation. Dev Biol 225:294–303

Sado T, Wang Z, Sasaki H et al (2001) Regulation of imprinted X-chromosome inactivation in mice by Tsix. Development 128:1275–1286

Sado T, Okano M, Li E et al (2004) De novo DNA methylation is dispensable for the initiation and propagation of X chromosome inactivation. Development 131:975–982

Sado T, Hoki Y, Sasaki H (2005) Tsix silences Xist through modification of chromatin structure. Dev Cell 9:159–165

Sarma K, Levasseur P, Aristarkhov A et al (2010) Locked nucleic acids (LNAs) reveal sequence requirements and kinetics of Xist RNA localization to the X chromosome. Proc Natl Acad Sci U S A 107:22196–22201

Schoeftner S, Sengupta AK, Kubicek S et al (2006) Recruitment of PRC1 function at the initiation of X inactivation independent of PRC2 and silencing. EMBO J 25:3110–3122

Shevchenko AI, Pavlova SV, Dementyeva EV et al (2009) Mosaic heterochromatin of the inactive X chromosome in vole Microtus rossiaemeridionalis. Mamm Genome 20:644–653

Silva J, Mak W, Zvetkova I et al (2003) Establishment of histone h3 methylation on the inactive X chromosome requires transient recruitment of Eed-Enx1 polycomb group complexes. Dev Cell 4:481–495

Simon JA, Kingston RE (2009) Mechanisms of polycomb gene silencing: knowns and unknowns. Nat Rev Mol Cell Biol 10:697–708

Simon MD, Wang CI, Kharchenko PV et al (2011) The genomic binding sites of a noncoding RNA. Proc Natl Acad Sci U S A 108:20497–20502

Simon MD, Pinter SF, Fang R et al (2013) High-resolution Xist binding maps reveal two-step spreading during X-chromosome inactivation. Nature 504:465–469

Soma M, Fujihara Y, Okabe M et al (2014) Ftx is dispensable for imprinted X-chromosome inactivation in preimplantation mouse embryos. Sci Rep 4:5181

Stavropoulos N, Rowntree RK, Lee JT (2005) Identification of developmentally specific enhancers for Tsix in the regulation of X chromosome inactivation. Mol Cell Biol 25:2757–2769

Sun BK, Deaton AM, Lee JT (2006) A transient heterochromatic state in Xist preempts X inactivation choice without RNA stabilization. Mol Cell 21:617–628

Sun S, Del Rosario BC, Szanto A et al (2013) Jpx RNA activates Xist by evicting CTCF. Cell 153:1537–1551

Takagi N, Sugawara O, Sasaki M (1982) Regional and temporal changes in the pattern of X-chromosome replication during the early post-implantation development of the female mouse. Chromosoma 85:275–286

Tavares L, Dimitrova E, Oxley D et al (2012) RYBP-PRC1 complexes mediate H2A ubiquitylation at polycomb target sites independently of PRC2 and H3K27me3. Cell 148:664–678

Tian D, Sun S, Lee JT (2010) The long noncoding RNA, Jpx, is a molecular switch for X chromosome inactivation. Cell 143:390–403

Tsai MC, Manor O, Wan Y et al (2010) Long noncoding RNA as modular scaffold of histone modification complexes. Science 329:689–693

Ulitsky I, Bartel DP (2013) lincRNAs: genomics, evolution, and mechanisms. Cell 154:26–46

Wang KC, Chang HY (2011) Molecular mechanisms of long noncoding RNAs. Mol Cell 43:904–914

Waterston RH, Lindblad-Toh K, Birney E et al (2002) Initial sequencing and comparative analysis of the mouse genome. Nature 420:520–562

Wilusz JE, Sunwoo H, Spector DL (2009) Long noncoding RNAs: functional surprises from the RNA world. Genes Dev 23:1494–1504
Wutz A (2011) Gene silencing in X-chromosome inactivation: advances in understanding facultative heterochromatin formation. Nat Rev Genet 12:542–553
Wutz A, Jaenisch R (2000) A shift from reversible to irreversible X inactivation is triggered during ES cell differentiation. Mol Cell 5:695–705
Wutz A, Rasmussen TP, Jaenisch R (2002) Chromosomal silencing and localization are mediated by different domains of Xist RNA. Nat Genet 30:167–174
Yen ZC, Meyer IM, Karalic S et al (2007) A cross-species comparison of X-chromosome inactivation in Eutheria. Genomics 90:453–463
Zhang LF, Huynh KD, Lee JT (2007) Perinucleolar targeting of the inactive X during S phase: evidence for a role in the maintenance of silencing. Cell 129:693–706
Zhao J, Sun BK, Erwin JA et al (2008) Polycomb proteins targeted by a short repeat RNA to the mouse X chromosome. Science 322:750–756
Zhao J, Ohsumi TK, Kung JT et al (2010) Genome-wide identification of polycomb-associated RNAs by RIP-seq. Mol Cell 40:939–953

Part V
Potential Outcomes for Clinical Medicine

Chapter 10
Regulation of pRB and p53 Pathways by the Long Noncoding RNAs *ANRIL*, *lincRNA-p21*, *lincRNA-RoR*, and *PANDA*

Yojiro Kotake and Masatoshi Kitagawa

Abstract Retinoblastoma protein (pRB) and p53 pathways play a key role in controlling the cell cycle and apoptosis in response to oncogenic insults and DNA damage. Disruption of these pathways deregulates the control of cell proliferation and represents a common event in the development of most types of human cancer. Recent studies have revealed that several long noncoding RNAs (lncRNAs) are involved in the regulation of pRB and p53 pathways, through transcriptional and translational control of target genes. In this chapter, we focus on four lncRNAs: *ANRIL*, *lincRNA-p21*, *lincRNA-RoR*, and *PANDA*. These lncRNAs are involved in the pRB and p53 pathways. *ANRIL* associates with and recruits polycomb proteins to repress the transcription of cyclin-dependent kinase (CDK) inhibitor *p15* and *p16* genes, resulting in the repression of pRB function. *lincRNA-p21*, *lincRNA-RoR*, and *PANDA* are induced by p53 in response to DNA damage and regulate apoptosis. We discuss the involvement of *ANRIL*, *lincRNA-p21*, *lincRNA-RoR*, and *PANDA* in cellular functions through the pRB and p53 pathways, and the molecular mechanisms by which these lncRNAs regulate the expression of target genes.

Keywords lncRNA • *ANRIL* • *lincRNA-p21* • *PANDA* • *lincRNA-RoR* • p53 • pRB

Y. Kotake (✉)
Department of Biological and Environmental Chemistry, Faculty of Humanity-Oriented Science and Engineering, Kinki University, 11-6 Kayanomori, Iizuka, Fukuoka 820-8555, Japan
e-mail: ykotake@fuk.kindai.ac.jp

M. Kitagawa
Department of Molecular Biology, Hamamatsu University School of Medicine, 1-20-1 Handayama, Higashi-ku, Hamamatsu, Shizuoka 431-3192, Japan

R. Kurokawa (ed.), *Long Noncoding RNAs*, DOI 10.1007/978-4-431-55576-6_10

10.1 Introduction

Retinoblastoma protein (pRB) and p53 tumor suppressor proteins play pivotal roles in the control of cell proliferation, acting as crucial gatekeepers for the cell cycle and apoptosis in response to oncogenic insults and DNA damage (Fig. 10.1) (Sherr and McCormick 2002; Campisi 2005). pRB is a negative regulator of the cell cycle and prevents transition from the G1 to the S phase by binding to and inhibiting E2F transcription in cells exciting mitosis, and in quiescent cells. For cells entering the cell cycle, extracellular mitogens induce the expression of cyclin Ds, which bind to and increase the kinase activity of cyclin-dependent kinase (CDK) 4/6. The

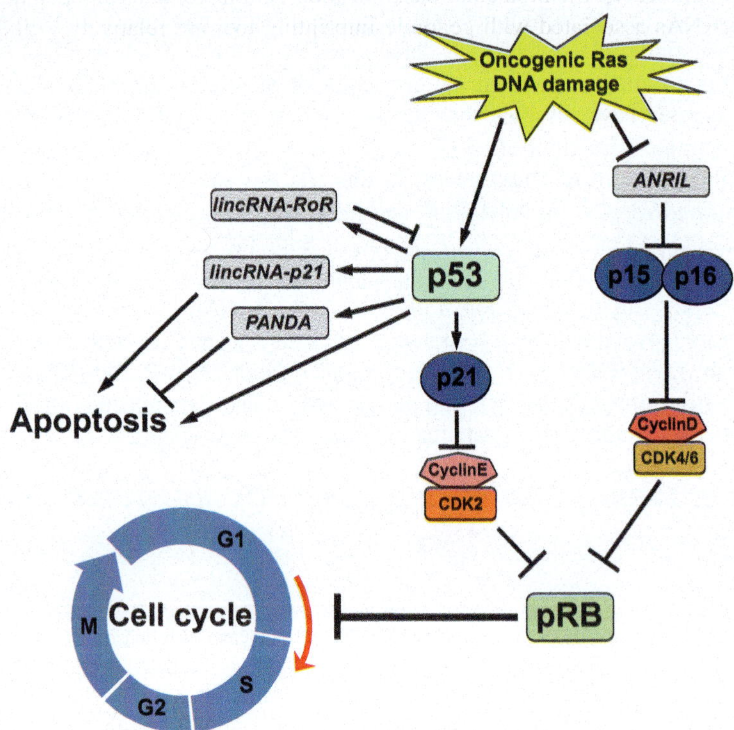

Fig. 10.1 Involvement of *ANRIL*, *lincRNA-p21*, *lincRNA-RoR*, and *PANDA* in the p53 and pRB pathways. The oncogenic Ras and DNA damage signals induce the expression of p15/p16 and p53, causing cell cycle arrest via pRB or apoptosis via p53, thus protecting cells from aberrant proliferation. *ANRIL* is involved in the p15/p16-pRB pathway. The oncogenic Ras signal represses the expression of *ANRIL*, leading to activation of *p15* and *p16* transcription. The *lincRNA-p21*, *lincRNA-RoR* and *PANDA* are induced by p53; *lincRNA-p21* mediates p53-induced apoptosis, while *PANDA* functions as a repressor of p53-mediated apoptosis. *lincRNA-RoR* represses p53 translation, leading to inhibition of p53-mediated cell cycle arrest and apoptosis

activated cyclin D–CDK4/6 complex phosphorylates pRB late in the G1 phase (Weinberg 1995; Kitagawa et al. 1996). At the G1–S boundary, the expression of cyclin Es and associated CDK2 kinase activity approaches maximum levels following cyclin D expression; the complex then further phosphorylates pRB (Hwang and Clurman 2005). Thus, cyclin D–CDK4/6 and cyclin E–CDK2 complexes cooperatively phosphorylate pRB, leading to functional inactivation of pRB and activation of E2F-mediated transcription, thereby promoting G1–S progression. CDK inhibitors (p15, p16, p18, p19, p21, p27, and p57) specifically bind to and inhibit their target cyclin–CDK complexes, leading to retention of pRB in the hypophosphorylated, growth-suppressive state, and preventing G1–S transition (Sherr and Roberts 1999). Disruption of the pRB pathway deregulates the control of G1–S progression, leading to cancer development (Sherr 1996). *ANRIL* (Antisense noncoding RNA in the *INK4* locus) represses the transcription of *p15* and *p16* by recruiting polycomb proteins to this locus, resulting in repression of the pRB pathway (Fig. 10.1) (Kotake et al. 2011; Yap et al. 2010; Kitagawa et al. 2013).

Transcription factor p53 mediates cellular responses to genotoxic and growth stresses induced by oncogenic insults and DNA damage (Laptenko and Prives 2006). Activated p53 regulates the transcription of a large number of target genes and is known to induce apoptosis, cell cycle arrest, and senescence to prevent tumorigenesis (Sherr and McCormick 2002; Vousden and Lu 2002; Vogelstein et al. 2000). Disruption of the p53-mediated checkpoint pathway is believed to be a necessary step for tumorigenesis. The p53 gene is frequently mutated in a wide range of human cancers (Hollstein et al. 1991; Levine et al. 1991). The p53 protein is stabilized and activated by post-translational modifications, such as phosphorylation and ubiquitination in response to DNA damage (Vousden and Lu 2002; Toledo and Wahl 2006); it is also regulated at transcriptional levels (Wang and El-Deiry 2006) and translational levels (Takagi et al. 2005). Recently, it was revealed that p53 induces a large number of long noncoding RNAs (lncRNAs), including *lincRNA-p21*, *lincRNA-RoR*, and *PANDA* (Fig. 10.1) (Hung et al. 2011; Huarte et al. 2010; Zhang et al. 2013; Subramanian et al. 2013). *lincRNA-p21* induces apoptosis by associating with and recruiting heterogeneous nuclear ribonucleoprotein-K (hnRNP-K) to target gene promoters, thereby repressing transcription (Huarte et al. 2010). In contrast to *lincRNA-p21*, *PANDA* represses apoptosis by associating with and inhibiting transcription factor NF-YA occupancy at apoptosis activator gene promoters, which results in the repression of transcription (Hung et al. 2011). *lincRNA-RoR* inhibits the translation of p53 mRNA by associating with hnRNP-I (Zhang et al. 2013), leading to inhibition of p53-mediated cell cycle arrest and apoptosis. In this chapter, we focus on *ANRIL*, *lincRNA-p21*, *lincRNA-RoR*, and *PANDA*, and we discuss their molecular mechanisms and functions with respect to regulation of pRB and p53 pathways.

10.2 *ANRIL*, a Long Noncoding RNA in the *INK4* Locus

10.2.1 ANRIL and INK4 Locus

ANRIL was discovered, by genetic analysis, within a 403 kb germ-line deletion of the melanoma-neural system tumor syndrome family (Pasmant et al. 2007). *ANRIL* spans over 126 kb of genomic sequence and contains 19 exons in human chromosome 9p21 (Fig. 10.2a). Expression analysis has revealed the existence of multiple *ANRIL* splicing variants, with some of them forming circular RNA structures (Folkersen et al. 2009; Burd et al. 2010). Exon 1 of *ANRIL* is located between

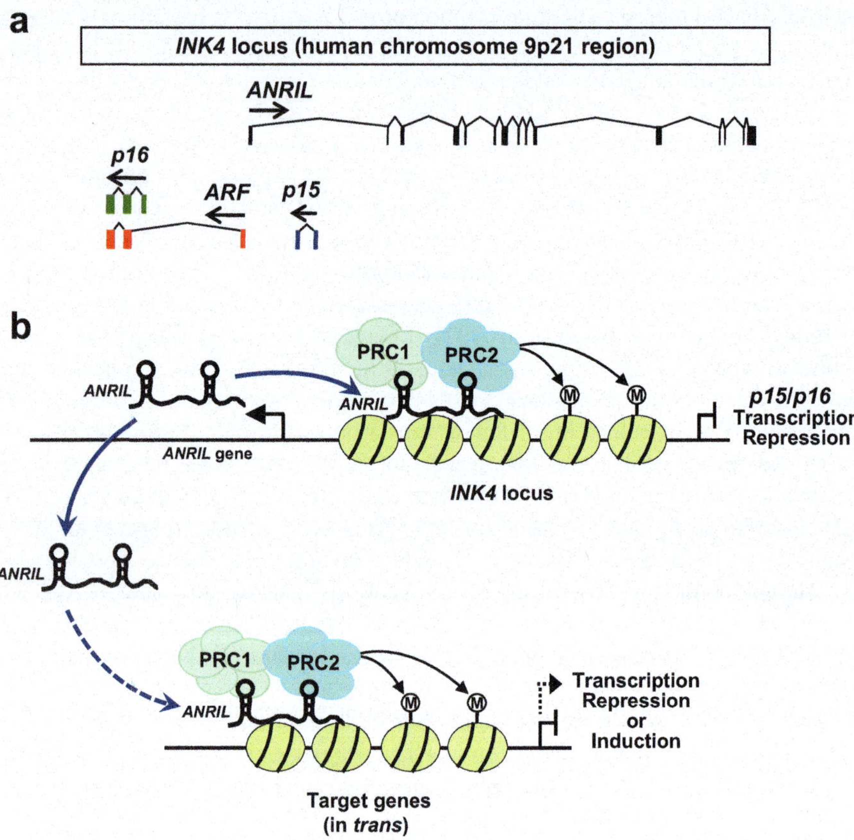

Fig. 10.2 *ANRIL*-mediated transcriptional regulation. (**a**) Genomic structures of the *ANRIL* and *INK4* loci. The exons encoding *ANRIL*, *p15*, *p16*, and *ARF* are shown. *Arrows* indicate the direction of the transcription. (**b**) *ANRIL* binds to and recruits PRC-1 and -2 on the *INK4* locus in a *cis*-acting manner. EZH2, a catalytic subunit of PRC-2, methylates histone H3K27 on the *INK4* locus, leading to the repression of *p15* and *p16* transcription. *ANRIL* also recruits PRC-1 and -2 to target gene promoters in a *trans*-acting manner and regulates their transcription

promoters of *ARF* and *p15* in the *INK4* locus, and is transcribed in the opposite direction to *p15* by RNA polymerase II.

The *INK4* locus encodes three tumor suppressor genes (*p15*, *p16*, and *ARF*) and is frequently deleted or mutated, or its expression is silenced, in a wide range of human cancers (Ruas and Peters 1998; Sharpless 2005). Both p15 and p16 function as inhibitors of CDK4 and CDK6, leading to increased growth-suppressive activities of the pRB family of proteins (Sherr 1996; Sherr and Roberts 1999). ARF binds to and inhibits the activity of MDM2 ubiquitin ligase, stabilizing and activating p53 (Pomerantz et al. 1998; Stott et al. 1998; Zhang et al. 1998). Genetic analyses using mice lacking each gene proved the tumor suppressor functions for these three genes (Serrano et al. 1996; Kamijo et al. 1997; Krimpenfort et al. 2001, 2007; Sharpless et al. 2001).

10.2.2 Transcriptional Regulation of the INK4 Locus by ANRIL

The antisense *p15* (*p15AS*) transcript that overlaps with *ANRIL* has been shown to induce heterochromatin formation and DNA methylation during *p15* silencing in a Dicer-independent manner (Yu et al. 2008). We and Yap et al. revealed that *ANRIL* is involved in the repression of *p15* and *p16* transcription through the recruitment of polycomb proteins (Yap et al. 2010; Kotake et al. 2011). The polycomb proteins form two complexes, *polycomb repression complex* (PRC)-1 and -2, which stably silence the transcription of target genes through histone modification (Cao et al. 2005; Wang et al. 2004a, b). In the hierarchical recruitment and gene silencing model, PRC-2-mediated histone H3 lysine 27 (H3K27) methylation is required for recruitment of PRC-1, which in turn causes H2A-K119 ubiquitination, leading to transcriptional repression of target genes. It has been shown that PRC-1 and -2 bind to and repress the *INK4* locus through H3K27 methylation (Kotake et al. 2007; Bracken et al. 2007; Zeng et al. 2011).

Yap et al. showed that *ANRIL* directly binds to CBX7, a PRC-1 component, in chromatin fractions (Yap et al. 2010). These findings suggest that *ANRIL* remains in the nucleus after transcription and affects chromatin. We also showed that SUZ12, a component of PRC-2, associates with *ANRIL* (Kotake et al. 2011). Inhibition of *ANRIL* disrupts the binding of PRC-1 and PRC-2 on the *INK4* locus, causing an increase in *p15* and *p16* mRNA levels (Kotake et al. 2011; Yap et al. 2010; Wan et al. 2013). Taken together, these data suggest that *ANRIL* binds to and recruits PRC-1 and -2 to the *INK4* locus, leading to the repression of *p15* and *p16* transcription (Fig. 10.2b). The inhibition of *ANRIL* limits cellular life span and induces senescence-associated beta-galactosidase in human fibroblasts, suggesting that *ANRIL* is involved in cellular senescence (Yap et al. 2010; Kotake et al. 2011).

10.2.3 *Trans*-acting Gene Regulation by *ANRIL*

Sato et al. showed that overexpressing an *ANRIL* splice variant suppressed the mRNA expression of several genes (*CEP290*, *EP300*, and *TCF7L1*) involved in the regulation of nuclear function (Sato et al. 2010). Congrains et al. showed that silencing *ANRIL* in human aortic vascular smooth muscle cells (HuAoVSMC) caused changes in the expression levels of several genes involved in the pathogenesis of atherosclerosis (Congrains et al. 2012). These results suggest that *ANRIL* regulates the *INK4* locus and other genes in a *trans*-acting manner. Interestingly, the genes affected by the silencing of *ANRIL* exons 1 or 19 differ (Congrains et al. 2012). *ANRIL* has several splicing variants (Burd et al. 2010; Folkersen et al. 2009), with part of them forming a circular structure (Burd et al. 2010). It might be that each splicing variant of *ANRIL* has a distinct function.

Recently, Holdt et al. reported that overexpressing *ANRIL* caused alterations in the expression levels of various genes. This leads to increased cell adhesion and promotion of cell growth and metabolic activity (Holdt et al. 2013). They also showed that the Alu sequence in *ANRIL* was required for its trans-regulation. *ANRIL* is involved in the transcriptional repression and activation of target genes in a *trans-acting* manner. This conflicting function of *ANRIL* suggests that it might also associate with additional transcriptional regulators and function as a scaffold for them on the promoters of target genes. Overexpressing *ANRIL* changes the distribution of SUZ12 and CBX7 on the promoters of *ANRIL* target genes (Holdt et al. 2013). Taken together, these data suggest that *ANRIL* migrates to and recruits PRC-1 and -2 to target gene promoters located in the *trans* region (Fig. 10.2b) along with lncRNAs, such as *HOTAIR* (Rinn et al. 2007; Gupta et al. 2010).

10.2.4 Regulation of *ANRIL*

The expression of *ANRIL* is transcriptionally regulated by E2F1 (Wan et al. 2013; Sato et al. 2010). Wan et al. showed that E2F1 is induced in an ATM-dependent manner after DNA damage and directly binds to the *ANRIL* promoter. This leads to activation of *ANRIL* transcription and results in the repression of *p15*, *p16*, and *ARF* (Wan et al. 2013). This pathway could be required for re-entry into the cell cycle after DNA repair. We demonstrated that enforced expression of oncogenic Ras (a constitutive active H-Ras mutant) decreased the expression of *ANRIL* (Kotake et al. 2011). In mouse and human fibroblasts, transcription of the *INK4* locus is induced by oncogenic Ras, causing stable cell cycle arrest (also known as premature senescence) (Brookes et al. 2002; Serrano et al. 1997). Reduction in the levels of *ANRIL* expression by oncogenic Ras might be required for the activation of the *INK4* locus and result in the induction of premature senescence to protect cells from hyperproliferative stimulation. However, the mechanisms by which oncogenic Ras represses *ANRIL* expression remain unclear.

10.2.5 Association of ANRIL with Disease

Genome-wide association studies have revealed that single-nucleotide polymorphisms (SNPs) on human chromosome 9p21 are associated with coronary artery disease (CAD) (Helgadottir et al. 2007; McPherson et al. 2007), ischaemic stroke (Gschwendtner et al. 2009; Matarin et al. 2008), type II diabetes (Saxena et al. 2007; Scott et al. 2007), and several cancers (Wrensch et al. 2009; Shete et al. 2009; Bishop et al. 2009). Recent studies have shown that SNPs in the *ANRIL* locus might also be associated with these diseases (Holdt et al. 2010; Cunnington et al. 2010; Broadbent et al. 2008; Iacobucci et al. 2011). One of the SNPs located in *ANRIL* is significantly associated with plexiform neurofibroma and correlates with reduced expression of *ANRIL* (Pasmant et al. 2011). Because *ANRIL* regulates the expression of the *INK4* locus and a large number of genes (Sato et al. 2010; Congrains et al. 2012; Holdt et al. 2013), disruption of its expression and functions can cause a wide variety of diseases.

10.3 lncRNAs Induced by p53

10.3.1 lincRNA-p21

Huarte et al. showed that several lncRNAs are induced in a p53-dependent manner (Huarte et al. 2010). One of these, *lincRNA-p21*, is involved in p53-mediated apoptosis; it is located about 15 kb upstream of the CDK inhibitor *p21* gene and is transcribed in the opposite direction of *p21* (Fig. 10.3). The promoter of *lincRNA-p21* has a highly conserved canonical p53-binding motif; p53 directly binds and activates the transcription of *lincRNA-p21* in response to DNA damage. Depletion of *lincRNA-p21* blocks DNA damage-associated apoptosis; conversely, overexpression of *lincRNA-p21* increases apoptosis induced by DNA damage. These results are indicators that *lincRNA-p21* is required for the induction of apoptosis in response to DNA damage. Huarte et al. demonstrated that *lincRNA-p21* functions as a repressor for certain p53 target genes. Pull-down assays using *lincRNA-p21* as bait identified hnRNP-K as a *lincRNA-p21*-associated protein, with *lincRNA-p21* associating with hnRNP-K through its 5′ terminal region. It is known that hnRNP-K is a component of the H1.2 complex that acts as a repressor in the p53 pathway (Kim et al. 2008). These previous findings indicate that *lincRNA-p21* associates with and recruits hnRNP-K to target gene promoters to repress transcription, resulting in the induction of apoptosis (Fig. 10.3). However, the mechanisms by which the *lincRNA-p21*/hnRNP-K complex recognizes the promoters of target genes in *trans* and represses transcription are unclear.

Yoon et al. revealed another function of *lincRNA-p21* in the regulation of target gene expression (Yoon et al. 2012). *lincRNA-p21* is located in the nucleus and

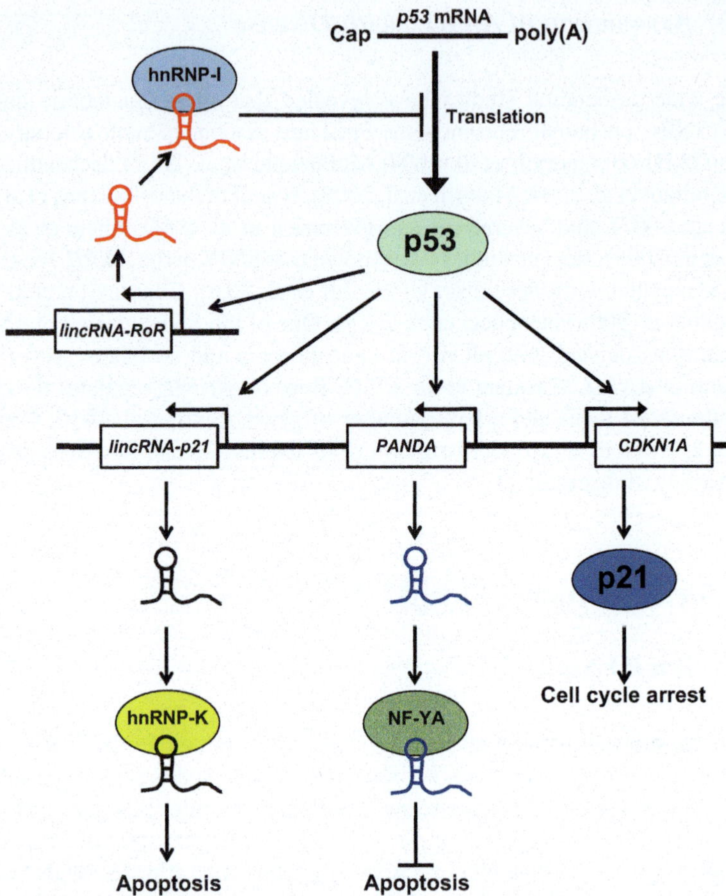

Fig. 10.3 The roles of lncRNAs induced by p53. *lincRNA-p21* and *PANDA* are transcribed upstream of the *CDKN1A* locus. *lincRNA-p21* associates with and recruits hnRNP-K to target gene promoters to repress transcription. This leads to induction of apoptosis. *PANDA* associates with NF-YA and inhibits binding to its target gene promoters, resulting in repression of the apoptotic response. *lincRNA-RoR* represses p53 translation through association with hnRNP-I, leading to the inhibition of apoptosis. Repression of p53 expression by *lincRNA-RoR* constitutes a feedback loop that controls the levels of p53 expression

cytoplasm; cytoplasmic *lincRNA-p21* associates with mRNAs encoding β-catenin and JunB, and represses their translation with cooperation from Rck. *lincRNA-p21* regulates the expression of target genes at the transcriptional and translational levels through association with different proteins.

10.3.2 PANDA

Hung et al. found 216 transcribed regions encoding putative lncRNAs in the promoters of cell cycle genes (Hung et al. 2011). There was no significant correlation between the expression of these putative lncRNAs and neighboring protein-coding genes, suggesting that most lncRNAs are not involved in the regulation of gene transcription in *cis*. *PANDA* (P21 associated ncRNA DNA damage activated) is a lncRNA induced by p53 in response to DNA damage response. *PANDA* is located about 5 kb upstream of the CDK inhibitor *p21*, with a 1.5 kb transcript that is transcribed in the opposite direction to *p21* (Fig. 10.3). Depletion of *PANDA* does not affect p21 expression and vice versa; however, *PANDA* depletion induces the expression of apoptosis activators, such as *APAF1*, *BIK*, *FAS*, and *LRDD*, in response to DNA damage. This results in increased sensitivity to apoptosis. *PANDA* was found to associate with NF-YA, a component of nuclear transcription factor Y (NF-Y). NF-Y forms a trimeric complex involving subunits A, B, and C (Mantovani 1999); it directly binds to the promoter of the apoptotic gene, *FAS*, and activates its transcription (Morachis et al. 2010). Most lncRNAs involved in transcriptional regulation recruit transcriptional regulators to the target locus; however, *PANDA* is not a recruiter but a repressor for NF-YA. Silencing *PANDA* increases NF-YA occupancy on the promoters of apoptotic genes, resulting in transcriptional activation (Hung et al. 2011). *PANDA* is induced by p53 and represses apoptosis in response to DNA damage by inhibiting NF-YA-mediated transcription (Fig. 10.3).

A gain-of-function p53 mutant observed in Li-Fraumeni syndrome retains *PANDA* induction. Selective induction of *PANDA* was observed in metastatic ductal carcinomas compared with that in normal breast tissue (Hung et al. 2011). Abnormal overexpression of *PANDA* might suppress apoptosis induced by DNA damage and overcome the p53 checkpoint, leading to carcinogenesis.

10.3.3 lincRNA-RoR

The *lincRNA-RoR* lncRNA is expressed at elevated levels in human-induced pluripotent stem cells (iPSCs) and modulates reprogramming of iPSCs (Loewer et al. 2010). The *lincRNA-RoR* gene is located on human chromosome 18; its transcript is 2.6 kb and contains four exons. Depletion of *lincRNA-RoR* results in diminished induction of iPSCs by reprogramming; in contrast, forced expression of *lincRNA-RoR* enhances the establishment of iPSCs during reprogramming. DNA microarray analysis revealed that silencing of *lincRNA-RoR* increases the expression of genes involved in the p53 pathway. Recently, it has been shown that *lincRNA-RoR* suppresses the translation of p53 (Zhang et al. 2013). Overexpressing *lincRNA-RoR* blocks DNA damage-induced apoptosis through the repression of p53 expression.

Silencing *lincRNA-RoR* increases p53 protein levels but not mRNA levels, indicating that *lincRNA-RoR* translationally represses p53 expression. Pull-down assays using deletion fragments of *lincRNA-RoR* as bait revealed a 500 bp region on exon 4 of *lincRNA-RoR* that associates with phosphorylated hnRNP-I through two potential hnRNP-I binding motifs. This binding is required for *lincRNA-RoR*-mediated repression of p53 translation. *lincRNA-RoR* itself is activated by p53, while p53 binds to the p53-response element on the *lincRNA-RoR* promoter. Thus, p53 and *lincRNA-RoR* form an autoregulatory feedback loop (Fig. 10.3).

10.4 Conclusion

The pRB and p53 pathways affect cell fate determination by regulating processes such as proliferation, senescence, apoptosis, and differentiation. We postulate that lncRNAs, which are involved in these pathways, might play pivotal roles in controlling cellular functions. Among the known lncRNAs, *ANRIL* is involved in cell proliferation and senescence via the p15/p16-pRB pathway. The *lincRNA-p21*, *PANDA*, and *lincRNA-RoR* lncRNAs are involved in the regulation of apoptosis and the derivation of iPSCs through the p53 pathway. What is required is the elucidation of the physiological significance of these lncRNAs. Since the disruption of pRB and p53 pathways leads to malignant transformation, these pathways must be strictly controlled by many factors, including multiple lncRNAs. It is possible that deregulation of these lncRNAs is associated with human cancer, and their investigation could provide new approaches to diagnosis and cancer therapies.

References

Bishop DT, Demenais F, Iles MM, Harland M, Taylor JC, Corda E, Randerson-Moor J, Aitken JF, Avril MF, Azizi E, Bakker B, Bianchi-Scarra G, Bressac-de Paillerets B, Calista D, Cannon-Albright LA, Chin AWT, Debniak T, Galore-Haskel G, Ghiorzo P, Gut I, Hansson J, Hocevar M, Hoiom V, Hopper JL, Ingvar C, Kanetsky PA, Kefford RF, Landi MT, Lang J, Lubinski J, Mackie R, Malvehy J, Mann GJ, Martin NG, Montgomery GW, van Nieuwpoort FA, Novakovic S, Olsson H, Puig S, Weiss M, van Workum W, Zelenika D, Brown KM, Goldstein AM, Gillanders EM, Boland A, Galan P, Elder DE, Gruis NA, Hayward NK, Lathrop GM, Barrett JH, Bishop JA (2009) Genome-wide association study identifies three loci associated with melanoma risk. Nat Genet 41(8):920–925, ng.411 [pii] 10.1038/ng.411

Bracken AP, Kleine-Kohlbrecher D, Dietrich N, Pasini D, Gargiulo G, Beekman C, Theilgaard-Monch K, Minucci S, Porse BT, Marine JC, Hansen KH, Helin K (2007) The polycomb group proteins bind throughout the INK4A-ARF locus and are disassociated in senescent cells. Genes Dev 21(5):525–530, 21/5/525 [pii] 10.1101/gad.415507

Broadbent HM, Peden JF, Lorkowski S, Goel A, Ongen H, Green F, Clarke R, Collins R, Franzosi MG, Tognoni G, Seedorf U, Rust S, Eriksson P, Hamsten A, Farrall M, Watkins H (2008) Susceptibility to coronary artery disease and diabetes is encoded by distinct, tightly linked SNPs in the ANRIL locus on chromosome 9p. Hum Mol Genet 17(6):806–814, ddm352 [pii] 10.1093/hmg/ddm352

Brookes S, Rowe J, Ruas M, Llanos S, Clark PA, Lomax M, James MC, Vatcheva R, Bates S, Vousden KH, Parry D, Gruis N, Smit N, Bergman W, Peters G (2002) INK4a-deficient human diploid fibroblasts are resistant to RAS-induced senescence. EMBO J 21(12):2936–2945. doi:10.1093/emboj/cdf289

Burd CE, Jeck WR, Liu Y, Sanoff HK, Wang Z, Sharpless NE (2010) Expression of linear and novel circular forms of an INK4/ARF-associated non-coding RNA correlates with atherosclerosis risk. PLoS Genet 6(12):e1001233. doi:10.1371/journal.pgen.1001233

Campisi J (2005) Senescent cells, tumor suppression, and organismal aging: good citizens, bad neighbors. Cell 120(4):513–522, S0092-8674(05)00111-X [pii] 10.1016/j.cell.2005.02.003

Cao R, Tsukada Y, Zhang Y (2005) Role of Bmi-1 and Ring1A in H2A ubiquitylation and Hox gene silencing. Mol Cell 20(6):845–854, S1097-2765(05)01817-4 [pii] 10.1016/j.molcel.2005.12.002

Congrains A, Kamide K, Katsuya T, Yasuda O, Oguro R, Yamamoto K, Ohishi M, Rakugi H (2012) CVD-associated non-coding RNA, ANRIL, modulates expression of atherogenic pathways in VSMC. Biochem Biophys Res Commun 419(4):612–616, S0006-291X(12)00283-5 [pii] 10.1016/j.bbrc.2012.02.050

Cunnington MS, Santibanez Koref M, Mayosi BM, Burn J, Keavney B (2010) Chromosome 9p21 SNPs associated with multiple disease phenotypes correlate with ANRIL expression. PLoS Genet 6(4):e1000899. doi:10.1371/journal.pgen.1000899

Folkersen L, Kyriakou T, Goel A, Peden J, Malarstig A, Paulsson-Berne G, Hamsten A, Hugh W, Franco-Cereceda A, Gabrielsen A, Eriksson P (2009) Relationship between CAD risk genotype in the chromosome 9p21 locus and gene expression. Identification of eight new ANRIL splice variants. PLoS One 4(11):e7677. doi:10.1371/journal.pone.0007677

Gschwendtner A, Bevan S, Cole JW, Plourde A, Matarin M, Ross-Adams H, Meitinger T, Wichmann E, Mitchell BD, Furie K, Slowik A, Rich SS, Syme PD, MacLeod MJ, Meschia JF, Rosand J, Kittner SJ, Markus HS, Muller-Myhsok B, Dichgans M (2009) Sequence variants on chromosome 9p21.3 confer risk for atherosclerotic stroke. Ann Neurol 65(5):531–539. doi:10.1002/ana.21590

Gupta RA, Shah N, Wang KC, Kim J, Horlings HM, Wong DJ, Tsai MC, Hung T, Argani P, Rinn JL, Wang Y, Brzoska P, Kong B, Li R, West RB, van de Vijver MJ, Sukumar S, Chang HY (2010) Long non-coding RNA HOTAIR reprograms chromatin state to promote cancer metastasis. Nature 464(7291):1071–1076, nature08975 [pii] 10.1038/nature08975

Helgadottir A, Thorleifsson G, Manolescu A, Gretarsdottir S, Blondal T, Jonasdottir A, Sigurdsson A, Baker A, Palsson A, Masson G, Gudbjartsson DF, Magnusson KP, Andersen K, Levey AI, Backman VM, Matthiasdottir S, Jonsdottir T, Palsson S, Einarsdottir H, Gunnarsdottir S, Gylfason A, Vaccarino V, Hooper WC, Reilly MP, Granger CB, Austin H, Rader DJ, Shah SH, Quyyumi AA, Gulcher JR, Thorgeirsson G, Thorsteinsdottir U, Kong A, Stefansson K (2007) A common variant on chromosome 9p21 affects the risk of myocardial infarction. Science 316(5830):1491–1493, 1142842 [pii] 10.1126/science.1142842

Holdt LM, Beutner F, Scholz M, Gielen S, Gabel G, Bergert H, Schuler G, Thiery J, Teupser D (2010) ANRIL expression is associated with atherosclerosis risk at chromosome 9p21. Arterioscler Thromb Vasc Biol 30(3):620–627, ATVBAHA.109.196832 [pii] 10.1161/ATVBAHA.109.196832

Holdt LM, Hoffmann S, Sass K, Langenberger D, Scholz M, Krohn K, Finstermeier K, Stahringer A, Wilfert W, Beutner F, Gielen S, Schuler G, Gabel G, Bergert H, Bechmann I, Stadler PF, Thiery J, Teupser D (2013) Alu elements in ANRIL non-coding RNA at chromosome 9p21 modulate atherogenic cell functions through trans-regulation of gene networks. PLoS Genet 9(7):e1003588, PGENETICS-D-13-00414 [pii] 10.1371/journal.pgen.1003588

Hollstein M, Sidransky D, Vogelstein B, Harris CC (1991) p53 mutations in human cancers. Science 253(5015):49–53

Huarte M, Guttman M, Feldser D, Garber M, Koziol MJ, Kenzelmann-Broz D, Khalil AM, Zuk O, Amit I, Rabani M, Attardi LD, Regev A, Lander ES, Jacks T, Rinn JL (2010) A large intergenic noncoding RNA induced by p53 mediates global gene repression in the p53 response. Cell 142(3):409–419, S0092-8674(10)00730-0 [pii] 10.1016/j.cell.2010.06.040

Hung T, Wang Y, Lin MF, Koegel AK, Kotake Y, Grant GD, Horlings HM, Shah N, Umbricht C, Wang P, Kong B, Langerod A, Borresen-Dale AL, Kim SK, van de Vijver M, Sukumar S, Whitfield ML, Kellis M, Xiong Y, Wong DJ, Chang HY (2011) Extensive and coordinated transcription of noncoding RNAs within cell-cycle promoters. Nat Genet 43(7):621–629, ng.848 [pii] 10.1038/ng.848

Hwang HC, Clurman BE (2005) Cyclin E in normal and neoplastic cell cycles. Oncogene 24(17):2776–2786, 1208613 [pii] 10.1038/sj.onc.1208613

Iacobucci I, Sazzini M, Garagnani P, Ferrari A, Boattini A, Lonetti A, Papayannidis C, Mantovani V, Marasco E, Ottaviani E, Soverini S, Girelli D, Luiselli D, Vignetti M, Baccarani M, Martinelli G (2011) A polymorphism in the chromosome 9p21 ANRIL locus is associated to Philadelphia positive acute lymphoblastic leukemia. Leuk Res 35(8):1052–1059, S0145-2126(11)00111-1 [pii] 10.1016/j.leukres.2011.02.020

Kamijo T, Zindy F, Roussel MF, Quelle DE, Downing JR, Ashmun RA, Grosveld G, Sherr CJ (1997) Tumor suppression at the mouse INK4a locus mediated by the alternative reading frame product p19ARF. Cell 91(5):649–659, S0092-8674(00)80452-3 [pii]

Kim K, Choi J, Heo K, Kim H, Levens D, Kohno K, Johnson EM, Brock HW, An W (2008) Isolation and characterization of a novel H1.2 complex that acts as a repressor of p53-mediated transcription. J Biol Chem 283(14):9113–9126, M708205200 [pii] 10.1074/jbc.M708205200

Kitagawa M, Higashi H, Jung HK, Suzuki-Takahashi I, Ikeda M, Tamai K, Kato J, Segawa K, Yoshida E, Nishimura S, Taya Y (1996) The consensus motif for phosphorylation by cyclin D1–CDK4 is different from that for phosphorylation by cyclin A/E–CDK2. EMBO J 15(24):7060–7069

Kitagawa M, Kitagawa K, Kotake Y, Niida H, Ohhata T (2013) Cell cycle regulation by long non-coding RNAs. Cell Mol Life Sci 70(24):4785–4794. doi:10.1007/s00018-013-1423-0

Kotake Y, Cao R, Viatour P, Sage J, Zhang Y, Xiong Y (2007) pRB family proteins are required for H3K27 trimethylation and polycomb repression complexes binding to and silencing p16INK-4alpha tumor suppressor gene. Genes Dev 21(1):49–54, 21/1/49 [pii] 10.1101/gad.1499407

Kotake Y, Nakagawa T, Kitagawa K, Suzuki S, Liu N, Kitagawa M, Xiong Y (2011) Long non-coding RNA ANRIL is required for the PRC2 recruitment to and silencing of p15(INK4B) tumor suppressor gene. Oncogene 30(16):1956–1962, onc2010568 [pii] 10.1038/onc.2010.568

Krimpenfort P, Quon KC, Mooi WJ, Loonstra A, Berns A (2001) Loss of p16Ink4a confers susceptibility to metastatic melanoma in mice. Nature 413(6851):83–86, 35092584 [pii] 10.1038/35092584

Krimpenfort P, Ijpenberg A, Song JY, van der Valk M, Nawijn M, Zevenhoven J, Berns A (2007) p15Ink4b is a critical tumour suppressor in the absence of p16Ink4a. Nature 448(7156):943–946, nature06084 [pii] 10.1038/nature06084

Laptenko O, Prives C (2006) Transcriptional regulation by p53: one protein, many possibilities. Cell Death Differ 13(6):951–961, 4401916 [pii] 10.1038/sj.cdd.4401916

Levine AJ, Momand J, Finlay CA (1991) The p53 tumour suppressor gene. Nature 351(6326):453–456. doi:10.1038/351453a0

Loewer S, Cabili MN, Guttman M, Loh YH, Thomas K, Park IH, Garber M, Curran M, Onder T, Agarwal S, Manos PD, Datta S, Lander ES, Schlaeger TM, Daley GQ, Rinn JL (2010) Large intergenic non-coding RNA-RoR modulates reprogramming of human induced pluripotent stem cells. Nat Genet 42(12):1113–1117, ng.710 [pii] 10.1038/ng.710

Mantovani R (1999) The molecular biology of the CCAAT-binding factor NF-Y. Gene 239(1):15–27, S0378-1119(99)00368-6 [pii]

Matarin M, Brown WM, Singleton A, Hardy JA, Meschia JF (2008) Whole genome analyses suggest ischemic stroke and heart disease share an association with polymorphisms on chromosome 9p21. Stroke 39(5):1586–1589, STROKEAHA.107.502963 [pii] 10.1161/STROKEAHA.107.502963

McPherson R, Pertsemlidis A, Kavaslar N, Stewart A, Roberts R, Cox DR, Hinds DA, Pennacchio LA, Tybjaerg-Hansen A, Folsom AR, Boerwinkle E, Hobbs HH, Cohen JC (2007) A common allele on chromosome 9 associated with coronary heart disease. Science 316(5830):1488–1491, 1142447 [pii] 10.1126/science.1142447

Morachis JM, Murawsky CM, Emerson BM (2010) Regulation of the p53 transcriptional response by structurally diverse core promoters. Genes Dev 24(2):135–147, gad.1856710 [pii] 10.1101/gad.1856710

Pasmant E, Laurendeau I, Heron D, Vidaud M, Vidaud D, Bieche I (2007) Characterization of a germ-line deletion, including the entire INK4/ARF locus, in a melanoma-neural system tumor family: identification of ANRIL, an antisense noncoding RNA whose expression coclusters with ARF. Cancer Res 67(8):3963–3969, 67/8/3963 [pii] 10.1158/0008-5472.CAN-06-2004

Pasmant E, Sabbagh A, Masliah-Planchon J, Ortonne N, Laurendeau I, Melin L, Ferkal S, Hernandez L, Leroy K, Valeyrie-Allanore L, Parfait B, Vidaud D, Bieche I, Lantieri L, Wolkenstein P, Vidaud M (2011) Role of noncoding RNA ANRIL in genesis of plexiform neurofibromas in neurofibromatosis type 1. J Natl Cancer Inst 103(22):1713–1722, djr416 [pii] 10.1093/jnci/djr416

Pomerantz J, Schreiber-Agus N, Liegeois NJ, Silverman A, Alland L, Chin L, Potes J, Chen K, Orlow I, Lee HW, Cordon-Cardo C, DePinho RA (1998) The Ink4a tumor suppressor gene product, p19Arf, interacts with MDM2 and neutralizes MDM2′s inhibition of p53. Cell 92(6):713–723, S0092-8674(00)81400-2 [pii]

Rinn JL, Kertesz M, Wang JK, Squazzo SL, Xu X, Brugmann SA, Goodnough LH, Helms JA, Farnham PJ, Segal E, Chang HY (2007) Functional demarcation of active and silent chromatin domains in human HOX loci by noncoding RNAs. Cell 129(7):1311–1323, S0092-8674(07)00659-9 [pii] 10.1016/j.cell.2007.05.022

Ruas M, Peters G (1998) The p16INK4a/CDKN2A tumor suppressor and its relatives. Biochim Biophys Acta 1378(2):F115–F177, S0304-419X(98)00017-1 [pii]

Sato K, Nakagawa H, Tajima A, Yoshida K, Inoue I (2010) ANRIL is implicated in the regulation of nucleus and potential transcriptional target of E2F1. Oncol Rep 24(3):701–707

Saxena R, Voight BF, Lyssenko V, Burtt NP, de Bakker PI, Chen H, Roix JJ, Kathiresan S, Hirschhorn JN, Daly MJ, Hughes TE, Groop L, Altshuler D, Almgren P, Florez JC, Meyer J, Ardlie K, Bengtsson Bostrom K, Isomaa B, Lettre G, Lindblad U, Lyon HN, Melander O, Newton-Cheh C, Nilsson P, Orho-Melander M, Rastam L, Speliotes EK, Taskinen MR, Tuomi T, Guiducci C, Berglund A, Carlson J, Gianniny L, Hackett R, Hall L, Holmkvist J, Laurila E, Sjogren M, Sterner M, Surti A, Svensson M, Tewhey R, Blumenstiel B, Parkin M, Defelice M, Barry R, Brodeur W, Camarata J, Chia N, Fava M, Gibbons J, Handsaker B, Healy C, Nguyen K, Gates C, Sougnez C, Gage D, Nizzari M, Gabriel SB, Chirn GW, Ma Q, Parikh H, Richardson D, Ricke D, Purcell S (2007) Genome-wide association analysis identifies loci for type 2 diabetes and triglyceride levels. Science 316(5829):1331–1336, 1142358 [pii] 10.1126/science.1142358

Scott LJ, Mohlke KL, Bonnycastle LL, Willer CJ, Li Y, Duren WL, Erdos MR, Stringham HM, Chines PS, Jackson AU, Prokunina-Olsson L, Ding CJ, Swift AJ, Narisu N, Hu T, Pruim R, Xiao R, Li XY, Conneely KN, Riebow NL, Sprau AG, Tong M, White PP, Hetrick KN, Barnhart MW, Bark CW, Goldstein JL, Watkins L, Xiang F, Saramies J, Buchanan TA, Watanabe RM, Valle TT, Kinnunen L, Abecasis GR, Pugh EW, Doheny KF, Bergman RN, Tuomilehto J, Collins FS, Boehnke M (2007) A genome-wide association study of type 2 diabetes in Finns detects multiple susceptibility variants. Science 316(5829):1341–1345, 1142382 [pii] 10.1126/science.1142382

Serrano M, Lee H, Chin L, Cordon-Cardo C, Beach D, DePinho RA (1996) Role of the INK4a locus in tumor suppression and cell mortality. Cell 85(1):27–37, S0092-8674(00)81079-X [pii]

Serrano M, Lin AW, McCurrach ME, Beach D, Lowe SW (1997) Oncogenic ras provokes premature cell senescence associated with accumulation of p53 and p16INK4a. Cell 88(5):593–602, S0092-8674(00)81902-9 [pii]

Sharpless NE (2005) INK4a/ARF: a multifunctional tumor suppressor locus. Mutat Res 576(1-2):22–38, S0027-5107(05)00151-X [pii] 10.1016/j.mrfmmm.2004.08.021

Sharpless NE, Bardeesy N, Lee KH, Carrasco D, Castrillon DH, Aguirre AJ, Wu EA, Horner JW, DePinho RA (2001) Loss of p16Ink4a with retention of p19Arf predisposes mice to tumorigenesis. Nature 413(6851):86–91, 35092592 [pii] 10.1038/35092592

Sherr CJ (1996) Cancer cell cycles. Science 274(5293):1672–1677

Sherr CJ, McCormick F (2002) The RB and p53 pathways in cancer. Cancer Cell 2(2):103–112, S1535610802001022 [pii]

Sherr CJ, Roberts JM (1999) CDK inhibitors: positive and negative regulators of G1-phase progression. Genes Dev 13(12):1501–1512

Shete S, Hosking FJ, Robertson LB, Dobbins SE, Sanson M, Malmer B, Simon M, Marie Y, Boisselier B, Delattre JY, Hoang-Xuan K, El Hallani S, Idbaih A, Zelenika D, Andersson U, Henriksson R, Bergenheim AT, Feychting M, Lonn S, Ahlbom A, Schramm J, Linnebank M, Hemminki K, Kumar R, Hepworth SJ, Price A, Armstrong G, Liu Y, Gu X, Yu R, Lau C, Schoemaker M, Muir K, Swerdlow A, Lathrop M, Bondy M, Houlston RS (2009) Genome-wide association study identifies five susceptibility loci for glioma. Nat Genet 41(8):899–904, ng.407 [pii] 10.1038/ng.407

Stott FJ, Bates S, James MC, McConnell BB, Starborg M, Brookes S, Palmero I, Ryan K, Hara E, Vousden KH, Peters G (1998) The alternative product from the human CDKN2A locus, p14(ARF), participates in a regulatory feedback loop with p53 and MDM2. EMBO J 17(17):5001–5014. doi:10.1093/emboj/17.17.5001

Subramanian M, Jones MF, Lal A (2013) Long non-coding RNAs embedded in the Rb and p53 pathways. Cancer (Basel) 5(4):1655–1675, cancers5041655 [pii] 10.3390/cancers5041655

Takagi M, Absalon MJ, McLure KG, Kastan MB (2005) Regulation of p53 translation and induction after DNA damage by ribosomal protein L26 and nucleolin. Cell 123(1):49–63, S0092-8674(05)00814-7 [pii] 10.1016/j.cell.2005.07.034

Toledo F, Wahl GM (2006) Regulating the p53 pathway: in vitro hypotheses, in vivo veritas. Nat Rev Cancer 6(12):909–923, nrc2012 [pii] 10.1038/nrc2012

Vogelstein B, Lane D, Levine AJ (2000) Surfing the p53 network. Nature 408(6810):307–310. doi:10.1038/35042675

Vousden KH, Lu X (2002) Live or let die: the cell's response to p53. Nat Rev Cancer 2(8):594–604, nrc864 [pii] 10.1038/nrc864

Wan G, Mathur R, Hu X, Liu Y, Zhang X, Peng G, Lu X (2013) Long non-coding RNA ANRIL (CDKN2B-AS) is induced by the ATM-E2F1 signaling pathway. Cell Signal 25(5):1086–1095, S0898-6568(13)00046-6 [pii] 10.1016/j.cellsig.2013.02.006

Wang S, El-Deiry WS (2006) p73 or p53 directly regulates human p53 transcription to maintain cell cycle checkpoints. Cancer Res 66(14):6982–6989, 66/14/6982 [pii] 10.1158/0008-5472.CAN-06-0511

Wang H, Wang L, Erdjument-Bromage H, Vidal M, Tempst P, Jones RS, Zhang Y (2004a) Role of histone H2A ubiquitination in polycomb silencing. Nature 431(7010):873–878, nature02985 [pii] 10.1038/nature02985

Wang L, Brown JL, Cao R, Zhang Y, Kassis JA, Jones RS (2004b) Hierarchical recruitment of polycomb group silencing complexes. Mol Cell 14(5):637–646, S1097276504002928 [pii] 10.1016/j.molcel.2004.05.009

Weinberg RA (1995) The retinoblastoma protein and cell cycle control. Cell 81(3):323–330, 0092-8674(95)90385-2 [pii]

Wrensch M, Jenkins RB, Chang JS, Yeh RF, Xiao Y, Decker PA, Ballman KV, Berger M, Buckner JC, Chang S, Giannini C, Halder C, Kollmeyer TM, Kosel ML, LaChance DH, McCoy L, O'Neill BP, Patoka J, Pico AR, Prados M, Quesenberry C, Rice T, Rynearson AL, Smirnov I, Tihan T, Wiemels J, Yang P, Wiencke JK (2009) Variants in the CDKN2B and RTEL1 regions are associated with high-grade glioma susceptibility. Nat Genet 41(8):905–908, ng.408 [pii] 10.1038/ng.408

Yap KL, Li S, Munoz-Cabello AM, Raguz S, Zeng L, Mujtaba S, Gil J, Walsh MJ, Zhou MM (2010) Molecular interplay of the noncoding RNA ANRIL and methylated histone H3 lysine

27 by polycomb CBX7 in transcriptional silencing of INK4a. Mol Cell 38(5):662–674, S1097-2765(10)00335-7 [pii] 10.1016/j.molcel.2010.03.021

Yoon JH, Abdelmohsen K, Srikantan S, Yang X, Martindale JL, De S, Huarte M, Zhan M, Becker KG, Gorospe M (2012) LincRNA-p21 suppresses target mRNA translation. Mol Cell 47(4):648–655, S1097-2765(12)00552-7 [pii] 10.1016/j.molcel.2012.06.027

Yu W, Gius D, Onyango P, Muldoon-Jacobs K, Karp J, Feinberg AP, Cui H (2008) Epigenetic silencing of tumour suppressor gene p15 by its antisense RNA. Nature 451(7175):202–206, nature06468 [pii] 10.1038/nature06468

Zeng Y, Kotake Y, Pei XH, Smith MD, Xiong Y (2011) p53 binds to and is required for the repression of Arf tumor suppressor by HDAC and polycomb. Cancer Res 71(7):2781–2792, 0008-5472.CAN-10-3483 [pii] 10.1158/0008-5472.CAN-10-3483

Zhang Y, Xiong Y, Yarbrough WG (1998) ARF promotes MDM2 degradation and stabilizes p53: ARF-INK4a locus deletion impairs both the Rb and p53 tumor suppression pathways. Cell 92(6):725–734, S0092-8674(00)81401-4 [pii]

Zhang A, Zhou N, Huang J, Liu Q, Fukuda K, Ma D, Lu Z, Bai C, Watabe K, Mo YY (2013) The human long non-coding RNA-RoR is a p53 repressor in response to DNA damage. Cell Res 23(3):340–350, cr2012164 [pii] 10.1038/cr.2012.164

Chapter 11
The Role of Androgen-Regulated Long Noncoding RNAs in Prostate Cancer

Ken-ichi Takayama and Satoshi Inoue

Abstract Recent transcriptome studies using next-generation sequencing have detected aberrant changes in the expression of long noncoding (lnc) RNA associated with cancer. Systematic analysis of transcription factor-binding sites and the regulated transcripts revealed that many lncRNAs are widely regulated at the transcriptional level. However, the functions of these transcripts have not been fully elucidated. In this study, using prostate cancer cells, we explored androgen receptor (AR)-regulated noncoding RNAs by a global transcriptome analysis. We found that the expression of a novel lncRNA (named *CTBP1-AS*) in the antisense region of *CTBP1* (carboxyl terminal binding protein 1) is rapidly induced by androgen treatment. *CTBP1-AS* is enriched in the nucleus of cancer cells and promotes androgen-dependent and castration-resistant tumor growth. We further presented the novel regulatory mechanism by which *CTBP1-AS* mediates epigenomic transcriptional control in the nucleus. *CTBP1-AS* interacts with an RNA-binding transcriptional and splicing factor, SFPQ/PSF, and repressed cell cycle regulators or AR coregulators including CTBP1. Thus, we showed that the expression of this novel lncRNA is induced by androgen treatment, and the lncRNA promotes prostate cancer growth.

Keywords Castration-resistant prostate cancer (CRPC) • Androgen receptor (AR) • *CTBP1-AS* • PSF • In situ hybridization (ISH) • RNA-fluorescence *in situ* hybridization (FISH) • Cap analysis of gene expression (CAGE) • Chromatin immunoprecipitation ChIP) • ChIP-sequence (ChIP-seq)

K. Takayama (✉) • S. Inoue
Department of Anti-Aging, Graduate School of Medicine, The University of Tokyo, Tokyo, Japan

Department of Geriatric Medicine, Graduate School of Medicine, The University of Tokyo, Tokyo, Japan
e-mail: ktakayama-tky@umin.ac.jp

R. Kurokawa (ed.), *Long Noncoding RNAs*, DOI 10.1007/978-4-431-55576-6_11

11.1 Introduction

11.1.1 The Emergence of Long Noncoding RNAs in Prostate Cancer Biology

Recent advances in transcriptome technology, especially next-generation sequencing, have revealed active transcription of more than 90 % of the human genome (Gutschner and Diederichs 2012). The worldwide ENCODE project has shown that only 2 % of transcripts are translated into proteins (Djebali et al. 2012). The noncoding RNAs (ncRNAs), the majority of transcripts in the nucleus, were initially thought of as the 'dark matter'. Generally ncRNAs are broadly divided into short (<200 nt) and long (>200 nt) transcripts. Short ncRNAs, particularly microRNAs (miRNAs), play important roles in cancer by post-transcriptionally modifying target mRNA or protein expression (Mercer et al. 2013). In this chapter, we summarize the function and clinical significance of the less understood long noncoding RNAs (lncRNAs) in prostate cancer. lncRNAs represent most of the transcribed ncRNA in the human genome. GENCODE v19 includes 13,870 human lncRNA-related genes, which produce 23,898 lncRNAs (Derrien et al. 2012). These lncRNAs exhibit a structure and biogenesis similar to those of the mRNAs. They are polyadenylated and may function in either nuclear or cytoplasmic fractions. However, only a few have been functionally characterized and experimentally validated. The evidence that lncRNAs are aberrantly expressed in numerous human diseases including cancer supports their importance (Du et al. 2013; Kan et al. 2010; Moran et al. 2012). A common mechanism seems to be their role as molecular scaffolds for targeting gene regulatory complexes to specific genomic loci. They can act as transcriptional regulators to modulate gene expression (Lee 2012). LncRNAs are uniquely equipped to function as locus-specific recruiters by tethering as a result of the formation of DNA–RNA heteroduplexes during transcription (Kung and Lee 2013).

11.1.2 Prostate Cancer-Associated Noncoding RNAs

NEAT1 and *MALAT1*: The mammalian nucleus is highly organized and contains distinct structural components comprising approximately ten types of nuclear bodies, including speckles and paraspeckles, which are thought to be involved in gene regulation. Some of these components contain specific lncRNAs that regulate nuclear body function. An essential architectural component of the paraspeckle structure, *nuclear enriched abundant transcript 1* (*NEAT1*), directly binds to SFPQ/PSF and represses the transcription of several genes. *NEAT1* is involved in several biological processes. Another paraspeckle long noncoding RNA, *metastasis-associated lung adenocarcinoma transcript 1 (MALAT1)* or *NEAT2*, is involved in gene regulation. In prostate cancer, *MALAT1* overexpression is correlated with poor prognosis and cancer progression (Ren et al. 2013). Although the mechanism of *MALAT1* upregulation in cancer cells is still unclear, multiple copy-number gains at the locus and chromosomal translocations have been reported (Paris et al. 2004).

PCAT-1: *Prostate cancer associated ncRNA transcript -1 (PCAT-1)* is a relatively small intergenic ncRNA located in the 8q24 gene desert, and it was identified through a global transcriptomic sequencing of prostate tumors (Prensner et al. 2011). *PCAT-1* expression is inversely correlated with the expression of the enhancer of zeste homolog 2 (EZH2), a histone methyltransferase that encodes components of Polycomb repressive complex 2 (PRC2), and a marker for prostate cancer progression; thus, PRC2 represses *PCAT*-1 expression. *PCAT-1* induces cell proliferation in vitro and has a predominantly repressive effect on gene expression, most notably on the expression of the tumor suppressor gene BRCA2 (Prensner et al. 2014). *PCAT-1* also interacts with the SUZ12 component of PRC2.

SCHLAP1: *Second chromosome locus associated with prostate 1 (SCHLAP1)* is an lncRNA transcribed from within an intergenic gene desert on 2q31.1. It was originally identified in an analysis of intergenic lncRNAs that are selectively upregulated in aggressive prostate cancer samples. *SCHLAP1* is highly expressed in 25 % of prostate tumors, with increased expression in metastatic cancer cells. *SCHLAP1* is involved in the regulation of the switch/sucrose non-fermenting (SWI/SNF) complex. This complex canonically controls transcription by using ATP hydrolysis to remodel chromatin and physically mobilize nucleosomes, particularly at gene promoters. *SCHLAP1* co-immunoprecipitates with SNF5 and prevents its genomic binding, thus antagonizing tumor-suppressive SWI/SNF-mediated gene regulation (Prensner et al. 2013).

PCGEM1: *Prostate cancer gene expression marker 1 (PCGEM1)* was originally identified as a prostate tissue-specific long ncRNA (Srikantan et al. 2000). It is involved in apoptosis inhibition by delaying p53 and p21 induction in an androgen-dependent manner (Petrovics et al. 2004). *PCGEM1* is overexpressed in at least half of prostate tumors. Recent reports have suggested that *PCGEM1* and another ncRNA, *prostate cancer noncoding RNA 1 (PRNCR1)*, are involved in AR-mediated gene transcription (Yang et al. 2013).

PCA3: *Prostate cancer gene 3 (PCA3)* is an lncRNA associated with prostate cancer and is a potentially useful biomarker (Bussemakers et al. 1999; Ferreira et al. 2013). It was originally discovered in 1999 by a differential display analysis of prostate tissues and cell lines. Its expression in 95 % of the prostate tumors is up to 100-fold higher than that in adjacent non-neoplastic tissue. Although its functional role is still unknown, urinary measurement of *PCA3* RNA levels can be helpful to detect prostate cancer, with tumor specificity superior to that of prostate-specific antigen (PSA) (Hessels and Schalken 2009; Lee et al. 2011).

11.1.3 Significance of Androgen Signaling in Prostate Cancer

11.1.3.1 The Structure and Mechanisms of Action of the Androgen Receptor (AR)

AR is a member of the nuclear receptor superfamily (Debes and Tindall 2004; Balk and Knudsen 2008). It is a key molecule for androgen signaling in its target organ, the prostate. Like other nuclear receptors, AR has a modular structure. It consists of an N-terminal domain (NTD)/transactivation domain, DNA-binding domain, and C-terminal ligand-binding domain (LBD) (Heemers and Tindall 2007; Jenster et al. 1991, 1995). The NTD is considered to be constitutively active and is important for transcriptional activation independent of ligand binding (Callewaert et al. 2006; Chamberlain et al. 1996; Dehm et al. 2007). The NTD contains the transcriptional activation function 1 (AF1) domain and mediates aberrant AR activity in castration-resistant prostate cancer cells. The LBD facilitates the binding of androgen to AR. Within the LBD, AF2 interacts with LXXLL-containing coregulators (Umesono and Evans 1989; Heery et al. 1997; Duff et al. 2006). Point mutations associated with prostate cancer have been mapped to the LBD and are associated with resistance to anti-androgens such as bicalutamide (Buchanan et al. 2001; Visakorpi et al. 1995; Bergerat and Ceraline 2009; Taplin et al. 1995, 2003). Testosterone is produced in the testes and is the most abundant circulating androgen (~90 %). Testosterone is able to diffuse into prostate cells and is converted to dihydrotestosterone (DHT) by the enzyme 5α-reductase (Schmidt and Tindall 2011). DHT directly binds to and activates AR. DHT binds the receptor more tightly than testosterone (Zhou et al. 1995). Before ligand binding, AR exists in the cytoplasm in a complex that includes molecular chaperones and co-chaperones from the heat shock protein (Hsp) family. After binding to androgen, a change in the conformation of the complex leads to AR nuclear translocation. In the nucleus, AR binds as a dimer to specific genomic sequences called androgen-responsive elements (AREs) in the promoter and enhancer regions of its target genes (Fig. 11.1).

11.1.3.2 The Roles of AR in Prostate Cancer Progression

AR also induces prostate cancer development and progression. Despite a favorable response to initial hormone therapy, most patients progress to lethal castration-resistant prostate cancer (CRPC) with elevation of AR expression (Chen et al. 2004), hypersensitivity to androgens (Waltering et al. 2009), intratumoral steroidogenesis (Locke et al. 2008), and AR variants (Sun et al. 2010). Thus, identification of AR downstream-signaling events is critical for understanding the progression to CRPC. PSA is a representative androgen-responsive gene in prostate cancer and is widely used as a clinical marker for the detection of the disease. Classic analyses of AR functions were performed using this gene as a model (Shang et al. 2002). However, the regulation of other AR target genes and their clinical relevance is not well understood.

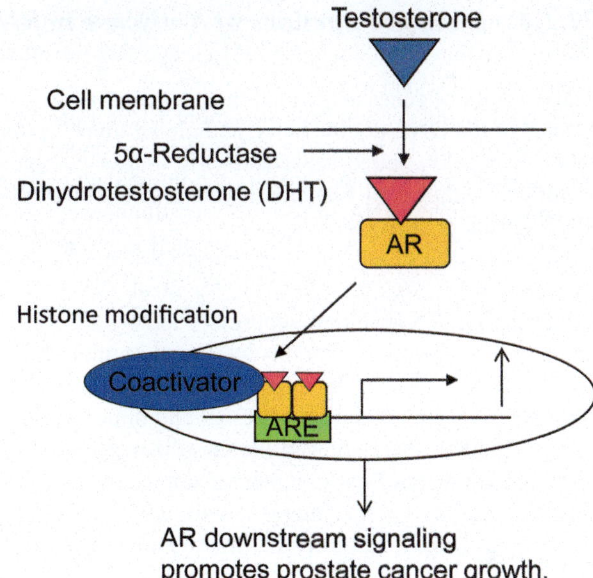

Fig. 11.1 AR functions as a nuclear receptor and ligand-dependent transcription factor. AR translocates to the nucleus and binds androgen response elements (*AREs*). AR epigenetically activates transcription by recruiting coactivators for histone modifications

11.1.3.3 Model Cells for Investigating AR Signaling in CRPC

Although AR overexpression in CRPC is commonly observed, the mechanism underlying altered AR expression is not fully understood. However, recent studies showed that both transcriptional and epigenomic changes are important for AR upregulation in prostate cancer. Cai et al. (2011) analyzed an androgen-dependent prostate cancer cell line and identified AR binding sites in the introns of the *AR* gene. Androgen treatment causes AR to bind the enhancer and recruit lysine-specific demethylase 1 (LSD1), which represses transcription by inhibiting histone H3K4 methylation, thus demonstrating the negative feedback loop that limits endogenous AR expression. However, if cells are incubated in castration levels of androgens, AR expression increases. Furthermore, low levels of androgens in CRPC are sufficient to activate AR target genes but insufficient to suppress the genes, including the *AR* gene itself.

To analyze AR involvement in the progression from hormone-sensitive to castration-resistant prostate cancer, several cell models have been developed. In particular, AR-positive prostate cancer models have been established by incubating LNCaP cells or VCaP cells in a hormone-depleted state (Cai et al. 2011; Culig et al. 1999; Kokontis et al. 1994; Chen et al. 2011). Such model cells were called LNCaP-abl (Chen et al. 2011) or LTAD (Takayama et al. 2012) cells. These cell lines express increased AR at both the mRNA and protein levels, suggesting hypersensitivity of AR signaling. The functions of AR-regulated genes have been analyzed using these cell models.

11.1.4 Diverse Functions of Antisense ncRNAs

The majority of genes in the mammalian genome can generate transcripts from both strands of the DNA double helix, and global transcriptome analyses have identified more than 1,000 paired sense/antisense transcripts, indicating that antisense transcription is important for gene regulation (Carninci et al. 2005; Katayama et al. 2005; Rosok et al. 2004).

One example of an antisense transcript is *cyclin dependent kinase (CDKN) 2B antisense RNA1 (CDKN2B-AS1)* (Yu et al. 2008), which is located within the CDKN2A/CDKN2B tumor suppressor locus on 9q21.3. CDK inhibitors p14, p15, and p16 are produced from this locus and regulate cell cycle progression at G1/S. *CDKN2B-AS1* specifically silences CDKN2B through heterochromatin formation, by increasing dimethylation of H3K9 and dimethylation of H3K4 at the gene promoter. Another study showed the coordinated upregulation of *CDKN2B-AS1* and chromobox7 (CBX7), a member of polycomb repressive complex 1 (PRC1), in prostate cancer. CBX7 interacts with *CDKN2B-AS1* at the gene promoter of CDKN2A/2B for histone H3K27 methylation (Yap et al. 2010).

11.1.5 Global Analysis of the AR-Mediated Transcriptional Program

Since the development of microarray technology, the androgen-mediated transcriptional program has been investigated in normal and cancerous prostate cells. The majority of large-scale gene expression studies have been performed in LNCaP cells (Wang et al. 2009), the most widely used model cell line for prostate cancer research. LNCaP cells are epithelial in origin, express AR, and exhibit androgen-sensitive growth and survival. Expression studies in LNCaP cells characterized the temporal program of transcription that reflects the cellular response to androgens and identified specific androgen-regulated genes or gene networks that participate in these responses.

The rapid development of technology to detect transcription factor-binding sites has revolutionized research on steroid hormone receptors. These technologies are based on chromatin immunoprecipitation (ChIP) analysis. After fixing the protein and chromatin associations by formaldehyde treatment, cells are lysed, and chromatin segmentation by ultrasound is performed. ChIP is performed using an antibody specific to the transcription factors of interest. Because DNA for ChIP analyses contains regions enriched for transcription factor-binding sites, there have been several attempts to analyze them.

For a comprehensive analysis of transcription factor-binding sites, ChIP-on-chip analyses that combine ChIP with genome tiling array technology (chip) have been used (Wang et al. 2007, 2009; Takayama et al. 2007, 2009; Jia et al. 2008; Massie et al. 2007). We first performed ChIP-on-chip using ENCODE (Takayama et al.

2007) and a chromosome 21 and 22 (Takayama et al. 2009) genome tiling array in LNCaP cells. The ENCODE array includes a representative sample (1 %) of the human genome. We have validated 10 AR binding sites and identified novel AR target genes, such as pepsinogen C (*PGC*), *UGT1A1*, and cadherin-2 (*CDH2*). CDH2 is known to be involved in prostate cancer progression. A high-affinity ARBS (AR binding site) was identified in intron 1 of the *CDH2* gene and validated by conventional ChIP analysis. Our study suggested that unbiased ARBSs are not located in the promoter regions and are far from the transcription start sites (TSSs) of RefSeq genes. Several other ChIP analyses have also been reported (Wang et al. 2009; Jia et al. 2008; Massie et al. 2007). Wang et al. (2009) mapped the ARBSs on chromosomes 21 and 22 in LNCaP cells. They expanded this to a genome-wide study comparing LNCaP cells and LNCaP-derived castration-resistant LNCaP-abl cells to identify direct AR-dependent target genes in the disease progression to CRPC[55]. They showed that AR is involved in CRPC via activation of cell cycle progression, mainly by inducing mitotic (M) phase-related genes.

More recently, high-throughput analysis of transcription factor-binding sites using highly developed sequencers, called ChIP-sequence (ChIP-seq) analyses, have been developed (Yu et al. 2010; Urbanucci et al. 2012; Tan et al. 2012). Several studies using ChIP-seq technology have been performed. Although ChIP-on-chip could not detect ARBSs in the regions in which probes could not be prepared, ChIP-seq was able to detect additional binding regions. ChIP-seq has demonstrated a genome-wide cell-based AR transcriptional program (Urbanucci et al. 2012). For instance, two sublines of LNCaP prostate cancer (PC) cell lines were used for ChIP-seq—one with twofold to threefold higher AR expression and the other with fourfold to fivefold higher AR expression than the control cells. Interestingly, the number of ARBSs and the AR binding strength were positively associated with the level of AR when cells were stimulated with low concentrations of androgens. These data demonstrated that the overexpression of AR sensitizes the receptor binding to chromatin, thus explaining how the AR signaling pathway is reactivated in CRPC cells. In another study (Tan et al. 2012), NKX3-1 and ARBSs across the prostate cancer genome were analyzed. *NKX3-1* is a homeobox gene required for prostate tumor progression, and mechanisms by which NKX3-1 controls the AR transcriptional network in prostate cancer were uncovered. NKX3-1 collaborates with AR and FOXA1 to mediate gene expression in advanced and recurrent prostate carcinoma.

11.1.6 Integrative Analysis to Discover the Androgen-Regulated Transcriptional Program

In addition to ChIP-seq, next-generation sequencing is used to analyze the transcriptome of prostate cancer cells. Whereas microarray detects the expression levels of transcripts that bind probes, sequence analysis measures the unbiased expression profiles of all transcripts. Therefore, combined analyses of genome-wide ARBSs and the androgen-regulated transcriptome have recently been reported. We have applied these techniques to the analysis of androgen-mediated transcriptional changes. Cap analysis of gene expression (CAGE) (Shiraki et al. 2003) is a high-throughput method to analyze gene expression and to profile TSSs, including those for promoter usage. CAGE is based on the sequencing of concatemers of DNA tags deriving from the initial 20 nucleotides at the 5′ ends of mRNAs. The frequency of CAGE tags is consistent with results from other analyses, such as microarrays. This analysis is high throughput, enabling an understanding of gene networks via the correlation between promoter usage and expression of gene transcription factors. We performed CAGE to determine androgen-regulated TSSs and ChIP-on-chip analysis to identify genome-wide ARBSs and histone H3 acetylated (AcH3) sites in the human whole genome (Takayama et al. 2011). Using CAGE, we identified 13,110 distinct, androgen-regulated TSSs. Cross-referencing with the gene expression database for prostate cancer (Oncomine), the majority of androgen-upregulated genes containing adjacent ARBSs and CAGE tag clusters in our study were previously confirmed as upregulated genes in prostate cancer. The integrated high-throughput genome analyses of CAGE and ChIP-on-chip provide useful information for elucidating the AR-mediated transcriptional network that contributes to the development and progression of prostate cancer. ncRNAs, including miRNAs, were also identified as androgen target transcripts in this study. We found many androgen-dependent TSSs widely distributed throughout the genome, including in the antisense (AS) direction of RefSeq genes. Several pairs of sense/antisense promoters were newly identified within single RefSeq gene regions, suggesting the involvement of AS noncoding RNA in transcriptional regulation.

A new technique, global nuclear run-on sequencing (GRO-seq), has been used to analyze sequential gene expression upon androgen treatment (Wang et al. 2011). For nuclear run-on reactions, cell nuclei are isolated after treatment with androgen for a specific amount of time. In the run-on step, RNA polymerases are allowed to run on about 100 bases in the presence of a ribonucleotide analog (5-bromouridine 5′-triphosphate [BrUTP]). BrU-containing RNA is selected through immunopurification with an antibody specific for the nucleotide analog. A cDNA library is then prepared for next-generation sequencing. Using this method, the authors discovered the production of enhancer-templated noncoding RNAs (eRNAs) including a unique class of enhancers that do not require nucleosome remodeling to induce specific enhancer-promoter looping and gene activation. GRO-seq data also suggest that AR induces both transcription initiation and elongation in a ligand-dependent manner. In combination with the AR binding and FOXA1 binding data, a large repository of active enhancers that can be dynamically tuned to elicit alternative gene expression programs was identified, and these may underlie many sequential gene expression events in prostate cancer progression.

11.2 Identification and Androgen-Regulation of Long Noncoding RNA, *CTBP1-AS*

11.2.1 An Overview of Our Study

Although AR signaling is important in development of prostate cancer, the roles of androgen-regulated lncRNAs have not been investigated. We investigated the functional role of a novel androgen-responsive long noncoding RNA, *CTBP1-AS* (Takayama et al. 2013). In this chapter, we focus on the mechanism of androgen-mediated cancer proliferation by this lncRNA and the clinical significance in prostate cancer progression.

Pairs of sense–antisense regulated tag clusters (TCs) with ARBSs were identified by a combinational study of ChIP-on-chip and CAGE analyses. Interestingly, CTBP1 was included among the androgen-regulated genes with antisense transcriptional activation. In addition, ARBS was overlapped with histone H3 acetylation chromatin status at the 3′-untranslated region (UTR) of CTBP1. This ARBS sequence included ARE motifs and was involved in androgen-mediated transcriptional activity (Fig. 11.2).

11.2.2 Cloning and Detection of CTBP1-AS

This androgen-regulated TC is located just downstream of ARBS and in the exon of AX747592 (registered in GENBANK), suggesting that a transcriptional variant of this transcript started from the TC. There was an upregulation of *CTBP1-AS* together with downregulation of CTBP1 in response to androgen. Multiple transcriptional termination sites in this antisense transcript (15–3 kb) were identified, although our northern blot analysis revealed that a ~5 kb transcript is the predominant isoform. Interestingly, expression of *CTBP1-AS* was more abundant in the nucleus relative to

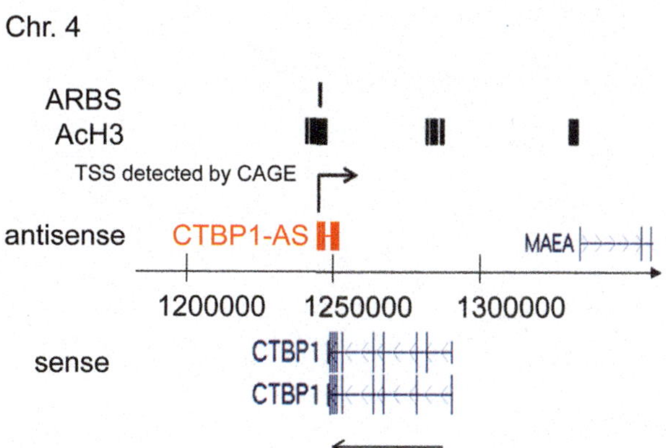

Fig. 11.2 Genomic location of *CTBP1-AS*. We identified an ARBS at the promoter region of *CTBP1-AS* by AR ChIP-chip or ChIP-seq analysis. Androgen-regulated transcriptional start sites (*TSSs*) were found in both sense and antisense promoters of *CTBP1* by CAGE analysis. Gene direction is indicated by *arrows*

the cytoplasm. Moreover, RNA-fluorescence *in situ* hybridization (FISH) analysis revealed that the expression of *CTBP1-AS* is induced diffusely throughout the nucleus by androgen.

The androgen-dependent decrease in CTBP1 protein levels was abolished by short interfering (si) RNA targeting *CTBP1-AS*. Taken together, these results indicated that *CTBP1-AS* is an antisense lncRNA regulated by AR in the nucleus and represses the sense *CTBP1*.

11.2.3 Clinical Significance of CTBP1-AS *in Prostate Cancer*

11.2.3.1 CTBP1-AS Is Upregulated in Prostate Cancer

Quantitative RT-PCR (qRT-PCR) and in situ hybridization are common methods for detecting lncRNA expression in clinical samples of prostate cancer. The upregulation of *CTBP1-AS* and downregulation of *CTBP1* are found to be associated with cancer by laser capture microdissection (LCM) and qRT-PCR analysis.

We did not detect *CTBP1-AS* expression in benign prostate tissues by *in situ* hybridization (ISH) study. However, *CTBP1-AS* expression was upregulated in the cancer samples, demonstrating an inverse correlation between *CTBP1-AS* and CTBP1 expression (Fig. 11.3b). Interestingly, *CTBP1-AS* expression increased with disease progression to metastasis. Taken together, our clinical studies indicate that this antisense lncRNA regulated by androgen is important for prostate cancer progression.

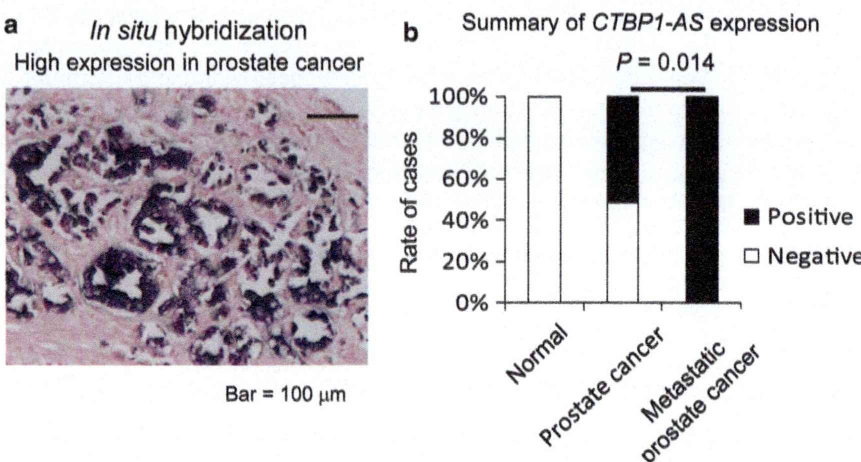

Fig. 11.3 Analysis of *CTBP1-AS* expression in clinical samples of prostate cancer. (**a**) We performed in situ hybridization using specific antisense probes for *CTBP1-AS* labeled with DIG. (**b**) *CTBP1-AS* expression is markedly increased in prostate tumors

11.2.3.2 Loss of CTBP1 Is Associated with Poor Prognosis of Prostate Cancer Patients

To investigate whether sense–antisense transcriptional regulation by *CTBP1-AS* is associated with prostate cancer progression, an immunohistochemical (IHC) analysis of CTBP1 expression for prostate cancer clinical specimens was performed. CTBP1 expression was downregulated in cancer samples compared with benign prostate samples. A Kaplan–Meier analysis showed poor cancer-specific survival in patients with lower CTBP1 expression than in those with higher CTBP1 expression. In addition, multivariate analysis demonstrated that CTBP1 downregulation was an independent prognostic factor.

11.2.4 CTBP1 Interacts with AR as a Co-repressor

CTBP1 functions as a co-repressor in the nucleus and is involved in tumor proliferation (Shi et al. 2003; Bergman et al. 2009; Chinnadurai 2007). We showed the function of CTBP1 as a co-repressor of AR in prostate cancer cells, suggesting the significance of the *CTBP1-AS*-mediated repression of CTBP1. Co-immunoprecipitation assays demonstrated ligand-dependent interaction of both exogenous and endogenous CTBP1 with AR. Androgen treatment induced ligand-dependent recruitment of CTBP1 to ARBS. Re-ChIP analysis showed colocalization of AR and CTBP1 at ARBSs; this colocalization decreased over time because of decreased expression of CTBP1. Induction of androgen-dependent target genes and promoter activity were upregulated by siCTBP1 transfection. In contrast, addition of exogenous CTBP1 repressed AR transcriptional activity. Demethylation of H3K9 is the representative histone modification in ARBSs induced by LSD1 (Metzger et al. 2005; Kahl et al. 2006; Wissmann et al. 2007). Progressive demethylation of histone H3K9 at the ARBS was observed by siCTBP1 treatment. CTBP1 interacts with the histone methyltransferase, G9a, even after androgen treatment, suggesting that H3K9 methylation by CTBP1 is probably mediated by G9a and that it presumably opposes LSD1 function. CTBP1 interaction with histone deacetylase (HDAC) is weak and becomes no longer apparent after androgen treatment. Taken together, CTBP1 functions as an AR co-repressor by inhibiting androgen-mediated demethylation.

We found that CTBP1 overexpression reduced cell proliferation with accompanying repression of androgen-regulated genes and that knocking down CTBP1 increased LNCaP cell proliferation. These results demonstrated the importance of CTBP1 in controlling cancer proliferation.

11.3 Potential Therapeutic Target of Androgen-Regulated Long Noncoding RNA in Prostate Cancer

11.3.1 CTBP1-AS *Activates AR Signaling*

On the basis of these findings, we further examined the effects of *CTBP1-AS* on AR transcriptional activity. First, we performed *CTBP1-AS* knockdown followed by androgen treatment. Microarray analysis demonstrated that transcriptional activation of androgen-induced genes was diminished by si*CTBP1-AS*. This inhibition of AR signaling is at least partially due to continuation of CTBP1 binding and inhibition of demethylation of H3K9 because si*CTBP1-AS* reversed androgen-mediated CTBP1 repression. Thus, our results indicate that *CTBP1-AS* positively regulates AR signaling.

11.3.2 Xenograft Model of Prostate Cancer

Next, we examined whether transcriptional regulation by *CTBP1-AS* is associated with tumor growth in prostate cancer. The xenograft model of prostate cancer cells is a common animal model used to evaluate tumor growth. Because *CTBP1-AS* is overexpressed in LTAD cells derived from LNCaP, a model of CRPC derived from LNCaP cells, these cells were injected subcutaneously into each side of male nude mice, which were castrated after tumor development. Each tumor was transfected with *CTBP1-AS* siRNA or control RNA.

Interestingly, si*CTBP1-AS* treatment inhibited LTAD castration-resistant tumor growth and induced CTBP1, suggesting that *CTBP1-AS* could be a target for treatment of CRPC. Taken together, our results indicate that *CTBP1-AS* overexpression can cause castration-resistant tumor growth by positively regulating AR signaling.

11.4 Investigation of Nuclear Function of Androgen-Regulated lncRNA

11.4.1 CTBP1-AS Induces Histone Deacetylation at CTBP1 Promoter for Transcriptional Repression

We were interested in the mechanism by which the CTBP1 sense–antisense transcript pair contributes to the transcriptional regulation of CTBP1. There are two potential mechanisms of sense-transcript repression: (1) post-transcriptional repression (*XIST* repression by *TSIX*) by degradation (Ogawa et al. 2008); and (2) epigenetic regulation by the recruitment of transcription factors (*CDKN2B-AS1*) (Yu et al. 2008). We first examined the possibility that CTBP1 is repressed via post-transcriptional degradation by antisense induction. However, the canonical antisense-mediated post-transcriptional degradation of sense mRNA was not observed in this CTBP1 pair after inhibition of transcription.

The interaction of lncRNAs with chromatin remodeling complexes induces heterochromatin formation in specific loci, leading to reduced target gene expression. For example, *HOX Antisense Intergenic RNA (HOTAIR)* (Gupta et al. 2010), a 2.2 kb ncRNA, is transcribed in the antisense direction from the HOXC gene cluster. *HOTAIR* functions *in trans* by interacting with and by recruiting polycomb repressive complex 2 (PRC2) to the HOXD locus, resulting in transcriptional silencing across a 40 kb region. In addition, *HOTAIR* interacts with a second histone modification complex, the LSD1/CoREST/REST complex, which coordinates the targeting of PRC2 and LSD1 to chromatin for coupled histone H3K27 methylation and K4 demethylation (Tsai et al. 2010).

The androgen-dependent repression of RNA polymerase II (pol II) recruitment is consistent with the androgen-induced CTBP1 repression in AR-regulated prostate cancer cells. Androgen treatment substantially reduced the chromatin markers required for activation, namely histone H3 acetylation and H3K4 methylation levels at the *CTBP1* promoter dependent on *CTBP1-AS* expression. This histone modification is induced by the recruitment of both HDAC and an HDAC-associated co-repressor, Sin3A, to the promoter region of *CTBP1*.

11.4.2 lncRNA Associated Protein

11.4.2.1 Identification of PSF as an Interacting Partner of *CTBP1-AS*

We suspected that the *CTBP1-AS*–Sin3A interaction is indirect and mediated by an RNA-binding protein. Sin3A forms complexes with the repressors PTB-associated splicing factor (PSF) or Non-POU domain-containing octamer-binding protein (NONO), which have both RNA- and DNA-binding domains, and binds to and represses gene promoter regions (Shav-Tal and Zipori 2002; Song et al. 2004). We performed knockdown experiments using siRNAs targeting PSF, NONO, and Sin3A to find that PSF is the major component associated with CTBP1 repression. In addition, PSF is recruited to the *CTBP1* promoter in a *CTBP1-AS*-dependent manner for inducing androgen-mediated deacetylation at the *CTBP1*-promoter. Moreover, the interaction between PSF and *CTBP1-AS* was confirmed by RNA immunoprecipitation (RIP) and RNA pull-down assays. By combining immunofluorescence and RNA-FISH, we showed that *CTBP1-AS* and PSF colocalized in the nucleus.

11.4.2.2 Identification of RNA-Associated Proteins by Mass Spectrometry

To identify proteins associated with lncRNA, mass spectrometry could be useful because it is a comprehensive method for determining RNA-binding proteins. In the analysis of *lncRNA-p21* (Huarte et al. 2010), RNA pull-down was performed.

Associated proteins were resolved on SDS-PAGE gel, and then specific bands were cut out for mass spectrometry analysis. In this study, hhnRNP-K was identified as the unique protein interacting with this lncRNA.

11.5 Functional Analysis of lncRNA

11.5.1 Cell Cycle Analysis

11.5.1.1 PSF Downstream Signals Regulate Prostate Cancer Cell Cycle Progression

Next, to analyze the impact of PSF on androgen-dependent transcriptional regulation, we performed siPSF knockdown followed by microarray analysis. Surprisingly, siPSF relieved the repression of more than 69.9 % of the androgen-repressed genes in LNCaP cells. In addition, most of the genes normally induced by androgen such as APP (Takayama et al. 2009) and TMPRSS2 were also repressed by siPSF, presumably because siPSF repressed AR activity by reversing CTBP1 repression. Pathway analysis has shown the enrichment of cell cycle-related genes among the PSF targets repressed by androgen.

Because gain- and loss-of-function experiments have shown that PSF promotes cell growth, we next examined whether PSF may regulate the cell cycle in prostate cancer. Flow cytometry analysis revealed that siPSF treatment inhibited cell cycle progression. Among the PSF target genes, we focused on p53 and SMAD3 because they are well-known cell cycle regulators that are repressed by androgen treatment (Song et al. 2010; Rokhlin et al. 2005) and are expressed at a low level in prostate cancer cells (Taylor et al. 2010; Schlomn et al. 2008). Interestingly, these two genes are also AR negative regulators (Shenk et al. 2001; Hayes et al. 2001). Androgen-dependent repression of both genes was abolished by siPSF treatment. Taken together, the present results indicate that *CTBP1-AS* and PSF cooperatively promote cell cycle progression by repressing cell cycle inhibitors.

11.5.2 Genome-Wide Transcriptional Program

11.5.2.1 *CTBP1-AS* and PSF Function Cooperatively to Modulate Global Androgen Signaling

Next, we compared *CTBP1-AS* and PSF target genes by microarray analysis. We showed that about 40 % of androgen-mediated repression was reversed by depletion of both *CTBP1-AS* and PSF, suggesting cooperative gene repression by both factors.

Using si*CTBP1-AS*, we confirmed that p53 and SMAD3 are also regulated by *CTBP1-AS*. We observed that PSF recruitment to the promoters was inhibited by si*CTBP1-AS* transfection.

11.5.2.2 ChIP-Sequence Analysis to Determine PSF-Binding Regions

We next explored AR- and PSF-binding sites by ChIP-seq analysis in LNCaP cells to investigate the PSF genomic function in prostate cancer. The distribution of binding regions was affected by androgen treatment. Notably, androgen-dependent PSF occupancy was observed in the vicinity of PSF-regulated genes such as *SMAD3* and *p53*. In the *CTBP1* genomic region, we identified two androgen-dependent PSF-binding sites around the *CTBP1* promoter. Androgen-dependent PSF binding in vitro was abrogated by *CTBP1-AS* knockdown, suggesting that *CTBP1-AS* influences the DNA-binding ability of PSF and subsequent transcriptional changes.

11.6 Summary and Future Plan

11.6.1 Summary of CTBP1-AS *Function*

We have described novel functions of prostate cancer-associated ncRNA *CTBP1-AS*, which was originally identified as an AR-regulated antisense transcript of the *CTBP1* gene locus by the combined study of AR ChIP-chip and CAGE analysis. *CTBP1-AS* is the first hormone-regulated natural antisense transcript identified that is directly associated with hormone-dependent cancer in vivo. We demonstrated the mechanism of *CTBP1-AS* by several experiments such as RNA pull-down, ISH, RNA-FISH combined with immunofluorescence, RIP assay, and ChIP-seq of interacting chromatin modifiers to investigate RNA function. We demonstrated that *CTBP1-AS* interacts with PSF, a transcriptional repressor, and recruits the HDAC–Sin3A complex to the *CTBP1* promoter for histone deacetylation. In addition, the release of the repressor CTBP1 from the regulatory regions of AR-regulated genes leads to transcriptional activation and the loss of repressing histone marks such as histone H3K9 methylation. In the *trans*-regulatory pathway, *CTBP1-AS* also guides PSF complexes to the regulatory regions of their endogenous target genes for transcriptional repression of genes that have suppressive functions for tumor growth and cell cycle progression (Fig. 11.4).

However, these studies are not sufficient to determine whole mechanisms of androgen-regulated lncRNAs. Additional genome-wide studies will be required to elucidate the biological significance of lncRNAs in prostate cancer carcinogenesis and to characterize lncRNA function. Many important aspects of lncRNA molecular interactions and chromatin binding sites have yet to be investigated.

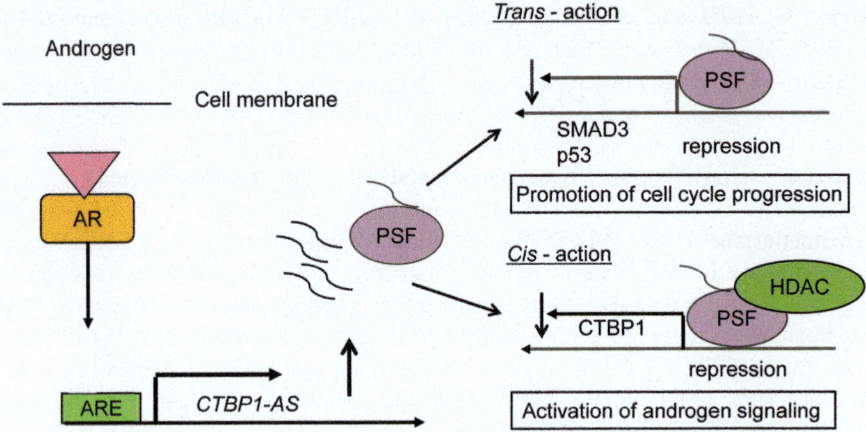

Fig. 11.4 Working model of *CTBP1-AS* for prostate cancer progression. Androgen-regulated lncRNA, *CTBP1-AS,* activates AR signaling and promotes the cell cycle progression for tumor growth

11.6.2 Usage of Androgen-Regulated lncRNAs for Therapeutic Targets and Biomarkers of CRPC

Long ncRNAs can be utilized as diagnostic and prognostic markers, and as novel specific therapeutic targets of CRPC. Current prostate cancer genomic data can be fully exploited if the noncoding regions are studied in detail. In addition, lncRNAs are interesting targets in cancer therapy because their cancer- and tissue- specific expression can be a major advantage over other therapeutic options. Several other studies have also demonstrated the clinical significance of lncRNAs (Gutschner and Diederichs 2012). However, before we can make use of these new therapeutic options (Leyten et al. 2014; Lin et al. 2013), many more functional and structural studies are necessary to fully understand lncRNA biology. An exponentially growing number of studies reporting new ncRNAs or androgen signaling molecules will contribute to progress in this field.

References

Balk SP, Knudsen KE (2008) AR, the cell cycle, and prostate cancer. Nucl Recept Signal 6:e001

Bergerat JP, Ceraline J (2009) Pleiotropic functional properties of androgen receptor mutants in prostate cancer. Hum Mutat 30:145–157

Bergman LM et al (2009) CtBPs promote cell survival through the maintenance of mitotic fidelity. Mol Cell Biol 29:4539–4551

Buchanan G et al (2001) Collocation of androgen receptor gene mutations in prostate cancer. Clin Cancer Res 7:1273–1281

Bussemakers MJ et al (1999) DD3: a new prostate-specific gene, highly overexpressed in prostate cancer. Cancer Res 59:5975–5979
Cai C et al (2011) Androgen receptor gene expression in prostate cancer is directly suppressed by the androgen receptor through recruitment of lysine-specific demethylase 1. Cancer Cell 20:457–471
Callewaert L et al (2006) Interplay between two hormone-independent activation domains in the androgen receptor. Cancer Res 66:543–553
Carninci P et al (2005) The transcriptional landscape of the mammalian genome. Science 309:1559–1563
Chamberlain NL et al (1996) Delineation of two distinct type 1 activation functions in the androgen receptor amino-terminal domain. J Biol Chem 271:26772–26778
Chen CD et al (2004) Molecular determinants of resistance to antiandrogen therapy. Nat Med 10:33–39
Chen Z et al (2011) Phospho-MED1-enhanced UBE2C locus looping drives castration-resistant prostate cancer growth. EMBO J 30:2405–2419
Chinnadurai G (2007) Transcriptional regulation by C-terminal binding proteins. Int J Biochem Cell Biol 39:1593–1607
Culig Z et al (1999) Switch from antagonist to agonist of the androgen receptor bicalutamide is associated with prostate tumour progression in a new model system. Br J Cancer 81:242–251
Debes JD, Tindall DJ (2004) Mechanism of androgen-refractory prostate cancer. N Engl J Med 351:1488–1490
Dehm SM et al (2007) Selective role of an NH2-terminal WxxLF motif for aberrant androgen receptor activation in androgen depletion-independent prostate cancer cells. Cancer Res 67:10067–10077
Derrien T et al (2012) The GENCODE v7 catalog of human long noncoding RNAs: analysis of their gene structure, evolution, and expression. Genome Res 22:1775–1789
Djebali S et al (2012) Landscape of transcription in human cells. Nature 489:101–108
Du Z et al (2013) Integrative genomic analyses reveal clinically relevant long noncoding RNAs in human cancer. Nat Struct Mol Biol 20:908–913
Duff J et al (2006) Structural dynamics of the human androgen receptor: implications for prostate cancer and neurodegenerative disease. Biochem Soc Trans 34:1098–1102
Ferreira LB et al (2013) PCA3 noncoding RNA is involved in the control of prostate-cancer cell survival and modulates androgen receptor signaling. BMC Cancer 12:507
Gupta RA et al (2010) Long non-coding RNA HOTAIR reprograms chromatin state to promote cancer metastasis. Nature 464:1071–1076
Gutschner T, Diederichs S (2012) The hallmarks of cancer: a long non-coding RNA point of view. RNA Biol 9:703–719
Hayes SA et al (2001) SMAD3 represses androgen receptor-mediated transcription. Cancer Res 61:2112–2118
Heemers HV, Tindall DJ (2007) Androgen receptor (AR) coregulators: a diversity of functions converging on and regulating the AR transcriptional complex. Endocr Rev 28:778–808
Heery DM et al (1997) A signature motif in transcriptional co-activatorsmediates binding to nuclear receptors. Nature 387:733–736
Hessels D, Schalken JA (2009) The use of PCA3 in the diagnosis of prostate cancer. Nat Rev Urol 6:255–261
Huarte M et al (2010) A large intergenic noncoding RNA induced by p53 mediates global gene repression in the p53 response. Cell 142:409–419
Jenster G et al (1991) Domains of the human androgen receptor involved in steroid binding, transcriptional activation, and subcellular localization. Mol Endocrinol 5:1396–1404
Jenster G et al (1995) Identification of two transcription activation units in the N-terminal domain of the human androgen receptor. J Biol Chem 270:7341–7346
Jia L et al (2008) Genomic androgen receptor-occupied regions with different functions, defined by histone acetylation, coregulators and transcriptional capacity. PLoS One 3:e3645

Kahl P et al (2006) Androgen receptor coactivators lysine-specific histone demethylase 1 and four and a half LIM domain protein 2 predict risk of prostate cancer recurrence. Cancer Res 66:11341–11347

Kan Z et al (2010) Diverse somatic mutation patterns and pathway alterations in human cancers. Nature 466:869–873

Katayama S et al (2005) Antisense transcription in the mammalian transcriptome. Science 309:1564–1566

Kokontis JM et al (1994) Increased androgen receptor activity and altered c-myc expression in prostate cancer cells after long-term androgen deprivation. Cancer Res 54:1566–1573

Kung JT, Lee JT (2013) RNA in the loop. Dev Cell 24:565–567

Lee JT (2012) Epigenetic regulation by long noncoding RNAs. Science 338:1435–1439

Lee GL et al (2011) Prostate cancer: diagnostic performance of the PCA3 urine test. Nat Rev Urol 8:123–124

Leyten GH et al (2014) Prospective multicentre evaluation of PCA3 and TMPRSS2-ERG gene fusions as diagnostic and prognostic urinary biomarkers for prostate cancer. Eur Urol 65:534–542

Lin DW et al (2013) Urinary TMPRSS2: ERG and PCA3 in an active surveillance cohort: results from a baseline analysis in the canary prostate active surveillance study. Clin Cancer Res 19:2442–2450

Locke JA et al (2008) Androgen levels increase by intratumoral de novo steroidogenesis during progression of castration-resistant prostate cancer. Cancer Res 68:6407–6415

Massie CE et al (2007) New androgen receptor genomic targets show an interaction with the ETS1 transcription factor. EMBO Rep 8:871–878

Mercer TR et al (2013) Structure and function of long noncoding RNAs in epigenetic regulation. Nat Struct Mol Biol 20:300–307

Metzger E et al (2005) LSD1 demethylates repressive histone marks to promote androgen-receptor-dependent transcription. Nature 437:436–439

Moran VA et al (2012) Emerging functional and mechanistic paradigms of mammalian long non-coding RNAs. Nucleic Acids Res 40:6391–6400

Ogawa Y et al (2008) Intersection of the RNA interference and X-inactivation pathways. Science 320:1336–1341

Paris PL et al (2004) Whole genome scanning identifies genotypes associated with recurrence and metastasis in prostate tumor. Hum Mol Genet 13:1303–1313

Petrovics G et al (2004) Elevated expression of PCGEM1, a prostate-specific gene with cell growth-promoting function, is associated with high-risk prostate cancer patients. Oncogene 23:605–611

Prensner JR et al (2011) Transcriptome sequencing across a prostate cancer cohort identifies PCAT-1, an unannotated lincRNA implicated in disease progression. Nat Biotechnol 29(8):742–749

Prensner JR et al (2013) The long noncoding RNA SChLAP1 promotes aggressive prostate cancer and antagonizes the SWI/SNF complex. Nat Genet 45(11):1392–1398

Prensner JR et al (2014) PCAT-1, a long noncoding RNA, regulates BRCA2 and controls homologous recombination in cancer. Cancer Res 74(6):1651–1660

Ren S et al (2013) Long noncoding RNA MALAT-1 is a new potential therapeutic target for castration resistant prostate cancer. J Urol 190:2278–2287

Rokhlin OW et al (2005) Androgen regulates apoptosis induced by TNFR family ligands via multiple signal pathway in LNCaP. Oncogene 24:6733–6784

Rosok O et al (2004) Systematic identification of sense–antisense transcripts in mammalian cells. Nat Biotechnol 22:104–108

Schlomn T et al (2008) Clinical significance of p53 alterations in surgically treated prostate cancers. Mod Pathol 21:1371–1378

Schmidt LJ, Tindall DJ (2011) Steroid 5 alpha-reductase inhibitors targeting BPH and prostate cancer. J Steroid Biochem Mol Biol 125:32–38

Shang Y et al (2002) Formation of the androgen receptor transcription complex. Mol Cell 9:601–610

Shav-Tal Y, Zipori D (2002) PSF and p54(nrb)/NonO–multi-functional nuclear proteins. FEBS Lett 531:109–114

Shenk JL et al (2001) p53 represses androgen-induced transactivation of prostate-specific antigen by disrupting hAR amino- to carboxyl-terminal interaction. J Biol Chem 276:38472–38479

Shi Y et al (2003) Coordinated histone modifications mediated by a CtBP co-repressor complex. Nature 422:735–738

Shiraki T et al (2003) Cap analysis gene expression for high-throughput analysis of transcriptional starting point and identification of promoter usage. Proc Natl Acad Sci U S A 100:15776–15781

Song X, Sui A, Garen A (2004) Binding of mouse VL30 retrotransposon RNA to PSF protein induces genes repressed by PSF: effects on steroidogenesis and oncogenesis. Proc Natl Acad Sci U S A 101:621–626

Song K et al (2010) DHT selectively reverses Smad3-mediated/TGF-beta-induced responses through transcriptional down-regulation of Smad3 in prostate epithelial cells. Mol Endocrinol 24:2019–2029

Srikantan V et al (2000) PCGEM1, a prostate-specific gene, is overexpressed in prostate cancer. Proc Natl Acad Sci U S A 97:12216–12221

Sun S et al (2010) Castration resistance in human prostate cancer is conferred by a frequently occurring androgen receptor splice variant. J Clin Invest 120:2715–2730

Takayama K et al (2007) Identification of novel androgen response genes in prostate cancer cells by coupling chromatin immunoprecipitation and genomic microarray analysis. Oncogene 26:4453–4463

Takayama K et al (2009) Amyloid precursor protein is a primary androgen target gene that promotes prostate cancer growth. Cancer Res 69:137–142

Takayama K et al (2011) Integration of cap analysis of gene expression and chromatin immunoprecipitation analysis on array reveals genome-wide androgen receptor signaling in prostate cancer cells. Oncogene 30(5):619–630

Takayama K et al (2012) TACC2 is an androgen-responsive cell cycle regulator promoting androgen-mediated and castration-resistant growth of prostate cancer. Mol Endocrinol 26:748–761

Takayama K et al (2013) Androgen-responsive long noncoding RNA *CTBP1-AS* promotes prostate cancer. EMBO J 32:1665–1680

Tan PY et al (2012) Integration of regulatory networks by NKX3-1 promotes androgen-dependent prostate cancer survival. Mol Cell Biol 32:399–414

Taplin ME et al (1995) Mutation of the androgen-receptor gene in metastatic androgen-independent prostate cancer. N Engl J Med 332:1393–1398

Taplin ME et al (2003) Androgen receptor mutations in androgen-independent prostate cancer: cancer and leukemia group B study 9663. J Clin Oncol 21:2673–2678

Taylor BS et al (2010) Integrative genomic profiling of human prostate cancer. Cancer Cell 18:11–22

Tsai MC et al (2010) Long noncoding RNA as modular scaffold of histone modification complexes. Science 329(5992):689–693

Umesono K, Evans RM (1989) Determinants of target gene specificity for steroid/thyroid hormone receptors. Cell 57:1139–1146

Urbanucci A et al (2012) Overexpression of androgen receptor enhances the binding of the receptor to the chromatin in prostate cancer. Oncogene 31:2153–2163

Visakorpi T et al (1995) In vivo amplification of the androgen receptor gene and progression of human prostate cancer. Nat Genet 9:401–406

Waltering KK et al (2009) Increased expression of androgen receptor sensitizes prostate cancer cells to low levels of androgens. Cancer Res 69:8141–8149

Wang Q et al (2007) A hierarchical network of transcription factors governs androgen receptor-dependent prostate cancer growth. Mol Cell 27:380–392

Wang Q et al (2009) Androgen receptor regulates a distinct transcription program in androgen-independent prostate cancer. Cell 138:245–256

Wang D et al (2011) Reprogramming transcription by distinct classes of enhancers functionally defined by eRNA. Nature 474:390–394

Wissmann M et al (2007) Cooperative demethylation by JMJD2C and LSD1 promotes androgen receptor-dependent gene expression. Nat Cell Biol 9:347–353

Yang L et al (2013) lncRNA-dependent mechanisms of androgen-receptor regulated gene activation programs. Nature 500:598–602

Yap KL et al (2010) Molecular interplay of the noncoding RNA ANRIL and methylated histone H3 lysine 27 by polycomb CBX7 in transcriptional silencing of INK4a Mol. Cell 38:662–674

Yu W et al (2008) Epigenetic silencing of tumour suppressor gene p15 by its antisense RNA. Nature 451:202–206

Yu J et al (2010) An integrated network of androgen receptor, polycomb, and TMPRSS2-ERG gene fusions in prostate cancer progression. Cancer Cell 17:443–454

Zhou ZX et al (1995) Specificity of ligand-dependent androgen receptor stabilization: receptor domain interactions influence ligand dissociation and receptor stability. Mol Endocrinol 9:208–218

Chapter 12
Macrophage Activation as a Model System for Understanding Enhancer Transcription and eRNA Function

Karmel A. Allison and Christopher K. Glass

Abstract Macrophages are innate immune cells that sense the presence of pathogens through conserved pattern recognition receptors, which include TLR4. Activation of TLR4 by bacterial lipopolysaccharide induces the expression of thousands of genes that function to initiate inflammation and coordinate innate and adaptive immune responses. Transcriptional activation of TLR4-responsive genes is mediated by signal-dependent transcription factors, such as NFκB, which bind to DNA regulatory elements termed enhancers. Recent findings indicate that macrophage enhancers are actively transcribed in concert with nearby genes. Similar observations have been reported for other cell types, raising the general question of whether enhancer transcription and/or the resulting enhancer RNAs (eRNAs) are of functional importance. Here, we review the use of macrophage activation as an experimental system for addressing these questions and highlight areas for future research.

Keywords Macrophage • Enhancer • Promoter • eRNA • mRNA • Transcription • TLR • Chromatin • Histone methylation • Histone acetylation • Nucleosome • NFkB

12.1 Macrophage Activation as a Model System

Macrophages are myeloid lineage cells, which play essential roles in innate and adaptive immune responses and contribute to diverse aspects of tissue homeostasis (Wynn et al. 2013). Importantly, many of the transcriptional programs required for

K.A. Allison
Department of Cellular and Molecular Medicine, School of Medicine, University of California, San Diego,
9500 Gilman Drive, La Jolla, CA 92093-0651, USA

C.K. Glass (✉)
Department of Medicine, School of Medicine, University of California, San Diego,
9500 Gilman Drive, La Jolla, CA 92093-0651, USA
e-mail: ckg@ucsd.edu

R. Kurokawa (ed.), *Long Noncoding RNAs*, DOI 10.1007/978-4-431-55576-6_12

appropriate responses to pathogens also contribute to the pathogenesis of numerous chronic inflammatory diseases, which include atherosclerosis, diabetes, arthritis, and cancer. The innate immune response to pathogens is triggered by the interaction of pathogen-associated molecular patterns with pattern recognition receptors, which include members of the toll-like receptor (TLR) family (Medzhitov and Horng 2009; Takeuchi and Akira 2010). TLR4 recognizes the lipopolysaccharide (LPS) component of gram-negative bacteria, and is arguably the most intensively studied pattern recognition receptor (Beutler 2000). TLR4 ligation results in activation of several latent, signal-dependent transcription factors, including NF-kappaB (NF-κB), interferon regulatory factors (IRFs), AP-1 factors, and STAT factors, which act in a combinatorial manner to both positively and negatively regulate the expression of thousands of genes (Medzhitov and Horng 2009; Smale 2012) (Fig. 12.1). This response has been intensively studied at the level of genomics (e.g., Escoubet-Lozach et al. 2011; Ghisletti et al. 2010; Kaikkonen et al. 2013; Ostuni et al. 2013), proteomics (e.g., Meissner et al. 2013), and lipidomics (e.g., Maurya et al. 2013) in macrophages.

The pathogen response of the mouse macrophage provides a powerful system for applying genomics and associated modeling approaches to the understanding of

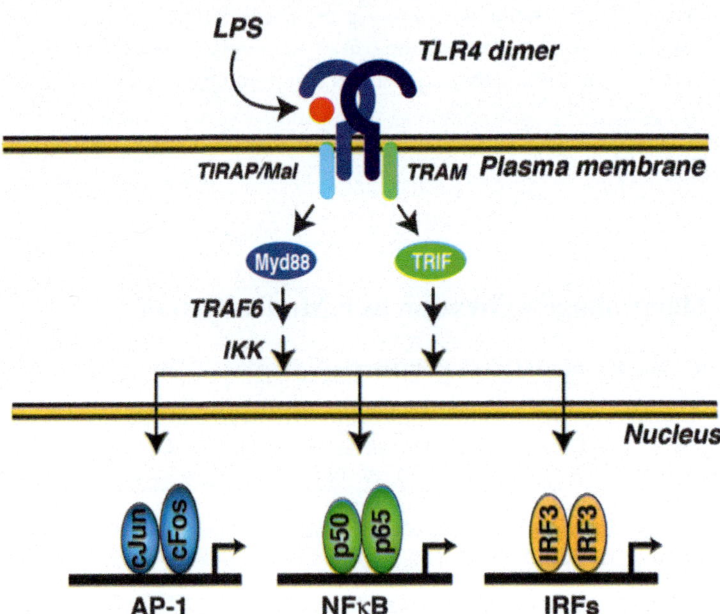

Fig. 12.1 TLR4-dependent gene expression. Toll-like receptor 4 (TLR4) is a pattern recognition receptor that is activated by lipopolysaccharide (*LPS*). Ligation of TLR4 results in transmission of an inflammatory signal through Myd88 and TRF, resulting in activation of latent, signal-dependent transcription factors, including NF-κB, interferon regulatory factors (*IRFs*), AP-1 factors, and STAT factors. These factors act in a combinatorial manner to both positively and negatively regulate the expression of thousands of genes

how transcriptional regulatory elements are selected from the genome and enable complex cell-specific programs of gene expression. First, mouse macrophages are exquisitely sensitive to LPS and other TLR4-specific analogues, such as Kdo2 lipid A (KLA). In addition to the large number of genes affected by TLR4 signaling, the dynamic range of transcriptional activation and repression exceeds three orders of magnitude for the most highly regulated genes. The response has both immediate–early and secondary phases based on the initial actions of latent transcription factors, which drive expression of cytokines such as TNF, and type I interferons, which induce expression and/or activities of a subsequent set of transcription factors. Although some genes, such as TNF and IL1b, are activated by TLR4 signaling in many cell types, a substantial component of the transcriptional response to TLR4 ligation is cell type specific. Therefore, the TLR4 response provides a powerful means of investigating the basis by which the genome is read to result in a signal-dependent, temporally orchestrated, and cell-specific response at the level of transcription.

Second, features of mouse macrophages themselves make them highly tractable for performing genomics analysis. As primary cells, they do not have the caveats associated with genomic alterations frequently associated with cell lines. Upward of 30 million thioglycollate-elicited macrophages (TGEMs) can be obtained from a single mouse. Importantly, more than 96 % of plated TGEMs express defining macrophage markers such as CD14 and CD11b, indicating a high degree of homogeneity. The relatively straightforward method for isolation has enabled simultaneous examination of multiple experimental conditions using all currently available and relevant genomics technologies, including ChIP-Seq, RNA-Seq, GRO-Seq, 5′-GRO-Seq, and HiC assays. Importantly, total and macrophage-specific loss-of-function alleles are available for many genes of interest, and siRNA knockdowns work well in TGEMs (e.g., Escoubet-Lozach et al. 2011; Heinz et al. 2010, 2013; Kaikkonen et al. 2013; Lam et al. 2013).

12.2 Studying Enhancers with High-Throughput Sequencing

As a result of this proclivity toward high-throughput studies, mouse macrophages are well suited for the investigation of the genome-wide phenomena such as enhancer selection and activation. Enhancers are regions of the genome that are able to regulate transcription from a distance, acting to increase or decrease transcriptional activity at target genes by looping into close proximity with promoters (Ong and Corces 2011). Enhancers act as platforms for the binding of many different transcription factors, as well as for the recruitment of the Mediator complex, cohesin, acetyltransferases, and RNA polymerase II (Pol II), all of which cooperate to regulate transcription at gene promoters.

With the advent of high-throughput sequencing, the ability to locate and characterize enhancers increased tremendously. Crucially, Heintzman et al. (2007) described a particular pattern of histone methylation that acts as a signature for enhancers genome wide, making it possible to map the enhancer landscapes of a variety of cells and organisms by looking for regions with high enrichment of H3K4 mono- and di-methylation (H3K4me1/2) and low enrichment of H3K4 tri-methylation (H3K4me3) (Fig. 12.2). While this histone signature does not guarantee that a given region is a functional enhancer, subsequent studies have shown that the majority of functional enhancers are indeed H3K4me1-high and H3K4me3-low (Heintzman et al. 2007; Heinz et al. 2010; Rada-Iglesias et al. 2011).

This genome-wide mapping of enhancers associated with specific chromatin signatures has led to the recognition that enhancers are distinct from cell type to cell type, even when gene expression is not (Heintzman et al. 2009). Further, recent studies have indicated that the cell type specificity of enhancers drives cell type-

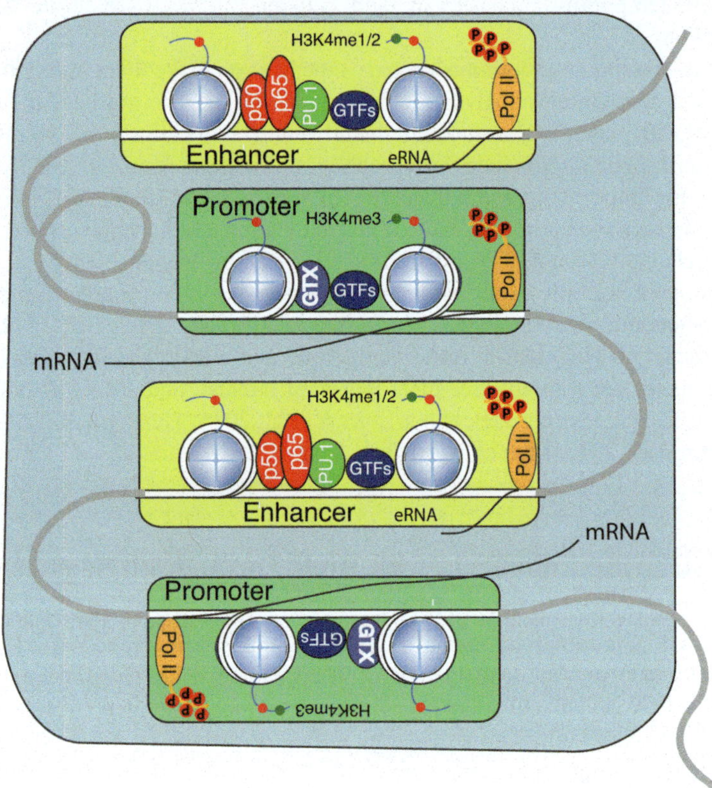

Fig. 12.2 Enhancer and promoter transcription. Enhancers (*yellow boxes*) are marked by mono- and dimethylation of H3K4, while gene promoters (*green boxes*) are marked by di- and trimethylation of H3K4. General transcription factors (*GTFs*) bind at both enhancers and promoters, whereas lineage-determining transcription factors (*LDTFs*) and signal-dependent transcription factors (*SDTFs*) bind primarily at enhancers, resulting in cell type-specific transcriptional activity. RNA polymerase II (*Pol II*) is recruited at both enhancers and promoters, generating mRNA along gene bodies and enhancer RNAs (*eRNAs*) at enhancers

specific gene expression (Heinz et al. 2010; Kaikkonen et al. 2013; Thurman et al. 2012; Visel et al. 2009). Investigations of how similar signaling cascades and transcription factors can establish distinct sets of enhancers in different cell types led to the formulation of a hierarchical model of enhancer selection and function. The hierarchical model relies on two classes of transcription factors—lineage-determining transcription factors (LDTFs) and signal-dependent transcription factors (SDTFs) (Garber et al. 2012; Mullen et al. 2011; Soufi et al. 2012; Trompouki et al. 2011). LDTFs are the relatively small set of transcription factors crucial to a particular lineage, and these factors collaborate in a combinatorial manner to compete with nucleosomes to bind DNA in a cell type-specific manner. The binding of these factors is proposed to establish open regions of chromatin that are associated with *cis-active* regulatory regions. SDTFs, which are activated in response to cell signals and are often shared across cell types, are then able to bind to the enhancers pre-established by LDTFs (Ghisletti et al. 2010; Heinz et al. 2010; Mullen et al. 2011; Trompouki et al. 2011). This multi-step model explains how SDTFs like NF-kappaB and the glucocorticoid receptor, which are widely expressed, can produce such different responses from cell type to cell type.

In the macrophage system, molecular and genetic studies of the process of differentiation from hematopoietic stem cells indicate essential roles of PU.1, C/EBPα/β and AP-1 family members as LDTFs (Heinz et al. 2010). In line with this, forced expression of PU.1 and C/EBPα in fibroblasts is sufficient to promote reprogramming to macrophage-like cells, albeit with low efficiency (Feng et al. 2008). Although PU.1, C/EBPs, and AP-1 factors function as LDTFs in macrophages, these lineage-determining functions are not exclusive to macrophages. For example, PU.1 is also a lineage-determining transcription factor for B cells (Klemsz et al. 1990; Scott et al. 1994), and C/EBPα is required for differentiation of other cell types, which include adipocytes (Cristancho and Lazar 2011; Herrera et al. 1989). Studies of the genome-wide locations of PU.1 in macrophages and B cells indicated cell-similar binding patterns at promoters but cell-specific binding patterns at distal inter- and intra-genic locations (Heinz et al. 2010). At these locations, PU.1 frequently co-localized with C/EBP and AP-1 factors in macrophages, and with B cell lineage-determining factors in B cells. Gain- and loss-of-function experiments demonstrated that co-binding of PU.1 with alternate lineage-determining factors was co-dependent. For example, loss of function of PU.1 resulted in loss of C/EBPα binding at sites where the two proteins co-bound within ~100 bp. Conversely, PU.1 binding at a subset of B cell-specific enhancers was dependent on co-expression and binding of early B cell factor (EBF). In contrast, the binding of PU.1 in macrophages was not dependent on the nearby binding of SDTFs, whereas the binding of SDTFs was dependent on the binding of PU.1 (Heinz et al. 2010). This co-dependence of LDTFs was shown to be in effect genome wide, as binding motifs for PU.1 that were disturbed by polymorphisms in multiple strains of inbred mice abrogated not just PU.1 binding but binding of C/EBP as well as the SDTF NF-kappaB (Heinz et al. 2013).

12.3 Transcription at Enhancers

In addition to a histone signature and transcription factor binding, recent studies in neurons and cancer cells revealed that widespread transcription characterizes enhancer regions. That is, in addition to driving transcription at gene promoters, enhancers themselves are transcribed. RNA Polymerase II (Pol II) was known to bind at several well-characterized enhancers including the beta-globin enhancer (Koch et al. 2008; Szutorisz et al. 2005), but these were thought to be idiosyncratic occurrences. However, Kim et al. showed widespread binding of Pol II to 12,000 neuronal enhancers marked by both H3K4me1 and the transcriptional co-activator CBP (Kim et al. 2010). High-throughput sequencing of total RNA showed that Pol II was active at these locations and generated nuclear bi-directional transcripts, called enhancer-RNAs (eRNAs), originating from the center of the CBP-identified enhancers. The identified eRNAs were further shown to positively correlate with expression of nearby genes, implying that eRNA synthesis occurs preferentially at active enhancers.

In parallel, De Santa et al. described the same phenomenon in macrophages subject to LPS stimulation, and found that 70 % of extragenic Pol II binding sites overlapped with histone marks indicative of enhancers (De Santa et al. 2010). The transcribed enhancers generated transcripts that were poly-adenylated but not spliced or exported from the nucleus. The eRNAs were also very low in abundance as compared with mRNAs; quickly induced, as eRNA synthesis preceded downstream gene synthesis at several loci investigated via RT-PCR; and transient, as they were highly susceptible to depletion via actinomycin D treatment.

This transience makes eRNAs difficult to capture with traditional RNA-Seq. However, the advent of global nuclear run-on sequencing (GRO-Seq) made possible the in-depth study of non-coding RNA kinetics across the whole genome (Core et al. 2008). Briefly, GRO-Seq takes advantage of a nuclear run-on reaction to tag nascent RNAs as they are assembled by Pol II. These tagged nascent transcripts are then sequenced, giving a real-time picture of transcription within the cell. Whereas RNA-Seq measures expression levels of stable, spliced RNA species, GRO-Seq returns data on rates of active transcription, of both coding and non-coding RNA species. Using GRO-Seq in coordination with the enhancer histone signature, it was estimated that ~18 % of transcripts in mouse macrophages were produced at enhancer-like regions, comprising almost a third of unannotated transcripts (Allison et al. 2014).

In MCF7 cells, Hah et al. used GRO-Seq to demonstrate extensive bidirectional transcription at enhancers (Hah et al. 2011), and showed further that the expression level of eRNAs was responsive to estrogen treatment. eRNAs in fact made up the largest class of transcripts that were initiated proximal to an estrogen receptor alpha binding site. Similarly, GRO-Seq was used in LNCaP cells to show widespread changes in enhancer transcription in response to FoxA1 and androgen receptor (AR) binding (Wang et al. 2011). As in Hah et al., expression levels in eRNAs were associated with transcription factor binding events and correlated with the expression levels of nearby genes.

The close association of eRNA production and gene expression levels raises the question of whether enhancer transcription or eRNAs themselves play a functional role. The act of transcription by Pol II has been associated with the conference of histone acetylation and the maintenance of open chromatin at extragenic regions (Travers 1999), as well as the deposition of methyl marks at histones (Gerber and Shilatifard 2003; Xiao et al. 2003), but initial reports of eRNAs did not assess the relevance of these effects at enhancers genome wide. In their study of LPS-treated macrophages, De Santa et al. found that transcription of an enhancer near the Ccl5 promoter was associated with increased acetylation in a manner sensitive to actinomycin D treatment (De Santa et al. 2010), but did not thoroughly establish the importance of transcription at this particular enhancer or enhancers generally in the recruitment of acetylation marks.

12.4 An Order of Events for Enhancer Transcription

Studies in primary mouse macrophages have helped elucidate the possible functions of enhancer transcription by taking advantage of the finding that TLR4 signaling induced the selection of thousands of "latent" or "de novo" enhancer-like regions in the genome. These de novo enhancers are defined by the new acquisition of H3K4me1 and/or H3K4me2 (Kaikkonen et al. 2013; Ostuni et al. 2013), marking enhancers that did not exist in the basal state but rather appeared upon cell activation. In the studies of Ostuni et al., "latent" enhancers were defined by (1) the lack of H3K4me1, H3K27Ac, and PU.1 in unstimulated cells; and (2) the presence of an LPS-induced H3K4me1 peak. There were ~500 such enhancers identified after 4 h of LPS stimulation, and ~1,000 after 24 h. The appearance of latent enhancers was not unique to LPS stimulation or TLR4 signaling, and several other stimuli, including TNFa and interleukin 1-beta (IL-1b), each induced a set of latent enhancers. Notably, Ostuni et al. did not find any histone mark or chromatin feature that allowed pre-identification of the latent enhancers in untreated conditions.

Kaikkonen et al. described a similar set of enhancers in macrophages by performing H3K4me2 ChIP-Seq of MNase-treated chromatin obtained following 0, 1, 6, 24, and 48 h of Kdo2-Lipid A (KLA) treatment (Kaikkonen et al. 2013). KLA, an LPS-analogue, induced ~32,000 inter- and intragenic locations marked by the enhancer histone signature prior to treatment, referred to as "pre-existing" enhancers. These regions were highly enriched for motifs recognized by PU.1, C/EBP, and AP-1 factors, consistent with previous findings (Ghisletti et al. 2010; Heinz et al. 2010), and were significantly correlated with the expression levels of nearby genes (Fig. 12.3). Notably, ~3,000 previously unmarked regions, termed "de novo" enhancers, gained H3K4me2 upon KLA stimulation. In contrast, ~1,000 regions lost this mark following KLA treatment. Gain and loss of H3K4me2 at enhancer-like regions was highly correlated with expression of nearby genes.

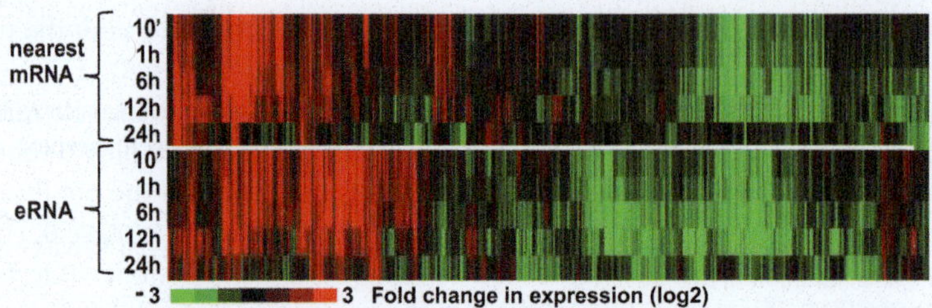

Fig. 12.3 Relationship of changes in eRNA expression to changes in mRNA expression. Heat map illustrating the relationship between signal-dependent changes in eRNAs and corresponding changes in expression of the nearest gene. Each column represents the amount of nascent RNA detected at a specific genomic location as a function of time following treatment of macrophages with a TLR4 agonist. Columns in the *bottom panel* correspond to locations of signal-regulated enhancers. Columns in the *upper panel* correspond to genomic locations of the nearest mRNA encoding gene. *Red* indicates upregulation of nascent RNA, and *green* represents downregulation. The pattern illustrates that eRNA expression is significantly correlated with expression of the nearest mRNA in stimulated macrophages (Reproduced from Kaikkonen et al. 2013, with permission)

The emergence of de novo enhancers in response to a temporally precise signal allowed Kaikkonen et al. to investigate the sequence of events that occur as closed chromatin transitions to active enhancer-like regions. ChIP-Seq experiments for histone acetylation (H3K2, H4K5, H4K8, and H4K12); for the LDTFs PU.1 and C/EBPα; and for the p65 (RelA) component of the SDTF NF-kappaB as a function of time following KLA treatment were conducted in parallel with GRO-Seq to define nascent RNA transcripts as a measure of transcriptional output. These experiments resulted in several important observations relevant to the general hierarchical model. First, p65 binding was primarily observed to occur at pre-existing enhancers characterized by high enrichment for PU.1, C/EBP, and AP-1 motifs. Co-binding of PU.1 and C/EBPα was confirmed at a high percentage of these sites by ChIP-Seq studies, though an increase in histone acetylation was associated with binding of p65.

In contrast to pre-existing enhancers, motif analysis of de novo H3K4me2-marked regions returned AP-1 and NF-kappaB motifs as the most highly enriched sequence elements. Importantly, however, C/EBP and PU.1 were also highly enriched. Consistent with these findings, while PU.1 or C/EBPα binding was absent under basal conditions, these factors were recruited to a large fraction of the de novo enhancer-like regions concomitantly with p65 within 1 h of KLA treatment. Loss-of-function studies indicated that in addition to the dependence of PU.1 binding on the nuclear entry of p65, DNA binding of p65 was dependent on PU.1 at locations where the two factors bound to closely spaced motifs. Therefore, at these locations, p65 acted as both a signal-dependent and collaborative transcription factor to facilitate the selection of new functional enhancers.

The collaborative binding of p65, PU.1, and C/EBP at de novo enhancers was temporally linked to the acquisition of histone acetylation and the initiation of enhancer transcription. Unexpectedly, these events substantially preceded the acquisition of H3K4me1 and H3K4me2. In addition to this temporal relationship, the length of eRNA transcripts was highly correlated with the distribution of H3K4me1 and H3K4me2 (Kaikkonen et al. 2013), suggesting that enhancer transcription was linked to the writing of these marks. To investigate this possibility, Kaikkonen et al. assessed the effects of inhibiting Pol II elongation on the H3K4me2 status after KLA treatment. Two different elongation inhibitors were used: the cyclin-dependent kinase (cdk) inhibitor flavopiridol, which at low concentrations preferentially inhibits the Cdk9 activity of P-TEFb (Chao and Price 2001); and IBET151, a selective inhibitor of BET (bromodomain and extra terminal domain) protein binding to acetylated histones, which disrupts the recruitment of P-TEFb complexes to acetylated histones that are dependent on BRD4 (Dawson et al. 2011; Nicodeme et al. 2010). Both drugs affected the elongation of KLA-induced nascent transcripts as evidenced by a decrease in the cumulative GRO-Seq tags beyond the transcription start site (TSS), with the effect of flavopiridol being more pronounced. Inhibition of eRNA elongation by IBET151 and flavopiridol was correlated with a decrease in the deposition of H3K4me2 at ~40 % and ~70 % of de novo enhancers. The effectiveness of drug treatment on reducing eRNA expression at individual enhancers was significantly correlated with a corresponding local reduction of H3K4me1 and H3K4me2. Similar effects were observed with three other inhibitors of Pol II. KLA-induced gain in H3K4me2 observed at many pre-existing enhances was blocked by inhibition of Pol II elongation.

Collectively, these findings suggest that enhancer H3K4me1/2 deposition is coupled to enhancer transcription, at least for de novo enhancers. Further, Kaikkonen et al. demonstrated that transcription-coupled H3K4 methylation at de novo enhancers was mediated by members of the Mll family of histone methyltransferases. These findings are consistent with the ability of Mlls to associate with the phosphorylated C terminal domain (CTD) of Pol II (Hughes et al. 2004; Krogan et al. 2003; MacConaill et al. 2006; Milne et al. 2005; Ng et al. 2003; Rana et al. 2011; Wood et al. 2003) and suggest that the progressive accumulation of H3K4 methylation at de novo enhancers results from their association with active forms of Pol II (Fig. 12.4). Whether this mechanism accounts for deposition of H3K4 methylation at pre-existing enhancers remains to be established, but these findings provide evidence for one functional consequence of enhancer transcription.

12.5 Enhancer Transcription as a Marker of Activity

The close temporal relationship between enhancer transcription, H3K4 methyl deposition, and gene transcription raises the question of how to interpret eRNA with respect to the enhancer signature derived from histone marks. While high levels of H3K4me1 with respect to H3K4me3 are considered characteristic of enhancer-like regions in the genome (Heintzman et al. 2007), this combination is not necessarily associated with enhancer activity. In the studies of Ostuni et al. the H3K4me1

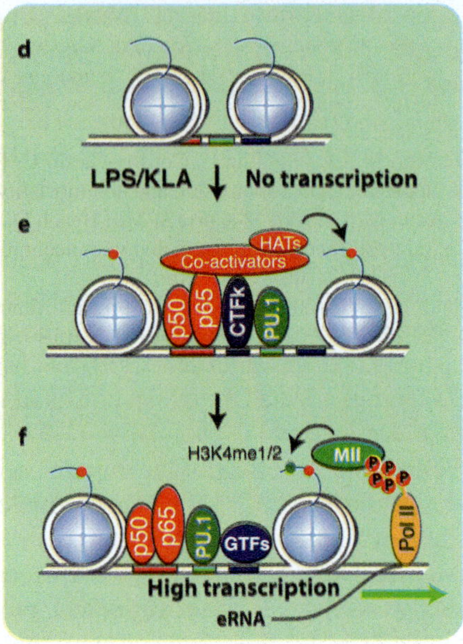

Fig. 12.4 Models for activation of pre-existing and de novo enhancers. (**a**) Pre-existing enhancers are bound in the basal state by lineage-determining transcription factors (*LDTFs*) and are marked by H3K4me1 and H3K4me2. Many of these pre-existing enhancers are transcribed at a low level by Pol II. (**b**) Upon initiation of an inflammatory signal (such as TLR4 stimulation with LPS), signal-dependent transcription factors (*SDTFs*) are recruited to the open chromatin at pre-existing enhancers. Co-activators such as histone acetyl transferases (*HATs*) are recruited to the enhancer with the SDTFs, resulting in increased Pol II activity. (**c**) Pol II transcription is associated with increased dimethylation of H3K4 at enhancers via the recruitment of methyltransferases. (**d**) In contrast to pre-existing enhancers, de novo enhancers lack LDTF binding and open chromatin in the basal state. (**e**) Upon stimulation, SDTFs and LDTFs are both recruited and coordinately open chromatin to form de novo enhancers. Co-activators and HATs acetylate surrounding histones. (**f**) The acetylated, bound de novo enhancers are then able to recruit Pol II and initiate transcription of eRNAs. Transcription is followed by recruitment of methyltransferases and the deposition of di-methyl at H3K4

signature associated with "latent" enhancers selected in response to LPS stimulation persisted after the transcriptional response to LPS subsided. These regions were associated with a more rapid response to subsequent stimulation, suggesting that the H3K4me1 mark provided a molecular memory that facilitated the second response (Ostuni et al. 2013). Kaikkonen et al. observed a similar persistence of H3Kme2 at de novo enhancers even after the transcriptional response of nearby genes had largely returned to baseline levels (Kaikkonen et al. 2013). Therefore, while both H3K4me1 and H3K4me2 are associated with enhancers, they do not necessarily reflect enhancer activity.

In contrast, Wang et al. found that enhancer activation was tightly coupled to eRNA production, as the gain and loss of AR binding was most closely correlated with eRNA synthesis rather than histone mark deposition or even the presence of the histone acetyltransferase p300 (Wang et al. 2011). Similarly, eRNA production as measured by GRO-Seq was highly correlated with nascent RNA production at the nearest mRNA encoding gene throughout the entire KLA time course (Kaikkonen et al. 2013). Providing further evidence for eRNA as a marker of active enhancers, Wu et al. studied tissue-specific RNA expression in mouse embryonic tissues (Wu et al. 2014). According to deep total RNA-Seq, previously validated enhancers were extensively transcribed, and eRNA marked a larger set of active enhancers than either H3K27Ac or p300. The enhancers marked by eRNA alone were subsequently tested for their ability to activate a lacZ reporter gene, and 8 out of 19 tested enhancers drove reporter expression in the predicted tissue-specific manner. Further, eRNAs have proven to be very sensitive indicators of transcription factor binding and activity. In addition to the correlation between AR binding and eRNA production seen by Wang et al., hundreds of eRNAs were observed to be responsive to p53 binding in p53-competent as compared with p53-null cells (Allen et al. 2014), and the effects of rosiglitazone treatment on adipocytes could be closely tracked via changes in eRNA expression levels at PPARgamma binding sites (Step et al. 2014).

In order to systematically determine the relationship between enhancer transcription and enhancer activity, Zhu et al. built a logistic regression model using 24 histone marks and p300 assayed by ChIP-Seq in conjunction with GRO-Seq from IMR90 cells (Zhu et al. 2013). The model revealed that the histone mark most predictive of eRNA synthesis was acetylation at H3K27 (H3K27Ac), which had previously been shown to be a mark of active enhancers (Creyghton et al. 2010). Models based on four histone marks achieved the highest area under the curve (AUC) value for predicting enhancer transcription, with 432 combinations of the 24 histone marks yielding AUC values within the top 5 % of all possible combinations. In addition to H3K27Ac, the activation marks H3K79me1, H3K9Ac, and H4K8Ac were positively associated with eRNA production. On the other hand, the repressive mark H3K27me3 was predictive of eRNA production with a negative coefficient. Despite the correlation of these histone marks with enhancer transcription, Zhu et al. found that eRNA was the single most predictive indicator of enhancer activity, with eRNA synthesis being more significantly associated with increased expression at nearby genes than histone marks in multiple cell types. These results are supported by a separate study by Pulakanti et al. where it was found that eRNA synthesis correlated with H3K27Ac deposition, hypomethylation, and occupancy of the DNA hydroxylase Tet1 in embryonic stem cells, all of which are traditional markers for transcriptional activity (Pulakanti et al. 2013).

12.6 The Functionality of eRNAs

These findings suggest that enhancer function is in some way linked to enhancer transcription. To directly study this possibility, Lam et al. investigated the mechanisms by which the Rev-Erbα and Rev-Erbβ nuclear receptors functioned to repress gene expression in macrophages. RevErbα/β are atypical members of the nuclear receptor family in that they constitutively interact with NCoR/HDAC3 co-repressor complexes but are unable to interact with nuclear receptor co-activators. As a consequence, they function as active transcriptional repressors upon binding to Rev-Erb recognition motifs (Yin and Lazar 2005; Zamir et al. 1996). Genome-wide location analysis of biotin tagged (BLRP) Rev-Erbα and Rev-Erbβ in RAW macrophages indicated that >80 % of their high confidence binding sites in the genome were at enhancer-like regions characterized by high H3K4me1/low H3K4me3 and occupied by macrophage LDTFs including PU.1 (Lam et al. 2013). Evaluation of Rev-Erb target genes, such as Cx3cr1, in Rev-Erbα/β double knockout macrophages using GRO-Seq revealed significant increases not only in the Cx3cr1 mRNA but also a corresponding increase in eRNA production from an enhancer-like region 28 kb downstream of the Cx3cr1 TSS that is occupied by Rev-Erbβ. Using a method to measure nascent RNA at the site of transcriptional initiation, termed 5′-GRO-Seq, the major sites of initiation were observed at the Cx3cr1 promoter and at the −28 kb enhancer. Lam et al. found that initiation from both locations was greatly suppressed by overexpression of BLRP-Rev-Erbα, suggesting that the consequences of Rev-Erb binding to its enhancer are direct suppression of enhancer transcriptional initiation and indirect suppression of initiation from the Cx3cr1 promoter.

However, the importance of the initiation of transcription still left open the question of whether the transcripts generated at enhancers are themselves important for enhancer function independent of the process of transcription. In order to gain an understanding of the significance of the products of enhancer transcription, Lam et al. designed siRNAs and antisense oligonucleotides (ASOs) to target the plus and minus strands of Cx3cr1 and Mmp9 eRNAs. Both methods were capable of reducing eRNA transcript levels on the basis of quantitative PCR analysis of nuclear RNA (Lam et al. 2013). Notably, reduction of eRNA expression was associated with decreased expression of nearby, but not distant, genes. This is exemplified by the ASO knockdown experiment presented in Fig. 12.5. Using a combination of 5′-GRO-Seq and conventional GRO-Seq to define the origin and length of the Cx3cr1 minus strand eRNA, an overlapping series of ASOs was synthesized and tested for ability to knockdown eRNA expression. The majority of these ASOs reduced eRNA expression, with the effects of two of the most potent ASOs illustrated at the bottom of Fig. 12.5. Both ASOs significantly reduced Cx3cr1 eRNA expression and Cx3cr1 mRNA expression. In contrast, neither ASO significantly affected expression of distant genes, such as the Mmp9 and Csrnp1 genes. Thus, at least in the case of the selected Rev-Erb target enhancers, the production of eRNAs was important for associated gene expression levels.

These results are supported by numerous studies outside macrophages. Melo et al. identified enhancers that produced eRNAs in a p53-dependent manner in immortalized human fibroblasts and MCF7 cells (Melo et al. 2013). The p53-bound enhancers were further shown to interact with distal p53-dependent gene promoters. Using siRNA, Melo et al. knocked down eRNAs at two p53 binding sites and demonstrated a concomitant loss of target mRNA as well as a reduction of Pol II at the promoters of the target genes. In primary human monocytes, Iiott et al. identified a set of LPS-inducible eRNAs whose expression correlated with that of nearby genes

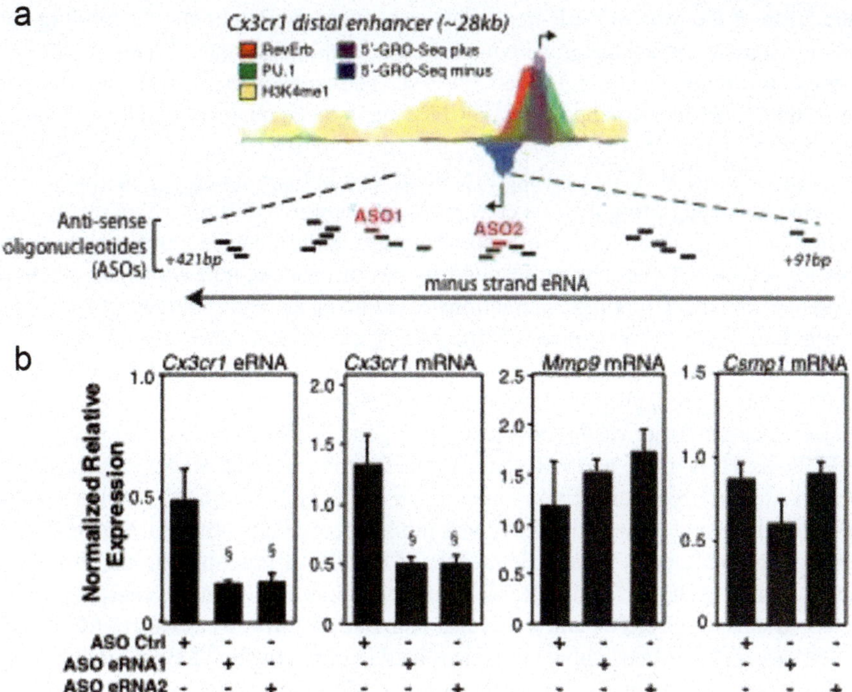

Fig. 12.5 Evidence for functional roles of eRNAs. (**a**) The *top panel* represents the Cx3cr1 28 kb distal enhancer and the experimental design for testing functional roles of eRNA. *Shaded regions* indicate locations of PU.1 binding, RevErb binding, and histone H3K4me1. *Directional arrows* represent eRNA transcription start sites defined by 5′-GRO-Seq. Antisense oligonucleotides (*ASOs*) were designed to tile eRNA generated along the minus strand of this enhancer region gene. (**b**) Two representative ASOs against the Cx3cr1 enhancer eRNA reduced expression of both the target eRNAs and the associated mRNA, as measured by qRT-PCR. The effect was specific to Cx3cr1 mRNA, such that Mmp9 mRNA and Csmp1 mRNA were unaffected by the ASOs (Adapted from Lam et al. 2013 with permission)

(NE et al. 2014). Many of these enhancers had NF-kappaB binding sites, suggesting they play an important role in transcriptional changes downstream of LPS stimulation via TLR4. To investigate the role of one particular enhancer near the highly induced IL1beta gene, Iiott et al. used locked nucleic acid (LNA)-based antisense inhibitors to suppress expression of the eRNA. The knockdown by LNAs (but notably not siRNAs) of the enhancer transcript resulted in reduced induction of IL1beta mRNA in response to LPS stimulation, demonstrating the importance of the eRNA at the IL1beta enhancer in the regulation of the target gene.

These loss-of-function studies demonstrated the importance of particular eRNAs in particular contexts but did not elucidate the mechanisms by which the eRNA was regulating target gene expression. One potential mechanism was elucidated by Li et al. in a study of estrogen receptor alpha (ER) binding in MCF7 cells (Li et al. 2013). Several enhancers that increased eRNA synthesis in response to E2 ligand treatment and subsequent ER binding were knocked down with siRNAs and LNAs.

Induction of the proximal ER-dependent genes was inhibited, but ER binding and methylations at the enhancer were unaltered. A high-throughput variant of chromosome conformation capture, termed 3D-DSL (Harismendy et al. 2011), was used to show that under normal conditions, ER binding induced qualitative and quantitative changes in promoter-enhancer looping. However, the presence of the eRNA-targeting siRNAs or LNAs resulted in alterations in the looping architecture at the targeted loci, suggesting that the eRNA transcripts were instrumental in the orchestration of ligand-dependent chromosome conformation changes. Further supporting this model, eRNA knockdown resulted in loss of cohesin, a protein with a role in promoter-enhancer looping interactions (Hadjur et al. 2009; Kagey et al. 2010; Schmidt et al. 2010), recruitment at the enhancers and associated promoters.

Similar results were obtained by Hsieh et al. in a study of androgen receptor (AR) enhancers in a prostate cancer cell line (Hsieh et al. 2014). eRNAs were produced at an enhancer of the AR-dependent gene KLK3 that was also marked by H3K27Ac, H3K4me1, and AR itself. Both the enhancer and KLK3 were induced upon androgen treatment. Upon knockdown of the eRNA with siRNA, expression of both KLK3 and the nearby KLK2 was inhibited, although other KLK genes in the locus were unaffected. To assess the role of the eRNA in regulating the two KLK genes, Hsieh et al. performed RNA immunoprecipitation with the sense and antisense strands of the eRNA and found that both pulled down AR as well as Mediator1 (Med1), which has been shown to be involved in chromosomal looping (Chen et al. 2011). 3C-qPCR demonstrated that either siRNA knockdown of the eRNA or knockdown of Med1 resulted in reduced looping of the KLK2 promoter to the enhancer locus. However, knockdown of the eRNA did not affect AR or Pol II occupancy at the enhancer itself, implying that the functional role of the eRNA in establishing enhancer-promoter looping is downstream of enhancer assembly. Complicating this result slightly, Hsieh et al. also found that knockdown of the eRNA at the KLK3 enhancer resulted in reduced expression of several AR target genes *in trans*, raising the question of what indirect effects the eRNA knockdown might have.

A second potential mechanism was highlighted by Mousavi et al. in an analysis of MyoD and MyoG binding activity in C2C12 skeletal muscle cells (Mousavi et al. 2013). At two enhancers, one proximal to MYOD1 and the other proximal to MYOG, knockdown of eRNAs with siRNA resulted in reduced recruitment of Pol II to both enhancers and their target genes. In line with this, binding of the transcription factor MyoD was reduced at the MYOG enhancer in response to knockdown of its eRNA, although this was not the case for the MYOD1 enhancer being studied. For both enhancers, knockdown of eRNAs resulted in reduced DNAse I accessibility at the target genes, although it was not clear how this effect was mediated, especially given that knockdown of the MYOD1 eRNA resulted in decreased accessibility at both MYOD1 and MYOG.

A third means by which eRNA might affect transcriptions was described in neurons responding to KCl-mediated membrane depolarization (Schaukowitch et al. 2014). Two enhancers proximal to the immediate early genes Arc and Gadd45b in neurons produce eRNA in response to stimulation. Strand-specific knockdown of the eRNAs with shRNA or LNAs resulted in reduced induction of their respective target genes, but, unlike the KLK3 locus, the looping of the enhancer to the promoter region was unaffected according to 3C assays. Similarly, recruitment of Med1 and members of the cohesion complex were unaffected. However, Schaukowitch et al. found that eRNA knockdown resulted in a marginally increased occupancy of the negative elongation factor (NELF) complex at the promoters of the target genes. NELF binds directly to Pol II and nascent RNA, and its release from the promoters of target genes is an important step in the unpausing of Pol II and subsequent mRNA elongation (Kwak and Lis 2013). In conjunction with increased NELF, there was a decrease in the elongating form of Pol II (phosphorylated at serine 5) with eRNA knockdown. Further, the eRNAs at Arc, Gadd45b, and c-fos were shown to pull down 1.5× to 2× more NELF in ultraviolet-crosslinked RNA immunoprecipitation assays, implying that the eRNA was binding directly to the RNA-binding domain of NELF. Thus, Schaukowitch et al. propose a model in which eRNA is able to destabilize the binding of NELF to nascent mRNAs, thereby allowing Pol II to elongate and continue to transcribe the mRNA.

12.7 Concluding Remarks

Enhancer transcription has emerged as a fascinating and quantitatively significant source of nuclear noncoding RNAs. Many questions remain regarding the general importance of enhancer transcription itself (apart from the eRNA product). At present, there is limited evidence that enhancer transcription is a major mechanism for deposition of H3K4me1/2 at enhancers other than the de novo enhancers studied by Kaikkonen et al. (2013). Although the proposed mechanism linking enhancer transcription to deposition of H3K4me1/2 via a PolII/MLL interaction is appealing, further studies of other classes of enhancers (e.g., those that are selected during transitions in cell differentiation) will be required to determine generality. In addition, while functional roles have been established for a small number of eRNAs, there is as yet no consistent picture of their mechanism(s) of action. Future studies are needed to examine questions such as the sequence determinants of activity and the identities of important interacting molecules, such as NELF. Macrophages are very likely to continue to be a robust experimental system for examination of these questions.

References

Allen MA, Andrysik Z, Dengler VL, Mellert HS, Guarnieri A, Freeman JA, Sullivan KD, Galbraith MD, Luo X, Kraus WL et al (2014) Global analysis of p53-regulated transcription identifies its direct targets and unexpected regulatory mechanisms. Elife 3:e02200

Allison KA, Kaikkonen MU, Gaasterland T, Glass CK (2014) Vespucci: a system for building annotated databases of nascent transcripts. Nucleic Acids Res 42:2433–2447

Beutler B (2000) Tlr4: central component of the sole mammalian LPS sensor. Curr Opin Immunol 12:20–26

Chao SH, Price DH (2001) Flavopiridol inactivates P-TEFb and blocks most RNA polymerase II transcription in vivo. J Biol Chem 276:31793–31799

Chen Z, Zhang C, Wu D, Chen H, Rorick A, Zhang X, Wang Q (2011) Phospho-MED1-enhanced UBE2C locus looping drives castration-resistant prostate cancer growth. EMBO J 30:2405–2419

Core LJ, Waterfall JJ, Lis JT (2008) Nascent RNA sequencing reveals widespread pausing and divergent initiation at human promoters. Science 322:1845–1848

Creyghton MP, Cheng AW, Welstead GG, Kooistra T, Carey BW, Steine EJ, Hanna J, Lodato MA, Frampton GM, Sharp PA et al (2010) Histone H3K27ac separates active from poised enhancers and predicts developmental state. Proc Natl Acad Sci U S A 107:21931–21936

Cristancho AG, Lazar MA (2011) Forming functional fat: a growing understanding of adipocyte differentiation. Nat Rev Mol Cell Biol 12:722–734

Dawson MA, Prinjha RK, Dittmann A, Giotopoulos G, Bantscheff M, Chan WI, Robson SC, Chung CW, Hopf C, Savitski MM et al (2011) Inhibition of BET recruitment to chromatin as an effective treatment for MLL-fusion leukaemia. Nature 478:529–533

De Santa F, Barozzi I, Mietton F, Ghisletti S, Polletti S, Tusi BK, Muller H, Ragoussis J, Wei CL, Natoli G (2010) A large fraction of extragenic RNA pol II transcription sites overlap enhancers. PLoS Biol 8:e1000384

Escoubet-Lozach L, Benner C, Kaikkonen MU, Lozach J, Heinz S, Spann NJ, Crotti A, Stender J, Ghisletti S, Reichart D et al (2011) Mechanisms establishing TLR4-responsive activation states of inflammatory response genes. PLoS Genet 7:e1002401

Feng R, Desbordes SC, Xie H, Tillo ES, Pixley F, Stanley ER, Graf T (2008) PU.1 and C/EBPalpha/beta convert fibroblasts into macrophage-like cells. Proc Natl Acad Sci U S A 105:6057–6062

Garber M, Yosef N, Goren A, Raychowdhury R, Thielke A, Guttman M, Robinson J, Minie B, Chevrier N, Itzhaki Z et al (2012) A high-throughput chromatin immunoprecipitation approach reveals principles of dynamic gene regulation in mammals. Mol Cell 47:810–822

Gerber M, Shilatifard A (2003) Transcriptional elongation by RNA polymerase II and histone methylation. J Biol Chem 278:26303–26306

Ghisletti S, Barozzi I, Mietton F, Polletti S, De Santa F, Venturini E, Gregory L, Lonie L, Chew A, Wei C-L et al (2010) Identification and characterization of enhancers controlling the inflammatory gene expression program in macrophages. Immunity 32:317–328

Hadjur S, Williams LM, Ryan NK, Cobb BS, Sexton T, Fraser P, Fisher AG, Merkenschlager M (2009) Cohesins form chromosomal cis-interactions at the developmentally regulated IFNG locus. Nature 460:410–413

Hah N, Danko CG, Core L, Waterfall JJ, Siepel A, Lis JT, Kraus WL (2011) A rapid, extensive, and transient transcriptional response to estrogen signaling in breast cancer cells. Cell 145:622–634

Harismendy O, Notani D, Song X, Rahim NG, Tanasa B, Heintzman N, Ren B, Fu XD, Topol EJ, Rosenfeld MG et al (2011) 9p21 DNA variants associated with coronary artery disease impair interferon-gamma signalling response. Nature 470:264–268

Heintzman ND, Stuart RK, Hon G, Fu Y, Ching CW, Hawkins RD, Barrera LO, Van Calcar S, Qu C, Ching KA et al (2007) Distinct and predictive chromatin signatures of transcriptional promoters and enhancers in the human genome. Nat Genet 39:311–318

Heintzman ND, Hon GC, Hawkins RD, Kheradpour P, Stark A, Harp LF, Ye Z, Lee LK, Stuart RK, Ching CW et al (2009) Histone modifications at human enhancers reflect global cell-type-specific gene expression. Nature 459:108–112

Heinz S, Benner C, Spann N, Bertolino E, Lin YC, Laslo P, Cheng JX, Murre C, Singh H, Glass CK (2010) Simple combinations of lineage-determining transcription factors prime cis-regulatory elements required for macrophage and B cell identities. Mol Cell 38:576–589

Heinz S, Romanoski CE, Benner C, Allison KA, Kaikkonen MU, Orozco LD, Glass CK (2013) Effect of natural genetic variation on enhancer selection and function. Nature 503(7477):487–492

Herrera R, Ro HS, Robinson GS, Xanthopoulos KG, Spiegelman BM (1989) A direct role for C/EBP and the AP-I-binding site in gene expression linked to adipocyte differentiation. Mol Cell Biol 9:5331–5339

Hsieh CL, Fei T, Chen Y, Li T, Gao Y, Wang X, Sun T, Sweeney CJ, Lee GS, Chen S et al (2014) Enhancer RNAs participate in androgen receptor-driven looping that selectively enhances gene activation. Proc Natl Acad Sci U S A 111:7319–7324

Hughes CM, Rozenblatt-Rosen O, Milne TA, Copeland TD, Levine SS, Lee JC, Hayes DN, Shanmugam KS, Bhattacharjee A, Biondi CA et al (2004) Menin associates with a trithorax family histone methyltransferase complex and with the hoxc8 locus. Mol Cell 13:587–597

Kagey MH, Newman JJ, Bilodeau S, Zhan Y, Orlando DA, van Berkum NL, Ebmeier CC, Goossens J, Rahl PB, Levine SS et al (2010) Mediator and cohesin connect gene expression and chromatin architecture. Nature 467:430–435

Kaikkonen MU, Spann NJ, Heinz S, Romanoski CE, Allison KA, Stender JD, Chun HB, Tough DF, Prinjha RK, Benner C et al (2013) Remodeling of the enhancer landscape during macrophage activation is coupled to enhancer transcription. Mol Cell 51:310–325

Kim TK, Hemberg M, Gray JM, Costa AM, Bear DM, Wu J, Harmin DA, Laptewicz M, Barbara-Haley K, Kuersten S et al (2010) Widespread transcription at neuronal activity-regulated enhancers. Nature 465:182–187

Klemsz MJ, McKercher SR, Celada A, Van Beveren C, Maki RA (1990) The macrophage and B cell-specific transcription factor PU.1 is related to the ets oncogene. Cell 61:113–124

Koch F, Jourquin F, Ferrier P, Andrau JC (2008) Genome-wide RNA polymerase II: not genes only! Trends Biochem Sci 33:265–273

Krogan NJ, Kim M, Tong A, Golshani A, Cagney G, Canadien V, Richards DP, Beattie BK, Emili A, Boone C et al (2003) Methylation of histone H3 by Set2 in Saccharomyces cerevisiae is linked to transcriptional elongation by RNA polymerase II. Mol Cell Biol 23:4207–4218

Kwak H, Lis JT (2013) Control of transcriptional elongation. Annu Rev Genet 47:483–508

Lam MT, Cho H, Lesch HP, Gosselin D, Heinz S, Tanaka-Oishi Y, Benner C, Kaikkonen MU, Kim AS, Kosaka M et al (2013) Rev-Erbs repress macrophage gene expression by inhibiting enhancer-directed transcription. Nature 498:511–515

Li W, Notani D, Ma Q, Tanasa B, Nunez E, Chen AY, Merkurjev D, Zhang J, Ohgi K, Song X et al (2013) Functional roles of enhancer RNAs for oestrogen-dependent transcriptional activation. Nature 498:516–520

MacConaill LE, Hughes CM, Rozenblatt-Rosen O, Nannepaga S, Meyerson M (2006) Phosphorylation of the menin tumor suppressor protein on serine 543 and serine 583. Mol Cancer Res 4:793–801

Maurya MR, Gupta S, Li X, Fahy E, Dinasarapu AR, Sud M, Brown HA, Glass CK, Murphy RC, Russell DW et al (2013) Analysis of inflammatory and lipid metabolic networks across RAW264.7 and thioglycolate-elicited macrophages. J Lipid Res 54:2525–2542

Medzhitov R, Horng T (2009) Transcriptional control of the inflammatory response. Nat Rev Immunol 9:692–703

Meissner F, Scheltema RA, Mollenkopf HJ, Mann M (2013) Direct proteomic quantification of the secretome of activated immune cells. Science 340:475–478

Melo CA, Drost J, Wijchers PJ, van de Werken H, de Wit E, Oude Vrielink JA, Elkon R, Melo SA, Leveille N, Kalluri R et al (2013) eRNAs are required for p53-dependent enhancer activity and gene transcription. Mol Cell 49:524–535

Milne TA, Dou Y, Martin ME, Brock HW, Roeder RG, Hess JL (2005) MLL associates specifically with a subset of transcriptionally active target genes. Proc Natl Acad Sci U S A 102:14765–14770

Mousavi K, Zare H, Dell'orso S, Grontved L, Gutierrez-Cruz G, Derfoul A, Hager GL, Sartorelli V (2013) eRNAs promote transcription by establishing chromatin accessibility at defined genomic loci. Mol Cell 51:606–617

Mullen AC, Orlando DA, Newman JJ, Loven J, Kumar RM, Bilodeau S, Reddy J, Guenther MG, DeKoter RP, Young RA (2011) Master transcription factors determine cell-type-specific responses to TGF-beta signaling. Cell 147:565–576

NE II, Heward JA, Roux B, Tsitsiou E, Fenwick PS, Lenzi L, Goodhead I, Hertz-Fowler C, Heger A, Hall N et al (2014) Long non-coding RNAs and enhancer RNAs regulate the lipopolysaccharide-induced inflammatory response in human monocytes. Nat Commun 5:3979

Ng HH, Robert F, Young RA, Struhl K (2003) Targeted recruitment of Set1 histone methylase by elongating Pol II provides a localized mark and memory of recent transcriptional activity. Mol Cell 11:709–719

Nicodeme E, Jeffrey KL, Schaefer U, Beinke S, Dewell S, Chung CW, Chandwani R, Marazzi I, Wilson P, Coste H et al (2010) Suppression of inflammation by a synthetic histone mimic. Nature 468:1119–1123

Ong CT, Corces VG (2011) Enhancer function: new insights into the regulation of tissue-specific gene expression. Nat Rev Genet 12:283–293

Ostuni R, Piccolo V, Barozzi I, Polletti S, Termanini A, Bonifacio S, Curina A, Prosperini E, Ghisletti S, Natoli G (2013) Latent enhancers activated by stimulation in differentiated cells. Cell 152:157–171

Pulakanti K, Pinello L, Stelloh C, Blinka S, Allred J, Milanovich S, Kiblawi S, Peterson J, Wang A, Yuan GC et al (2013) Enhancer transcribed RNAs arise from hypomethylated, Tet-occupied genomic regions. Epigenetics 8:1303–1320

Rada-Iglesias A, Bajpai R, Swigut T, Brugmann SA, Flynn RA, Wysocka J (2011) A unique chromatin signature uncovers early developmental enhancers in humans. Nature 470:279–283

Rana R, Surapureddi S, Kam W, Ferguson S, Goldstein JA (2011) Med25 is required for RNA polymerase II recruitment to specific promoters, thus regulating xenobiotic and lipid metabolism in human liver. Mol Cell Biol 31:466–481

Schaukowitch K, Joo JY, Liu X, Watts JK, Martinez C, Kim TK (2014) Enhancer RNA facilitates NELF release from immediate early genes. Mol Cell 56:29–42

Schmidt D, Schwalie PC, Ross-Innes CS, Hurtado A, Brown GD, Carroll JS, Flicek P, Odom DT (2010) A CTCF-independent role for cohesin in tissue-specific transcription. Genome Res 20:578–588

Scott EW, Simon MC, Anastasi J, Singh H (1994) Requirement of transcription factor PU.1 in the development of multiple hematopoietic lineages. Science 265:1573–1577

Smale ST (2012) Transcriptional regulation in the innate immune system. Curr Opin Immunol 24:51–57

Soufi A, Donahue G, Zaret KS (2012) Facilitators and impediments of the pluripotency reprogramming factors' initial engagement with the genome. Cell 151:994–1004

Step SE, Lim HW, Marinis JM, Prokesch A, Steger DJ, You SH, Won KJ, Lazar MA (2014) Anti-diabetic rosiglitazone remodels the adipocyte transcriptome by redistributing transcription to PPARgamma-driven enhancers. Genes Dev 28:1018–1028

Szutorisz H, Dillon N, Tora L (2005) The role of enhancers as centres for general transcription factor recruitment. Trends Biochem Sci 30:593–599

Takeuchi O, Akira S (2010) Pattern recognition receptors and inflammation. Cell 140:805–820

Thurman RE, Rynes E, Humbert R, Vierstra J, Maurano MT, Haugen E, Sheffield NC, Stergachis AB, Wang H, Vernot B et al (2012) The accessible chromatin landscape of the human genome. Nature 489:75–82

Travers A (1999) Chromatin modification by DNA tracking. Proc Natl Acad Sci U S A 96:13634–13637

Trompouki E, Bowman TV, Lawton LN, Fan ZP, Wu DC, DiBiase A, Martin CS, Cech JN, Sessa AK, Leblanc JL et al (2011) Lineage regulators direct BMP and Wnt pathways to cell-specific programs during differentiation and regeneration. Cell 147:577–589

Visel A, Blow MJ, Li Z, Zhang T, Akiyama JA, Holt A, Plajzer-Frick I, Shoukry M, Wright C, Chen F et al (2009) ChIP-seq accurately predicts tissue-specific activity of enhancers. Nature 457:854–858

Wang D, Garcia-Bassets I, Benner C, Li W, Su X, Zhou Y, Qiu J, Liu W, Kaikkonen MU, Ohgi KA et al (2011) Reprogramming transcription by distinct classes of enhancers functionally defined by eRNA. Nature 474:390–394

Wood A, Schneider J, Dover J, Johnston M, Shilatifard A (2003) The Paf1 complex is essential for histone monoubiquitination by the Rad6-Bre1 complex, which signals for histone methylation by COMPASS and Dot1p. J Biol Chem 278:34739–34742

Wu H, Nord AS, Akiyama JA, Shoukry M, Afzal V, Rubin EM, Pennacchio LA, Visel A (2014) Tissue-specific RNA expression marks distant-acting developmental enhancers. PLoS Genet 10:e1004610

Wynn TA, Chawla A, Pollard JW (2013) Macrophage biology in development, homeostasis and disease. Nature 496:445–455

Xiao T, Hall H, Kizer KO, Shibata Y, Hall MC, Borchers CH, Strahl BD (2003) Phosphorylation of RNA polymerase II CTD regulates H3 methylation in yeast. Genes Dev 17:654–663

Yin L, Lazar MA (2005) The orphan nuclear receptor Rev-erbalpha recruits the N-CoR/histone deacetylase 3 corepressor to regulate the circadian Bmal1 gene. Mol Endocrinol 19:1452–1459

Zamir I, Harding HP, Atkins GB, Horlein A, Glass CK, Rosenfeld MG, Lazar MA (1996) A nuclear hormone receptor corepressor mediates transcriptional silencing by receptors with distinct repression domains. Mol Cell Biol 16:5458–5465

Zhu Y, Sun L, Chen Z, Whitaker JW, Wang T, Wang W (2013) Predicting enhancer transcription and activity from chromatin modifications. Nucleic Acids Res 41:10032–10043

Chapter 13
Long Noncoding RNA Functions as a Regulator for Steroid Hormone Receptor-Related Breast and Prostate Cancers

Chunyu Jin and Michael G. Rosenfeld

Abstract Steroid hormone receptors have essential roles in various biological processes, including pathogenesis, invasion, and metastatic behavior of breast and prostate cancer. The precise and dynamic regulation of steroid hormone receptors, members of the nuclear receptor (NR) gene family, remains an essential, evolving question. Here we summarize several recent studies that have uncovered the roles of a new layer of NR regulators, long noncoding RNAs (ncRNAs). We present representative examples to demonstrate the diverse roles of ncRNAs, including long ncRNAs (lncRNAs) and enhancer RNAs (eRNAs), in the context of the NR functional network, focusing on the molecular mechanism, biological significance, and clinical relevance of these lncRNA regulators. Several breast/prostate cancer-associated lncRNA regulators not yet directly linked to NRs are also included in this chapter.

Keywords lncRNA • eRNA • Nuclear receptor • Steroid hormone • Breast cancer - Prostate cancer

13.1 Introduction

Nuclear receptors (NRs), characterized as hormone NRs, metabolic NRs, and orphan NRs (Gadaleta and Magnani 2014), play critical roles in differentiation, development, homeostasis, and metabolism (Mangelsdorf et al. 1995). The common structure of NRs consists of a C-terminal domain that is often a ligand binding domain (LBD), DNA binding domain (DBD), N-terminal domain, and hinge region between LBD and DBD; their structure and function have been extensively reviewed (Bain et al. 2007; Jin and Li 2010; Ribeiro et al. 1995).

Estrogen receptor (ER), androgen receptor (AR), progesterone receptor (PR), and glucocorticoid receptor (GR) all belong to the steroid nuclear receptor family,

C. Jin (✉) • M.G. Rosenfeld (✉)
Howard Hughes Medical Institute, Department of Medicine, University of California San Diego, La Jolla, CA 92093, USA
e-mail: c3jin@ucsd.edu; mrosenfeld@ucsd.edu

R. Kurokawa (ed.), *Long Noncoding RNAs*, DOI 10.1007/978-4-431-55576-6_13

which usually reside in cytoplasm in the absence of ligand and translocate into the nucleus in the presence of ligand, binding to DNA on their cognate binding element and regulating target gene expression. A bunch of transcription factors (TFs) bind to NRs to form a huge protein complex of about a megadalton and cooperatively control the gene expression program (Liu et al. 2014). Dysregulation of nuclear receptors is associated with various diseases, including breast cancer and prostate cancers (Auchus and Fuqua 1994). NRs also interact with and require the actions of a large number of so-called co-activators or co-repressors as mediators of the epigenomic changes that coordinate their activity, to achieve precise regulation of gene expression (Beato et al. 1995; Dennis and O'Malley 2005; Garcia-Bassets et al. 2007; Glass et al. 1997; Korzus et al. 1998; Malik and Roeder 2005; Perissi et al. 2004; Rosenfeld et al. 2006; Spiegelman and Heinrich 2004; Xu et al. 1999). Dysregulation of coregulators has been implicated in various pathological states (Lonard and O'Malley 2012). The identification of regulatory lncRNAs as NR coregulators expands the coregulators from proteins to lncRNAs.

Indeed, with over 100 years of study of hormone regulation (Tata 2005), the linkage between noncoding RNA and hormone receptor regulation has only emerged in the past two decades. In this chapter, we focus on the nuclear receptor regulatory noncoding RNAs, including eRNAs, and will also briefly review other functional ncRNAs that are not yet recognized as nuclear receptor regulators but appear to be of biological significance in breast or prostate cancer development and diagnosis. The goal is to provide a general picture of the major findings for each lncRNA discussed, the working model of their actions, and their clinical relevance.

13.2 Steroid Hormone Receptor Associated lncRNAs

13.2.1 SRA

SRA (Steroid Receptor RNA Activator) was the first lncRNA identified that interacts with NRs and regulates their activities. It was initially found as a PR cofactor by a yeast two-hybrid screening and was further validated by a CMV-hPR luciferase reporter assay (Lanz et al. 1999). Various approaches failed to detect an encoded protein, and the protein synthesis inhibitor cycloheximide did not diminish the transactivation effect of SRA on steroid hormone receptors, indicating that SRA is a noncoding RNA. SRA lncRNA was found to interact with a nuclear receptor coactivator SRC-1, which is recruited by nuclear receptors (Lanz et al. 1999), including PR, GR, ER (Deblois and Giguere 2003), retinoic acid receptor (RAR) (Zhao et al. 2004), thyroid hormone receptor (TR) (Xu and Koenig 2004), AR (Agoulnik and Weigel 2009), steroidogenic factor 1 (SF-1) (Xu et al. 2009), and peroxisome proliferator-activated receptor-gamma (PPARγ) (Xu et al. 2010). To identify functional domains within SRA, a series of deletion mutants were generated to detect the PR coactivation function. The results indicated that the transactivation function of SRA was contributed from all over, rather than from a special region. Thermological modeling of SRA identified several stem-loop motifs, in which six motifs are important for coactivation, verified by site mutations (Lanz et al. 2002).

As more SRA isoforms were identified, some were found to be potentially capable of encoding a protein, the steroid receptor RNA activator protein (SRAP) (Emberley et al. 2003). Alternative splicing of the first intron appeared to affect the generation of coding or noncoding isoforms (Hube et al. 2006). The balance between fully spliced and intron-1-containing SRA varies in breast tumors and affected cancer cell growth. Mutating the protein translational start site ATG to TTG, which impaired protein-coding capability, also impaired the transactivation activity of SRA to ER responsive element (ERE), indicating that SRAP, besides SRA RNA, contributed to ER activation (Chooniedass-Kothari et al. 2010b). Both SRA RNA and protein are regulators of steroid hormone receptor (Leygue 2007). However, the activation or repression role of NRs is variable in different contexts. Using an SRA motif STR7, a stem-loop structure that is essential for coactivator function (Lanz et al. 2002), as a bait for a yeast three-hybrid screen, an RNA binding domain-containing protein, SLIRP1 (SRA stem-loop interacting RNA binding protein) was identified to interact with SRA (Hatchell et al. 2006). SLIRP1 is conserved in amino acid sequence among human, rat, and mouse, and shares substantial homology with SHARP (SMRT/HDAC1 associated repressor protein), an NR corepressor that interact with SRA (Shi et al. 2001). SRA-SLIRP1 interaction is required for the repression role of SLIRP1 to NRs (Hatchell et al. 2006). Similarly, SRAP was also capable of repression detected by luciferase reporter assay; endogenous SRAP physically interacts with multiple transcription factors, suggesting a repressive role of SRAP in various pathways (Chooniedass-Kothari et al. 2004, 2010a).

SRA was also found to interact with other partners and in turn to be involved in various biological processes (Xu et al. 2010). SRA also elevates insulin signaling and glucose uptake in differentiated adipocytes, at least in part because of the interaction with PPARγ (Xu et al. 2010). SRA was found to interact with CCCTC binding factor (CTCF) together with DEAD-box RNA helicase p68 (DDX5) and is required for the insulator function (Yao et al. 2010). SRA interacts with MyoD, a master transcription factor that controls myoblast differentiation, and potentiates MyoD activity with cooperative regulation of p68 and p72, thereby influencing skeletal muscle differentiation (Caretti et al. 2006). Interestingly, SRAP interacts with its RNA counterpart, SRA RNA, and prevents its activation of MyoD activity, and the ratio of SRAP relative to SRA RNA is changing during myogenic differentiation, suggesting a sophisticated regulation model of the SRA gene in muscle differentiation (Hube et al. 2011).

To establish the transcriptome consequences altered by depleting SRA, a microarray analysis was performed in Hela cells with respect to GR function and in MCF7 cells for ER function (Foulds et al. 2010). Consistent with previous studies, the majority of significantly changed genes were downregulated upon SRA siRNA knockdown, suggesting a coactivation function of SRA; however, in MCF7 cells, only a small proportion of these are supposed ER target genes in the presence of estrogen. Meanwhile, well-characterized GR target genes were affected by SRA knockdown, but these were not robust enough to call significant differences in dexamethasone (Dex)-treated Hela cells, indicating that the NR coactivation function of SRA was not that specific (Foulds et al. 2010). Knockdown of SRA in triple-negative breast cancer MDA-MB-231 cell lines impaired the invasiveness, indicating that the dependency of NRs in SRA function may be less than originally thought (Foulds et al. 2010).

Aberrant SRA transcription was observed in a wide range of tumors. It is noted to be upregulated in tumors compared with normal tissues, such as breast cancer (Murphy et al. 2000) and ovarian cancer (Hussein-Fikret and Fuller 2005). Particularly, SRA expression levels in ER+/PR+ and ER−/PR− breast tumor samples were lower than those in ER+/PR− and ER−/PR+ samples, and the expression of a new isoform with deletion of 203 bp was correlated with the tumor grade (Leygue et al. 1999). By tissue-microarray analysis of 372 breast tumors, SRAP levels significantly correlated with ER+, PR+, and older (age >64 years) patients, and higher SRAP expression in ER+, PR+, or younger (age ≤64 years) patients correlated with a worse survival rate (Yan et al. 2009). However, another study showed that patients with SRAP-positive primary tumors had a significantly higher survival likelihood from recurrent disease than patients with SRAP-negative samples, suggesting a protective role of SRAP (Chooniedass-Kothari et al. 2006). More studies seem to be required for revealing the correlation of SRA and SRAP with diseases in different contexts.

Prostate cancer studies showed that SRA was required for expression of some AR target genes in the presence of the AR ligand dihydrotestosterone (DHT), but not in all of the several genes that were tested. Knockdown of SRA reduced proliferation of prostate cancer LNCaP (AR+) and DU145 (AR−) cells but not PC-3 (AR−) cells, suggesting that the biological function of SRA is not so tightly correlated with AR function (Agoulnik and Weigel 2009). SRAP was also expressed in prostate cancer cells and was reported to be necessary for AR-activated transcription (Kawashima et al. 2003; Kurisu et al. 2006).

The in vivo role of SRA was assessed in a transgenic mouse model that robustly expressed human SRA. While overexpression of SRA indeed elevated estrogen-controlled progesterone receptor (PGR) gene expression and promoted cellular proliferation and differentiation, no alterations progressed to malignancy, indicating that overexpression alone was not sufficient to induce tumorigenesis (Lanz et al. 2003).

SRA also contributed to repression of hormone-induced gene expression in the absence of hormone, acting as scaffold for a complex containing HP1γ, LSD1, HDAC1/2, and CoREST (Beato and Vicent 2013). SRA has been shown to harbor pseudouridine, as a substrate of the mammalian pseudouridine synthase 1 (hPus1p) (Huet et al. 2014). The functional meaning of this modification is as yet unclear.

13.2.2 Gas5

A series of Gas (Growth-Arrest-Specific) genes were nominated by a subtraction cDNA library enriched for RNAs preferentially expressed in growth-arrested cells (Schneider et al. 1988). A yeast two-hybrid screen with the GR DBD as bait resulted in 2 out of 118 positive clones being mapped to Gas5, indicating the interaction between Gas5 and GR (Kino et al. 2010). Considering *GAS5* as a noncoding multiple small nucleolar RNA (snoRNA) host gene (Smith and Steitz 1998), further experiments were conducted to clarify that Gas5 interacts with GR as lncRNA rather than as snoRNA expressed from a Gas5 intronic region. This interaction was enhanced by the GR agonist, Dex. A reporter assay in which Gas5 and GR were overexpressed with glucocorticoid-responsive mouse mammary tumor virus (MMTV) promoter/reporter indicated that Gas5 can significantly suppress Dex-stimulated GR transcriptional activity (Kino et al. 2010). Gas5 was observed to translocate to the nucleus in response to Dex, accompanying GR (Kino et al. 2010). The interaction of Gas5 and GR reduces GR binding to GR binding elements (GREs) and suppresses its target gene expression, assessed by a cellular inhibitor of apoptosis 2 (cIAP2) gene construct, indicating a "decoy" function of Gas5 and GR association, and this effect was confirmed using other GR target genes' reporter construct and endogenous gene expression (Kino et al. 2010). Mapping results identified that 400–598 nt of Gas5 is responsible for binding to GR, which contains two GRE-like sequences. The dissociation constant (K_d) of Gas5:GR DBD interaction was ~30 nM, comparable to that of GRE DNA to GR DBD (Rundlett and Miesfeld 1995). This led to the suggestion of a competing role of *GAS5* with GRE DNA for GR binding (Kino et al. 2010). Gas5 also binds to DBD of the mineralocorticoid receptor (MR), PR, and AR, and inhibits their transcriptional activity in a ligand-dependent manner, but this is not the case for ERα (Kino et al. 2010). Gas5 was found to be downregulated in breast cancer (Mourtada-Maarabouni et al. 2009), renal cell carcinoma (Qiao et al. 2013), pancreatic cancer (Lu et al. 2013), bladder cancer cells (Liu et al. 2013), non-small cell lung cancer (Shi et al. 2013), gastric cancer (Sun et al. 2014), castration-resistant prostate cancer (Pickard et al. 2013), and colorectal cancer (Yin et al. 2014), and has been reported to inhibit proliferation or promote apoptosis in cells of these cancers. Besides, Gas5 also plays an important role in T-cell differentiation (Mourtada-Maarabouni et al. 2008, 2010).

A UPF1-mediated RNA degradation pathway contributed to the regulation of Gas5 transcript level (Tani et al. 2013). miR-21 and Gas5 were present in the Gas5-RISC complex and negatively regulate each other; a functional miR-21 binding site was found in exon4 of Gas5, suggesting a direct regulation mechanism (Tani et al. 2013).

13.2.3 PCGEM1 and PRNCR1

PCGEM1 (Prostate Cancer Gene Expression Marker 1), located on chromosome 2q32, was initially identified by differential display analysis of paired normal and prostate cancer tissues (Srikantan et al. 2000). It is a prostate tissue-specific transcript, overexpressed in prostate cancer patients, which lacks protein-coding capacity and is therefore considered as a noncoding RNA (Srikantan et al. 2000). Elevated expression of PCGEM1 is associated with a high risk of prostate cancer and promotes cancer cell growth (Petrovics et al. 2004). *PCGEM1* overexpression in LNCaP cells inhibits apoptosis induced by the DNA brake-inducing drug doxorubicin, while attenuation of the apoptotic response appears to be androgen-dependent (Fu et al. 2006). PRNCR1 (prostate cancer noncoding RNA 1) was found by SNP (single nucleotide polymorphism) association with prostate cancer susceptibility in the chromosome 8q24 "gene" desert region, in which two SNPs, rs1456315 and rs7463708, showed most significantly an association with prostate cancer susceptibility (Chung et al. 2011). Chromosome 8q24 exhibits multiple susceptibility loci on prostate, breast, and colon cancers (Ahmadiyeh et al. 2010; Al Olama et al. 2009; Gudmundsson et al. 2007; Schumacher et al. 2007), and is frequently amplified in prostate cancers (Sato et al. 1999), suggesting a correlation of *PRNCR1* with prostate cancer. This gene locus also harbors SNPs that are associated with colorectal cancer (Li et al. 2013a). Knockdown of PRNCR1 attenuates the transactivation activity of AR and the viability of prostate cancer cells (Chung et al. 2011).

The biotinylated RNA pull-down assay followed by mass spectrometry permitted the identification of protein partners for these two lncRNAs, PCGEM1 and PRNCR1, resulting in AR, pygopus homolog2 (Pygo2), and beta-catenin interacting with PCGEM1 and DOT1-like histone H3 methyltransferase (DOT1L) and AR interacting with PRNCR1 (Yang et al. 2013). The interactions of PCGEM1 and PRNCR1 with AR were also confirmed by a reciprocal method, RNA immunoprecipitation (RIP) (Yang et al. 2013). The DHT-treated time-course RIP assay indicated a sequential binding event of AR with PRNCR1 prior to PCGEM1. With inhibitors of demethylase, deacetylase, and phosphatase during the entire procedure to fully preserve the existed post-translational modifications, several modifications with pulled-down AR were identified, including methylation, acetylation, and phosphorylation, with varying robustness of the signal. Flag-tagged AR lysine to Arginine (K-R) mutation constructs for modification sites that were suggested by mass spectrometry, such as methylation on K143, K237, K291, K318, and K349, and acetylation on K631 and K634, were generated to test the necessity of the modifications for interaction; anti-Flag RIP assays in transfected LNCaP cells suggested that K349 and K631/634 were essential sites for PCGEM1-AR and PRNCR1-AR interactions, respectively (Yang et al. 2013). K631/634 acetylation has been reported to be induced by DHT and facilitated AR transactivation (Fu et al. 2000, 2003). One of the partners of PRNCR1, DOT1L, a methyltransferase with a known substrate of H3K79 (Feng et al. 2002), was capable of AR K349 methylation detected by an in vitro methyltransferase assay. The binding sites of AR to PRNCR1 and PCGEM1 were mapped to AR DBD and N-terminus, respectively. However, another study failed to detect the interaction of AR to these two lncRNAs, possibly because AR modifications were poorly preserved during the experimental process, as none of the previously reported AR medications were successfully captured (Prensner et al. 2014b).

Knockdown either of these two lncRNAs attenuates canonical AR target gene expression in a DHT-dependent manner, including *FKBP5*, *TMPRSS2*, *NKX3.1*, *KLK2*, and *KLK3*. Global run-on sequencing (GRO-Seq) confirmed this effect genome widely on 617 DHT-upregulated genes with AR-bound enhancer within

200 kb of promoter (Yang et al. 2013). Knockdown of either of these two lncRNAs also decreased expression of canonical AR target gene expression in CWR22Rv1 cells, a castration-resistant prostate cell line that expresses the constitutive-activated AR (Tepper et al. 2002). The genomic binding loci mapped by chromatin isolation by RNA purification (ChIRP) revealed ~82 % of *PGCEM1* co-localizing with AR-bound sites, of which ~70 % corresponded to AR-bound, enhancer mark H3K4me[1]-marked loci (Yang et al. 2013). By ChIP-3C (ChIP-chromatin conformation capture) experiments, these two lncRNAs were found to be required for AR-bound enhancer–promoter looping (Yang et al. 2013).

Using CWR22Rv1 cells as a model, in vitro cell proliferation and in vivo mouse xenograft tests demonstrated that depletion of either of these two lncRNAs significantly reduces the cancer cell growth both in vitro and in vivo, suggesting clinical potential as a therapeutic target (Yang et al. 2013). Like miR-21 and Gas5, similar effects exist in miR-145 and PCGEM1, which negatively regulate each other reciprocally, but whether it is a direct effect is still unclear. Knockdown of PCGEM1 or overexpression of miR-145 inhibits LNCaP cell migration, invasion, and progression of tumor xenograft, demonstrating that both PCGEM1 and its regulator mir-145 are functional in prostate cancer biology (He et al. 2014).

13.3 eRNA

Enhancer RNAs (eRNAs) are the bidirectional RNAs transcribed from enhancers, exemplified in GRO-seq analysis of neuronal enhancers (Kim et al. 2010). In the AR system, AR-activated enhancers marked by increased eRNA are responsible for activation of nearby coding gene expression (Wang et al. 2011). Similar observations were reported in the ER system (Hah et al. 2011; Li et al. 2013b). To determine whether eRNA might be functional, siRNA and locked nucleic acid antisense oligonucleotide (LNAs)-mediated eRNA knockdown were performed on several eRNAs, including the eRNAs of the canonical ER target genes *TFF1*, *FOXC1*, and *CA12*. As a result, the coding gene expression level was correlated with the cognate eRNA level, indicating that the presence of eRNAs was of functional significance in ERα-regulated gene expression (Li et al. 2013b). Furthermore, a GAL4-BoxB-tethering-based reporter assay of FOXC1 eRNA suggested that the sequence-specific eRNA transcript *per se*, rather than merely the process of enhancer transcription, was required for activating its cognate coding target gene (Li et al. 2013b). 3D-DSL experiments demonstrated that eRNA knockdown disrupts ligand-induced enhancer–promoter interaction, detected by *NRIP1* and *GREB1* gene loci. Some protein complex have been identified to mediate enhancer–promoter looping, including mediator and cohesin (Kagey et al. 2010), and similar results were observed for AR-induced enhancers (Hsieh et al. 2014). To provide mechanism insights into how eRNAs affect enhancer–promoter interaction, which remains poorly understood, eRNA pull-down revealed an interaction between eRNA with SMC3 and RAD21, subunits of the cohesin complex. Knockdown of eRNA resulted in a decrease of cohesin recruitment to enhancers in response to ER ligand 17β-oestradiol (E2). A 3C assay for *NRIP1* and *GREB1* loci demonstrated that depletion of RAD21 led to loss of enhancer–promoter interactions, indicating that eRNAs, in response to an E2 signal, regulate their adjacent genes' expression dependent on cohesin-mediated enhancer–promoter looping events (Li et al. 2013b).

13.4 Other Functional Breast/Prostate Cancer Related lncRNAs

13.4.1 PCA3

PCA3 (Prostate Cancer Antigen 3, also called DD3) is a prostate-specific lncRNA, highly overexpressed in prostate cancer (Bussemakers et al. 1999), and probably the first well-identified lncRNA to serve as a specific marker for indicating prostate cancers. A PCA3 urine assay has been developed to be a potential diagnostic method for prostate cancer diagnosis (Groskopf et al. 2006; Kirby 2007; Marks et al. 2007; van Gils et al. 2007). Urinary PCA3 could be superior to serum prostate-specific antigen (PSA) determination for predicting the biopsy outcome (Marks et al. 2007). Follow-up of men with an elevated PCA3 score indicated that the PCA3 score was able to predict prostate cancer in men with one or two previous negative repeat biopsies (Remzi et al. 2010). However, the PCA3 score seemed not to serve as an independent predictor for tumor volume or for non-organ-confined disease, so it fails to predict aggressive prostate cancers (Augustin et al. 2013). PCA3 is encoded from the intron of the *BMCC1* gene, but these two genes do not appear to be co-regulated, suggesting that PCA3 is an independent transcript, albeit that the possibility of PCA3 being a *cis*-regulator of BMCC1 has not been ruled out (Salagierski et al. 2010). Despite hundreds of studies on PCA3 as a prostate cancer marker, there are few reports on the mechanism. It has been reported that *PCA3* knockdown inhibits cell growth and viability, probably through the AR signaling pathway. AR target genes including *PSA* and *NDRG1* are downregulated upon *PCA3* siRNA knockdown (Ferreira et al. 2012); the precise functional mechanisms remain unclear.

13.4.2 HOTAIR

HOTAIR (HOX Antisense Intergenic RNA) is a 2,158 nt noncoding RNA transcribed from the *HoxC* locus. Depletion of HOTAIR led to a 40 kb transcriptional activation around the *HoxD* locus (Rinn et al. 2007). HOTAIR binds to the PRC2 complex (detected by components Suz12 and Ezh2) and mediates the formation of H3K27me3 on the *HoxD* locus, whereas HOTAIR-depleted cells substantially lost H3K27me3 occupancy of the *HoxD* locus; therefore, it regulated transcription as a scaffold of the epigenetic control complex (Rinn et al. 2007). Biochemical experiments demonstrated that the Ezh2-EED heterodimer was necessary and sufficient for binding to HOTAIR, utilizing an 89 nt motif (Wu et al. 2013). The HOTAIR expression level was significantly higher in breast tumors than in normal breast epithelia, and a high HOTAIR level was a significant predictor of metastasis and death (Gupta et al. 2010). In vivo experiments indicated that overexpression of HOTAIR dramatically promoted metastasis of the MDA-MB-231 xenograft model. HOTAIR overexpression caused 854 genes to gain Suz12 and H3K27me3 occupancy, with downregulated expression that was linked to breast tumor aggressiveness (Gupta et al. 2010). *HOTAIR* itself is transcriptionally induced by E2 (Bhan et al. 2013), providing another link to breast cancer. Besides breast cancer, *HOTAIR* could also promote the metastasis of esophageal squamous cell carcinoma (Chen et al. 2013), gastric carcinoma (Emadi-Andani et al. 2014; Lee et al. 2014), non-small cell lung cancer (Nakagawa et al. 2013), and epithelial ovarian cancer (Qiu et al. 2014), and serve as a molecular marker for ER-positive primary breast cancer metastasis (Sorensen et al. 2013) and nasopharyngeal carcinoma progression and survival (Nie et al. 2013). *HOTAIR* was recently found to facilitate ubiquitination of E3 ubiquitin ligase to its substrates (Yoon et al. 2013).

13.4.3 PCAT-1

PCAT-1 (prostate cancer associated transcript 1) is located in the chromosome 8q24 gene desert ~725 kb upstream of the c-MYC oncogene. *PCAT-1* and other PCAT series lncRNAs were identified by systematical RNA-seq of a cohort of 102 prostate tissues and cell lines (Prensner et al. 2011). It was found that 19.8 % of transcripts were unannotated intronic and intergenic transcripts with conservation and lack of a high-quality open reading frame (ORT), indicating that they are indeed ncRNAs. Further characterization of these transcripts identified that *PCAT-1* and *PCAT-14* showed cancer-specific upregulation in tumor samples compared with matched normal samples. Although the 8q24 region is frequently amplified (Beroukhim et al. 2010), high expression of *PCAT-1* in localized tumors is not due to the 8q24 amplification (Prensner et al. 2014a). *PCAT-1* and Ezh2 expression are mutually exclusive in cancer tissue samples, but a core PRC2 component, SUZ12, binds the *PCAT-1* promoter as well as the *PCAT-1* transcript. Overexpression of *PCAT-1* in RWPE benign immortalized prostate cells accelerated cell proliferation, and knockdown of *PCAT-1* in LNCaP cells decreased proliferation (Prensner et al. 2011). Increased *PCAT-1* expression level is also found in colorectal cancer (CRC) tissue samples, compared with matched normal tissues, and CRC patients with *PCAT-1* overexpression showed a poorer survival rate, suggesting that it may also serve as a biomarker for colorectal cancer (Ge et al. 2013). *PCAT-1* and *BRCA2* expression are negatively correlated in prostate cancer specimens, and knockdown or overexpression of *PCAT-1* caused up- or downregulated *BRCA2* expression, respectively. Overexpression of *PCAT-1* increased γ-H2AX foci, suggesting that *PCAT-1* impairs the repair of double-strand DNA breaks, enabling increased cell death following genotoxic stress. *PCAT-1* regulates *BRCA2* mRNA decay analogous to the effects of miRNA, revealed by *BRCA2* 3′UTR-luc activity, a function subserved by the 5′ end of PCAT-1 (Prensner et al. 2014a).

13.4.4 SChLAP1

The *SChLAP1* (Second Chromosome Locus Associated with Prostate-1) gene encodes multiple isoforms of transcripts and spans nearly 200 kb on chromosome 2q31.1. Like *PCAT-1*, *SChLAP1* was initially found by an unannotated transcript presented in prostate cancer tissue RNA-seq (Prensner et al. 2011). In localized prostate tumors, stratifying 235 individual samples with localized prostate cancer, *SChLAP1* expression was correlated with higher Gleason scores, a grading system of prostate cancer based on microscopic appearance. *SChLAP1* expression correlates with prostate cancer-specific mortality, with a shorter median time for biochemical recurrence, and serves as a single-gene predictor of aggressive prostate cancer (Prensner et al. 2013). Knockdown of *SChLAP1* impaired cell invasion and proliferation in vitro. As an in vivo test, intracardiac injection of CB-17 SCID mice with 22Rv1 cells stably knocking down *SChLAP1* reduces metastasis in both tumor sites and size. RNA-seq revealed that the affected genes upon knockdown of *SChLAP1* were inversely correlated with the SWI/SNF complex-regulated genes. Indeed *SChLAP1* was reported to be physically associated with SNF5, a core component of SWI/SNF complex. Overexpression of *SChLAP1* attenuates SNF5 binding to genomic loci, in turn agonizing its target genes expression (Prensner et al. 2013).

13.4.5 BCAR4

Tamoxifen is an antiestrogen drug that is widely used as endocrine therapy for ERα-positive breast cancer. BCAR4 (breast cancer antiestrogen resistance 4) was initially identified by cDNA libraries screening for 4-hydroxy-tamoxifen (OH-TAM) resistance (Meijer et al. 2006). Ectopic expression of *BCAR4* induces OH-TAM resistance and anchorage-independent ZR-75-1 cell growth (Meijer et al. 2006). Subsequent studies demonstrated that high BCAR4 mRNA levels are an independent predictive factor for poor progression-free survival after starting tamoxifen treatment (Godinho et al. 2010, 2012). Injection of BCAR4-expressing ZR-75-1 cells into nude mice resulted in rapidly growing tumors (Godinho et al. 2011). Tissue specificity analysis suggested that *BCAR4* mRNA is highly expressed in the human placenta and oocytes but is absent in other normal tissues (Godinho et al. 2011). Although *BCAR4* is theoretically capable of encoding a protein (Angulo et al. 2013; Meijer et al. 2006), its noncoding function appears to be critical for breast cancer metastasis, acting as a downstream effector of chemokine CCL21

(Xing et al. 2014). CCL21-induced binding of BCAR4 and SNIP1 releases the inhibitory role of SNIP1 to P300-dependent histone acetylation activity, which, in turn enables BCAR4-recruited PNUTS to bind H3K18ac, activating PP1 phosphatase and modulating RNA polymerase II Ser5 phosphorylation that initiates transcription at Gli2 target gene promoters, controlling a group of genes' expression (Xing et al. 2014). shRNAs or LNA-mediated knockdown of *BCAR4* in MDA-MB-231-derived LM2 cells lead to a significant decrease of metastasis tested by a mouse model, suggesting its therapeutic potential (Xing et al. 2014).

13.5 Perspective

Nuclear receptors have both activation and repression functional effects on their target genes' expression, which may largely depend on their cofactors. The exchange of cofactors, from co-repressors to co-activators, in response to ligands, switches the target genes from transcriptional repressive to transcriptional active (Perissi et al. 2004). Over decades of studies, co-repressors or co-activators have usually been thought to be proteins, exemplified by histone acetyltransferase containing complex or histone deacetylase complex. The discovery of functional noncoding RNA extends the insight of co-activators and co-repressors to a new layer. So far, we realize that lncRNAs are capable of binding nuclear receptors, even on a nanomolar scale, which is comparable to DNA–protein binding affinity. This RNA–protein binding event causes various consequences: the RNA could be a "decoy" function by competing with a DNA binding element for the nuclear receptor; it could be an "attractive" function to recruit proteins to a specific place; and it also could be a "scaffold" function to stabilize a protein complex (Fig. 13.1). The working models for a few lncRNAs with their NR partners have been finely described, but not many, making it still difficult to answer these questions: do the nuclear receptor binding lncRNAs have a common feature; does a "RNA–protein interaction code" exist that could allow us to quickly identify every RNA's interacting partners, at least at the domain level; and what is the determinant of the RNA–protein interaction, nucleotide sequences or secondary structure, or higher level structure? We also want to know whether the relatively small number of lncRNAs play a crucial role in regulating nuclear receptor activity, or is the number far beyond what we have found, in a tissue-, spatial-, temporal-specific manner?

Regarding the consequence of lncRNAs as a regulator of NRs, the roles of lncRNAs, as far as we have found, are focused on NR target gene transcription. However, except for PCGEM1 and PRNCR1, which have been studied genome widely, most studies have only been done on a few cases, in which the global effects are still unclear. Therefore, it is possible that global effects are not fully consistent with specific gene cases, exemplified by the SRA, which was initially reported as a coactivator of NRs but ultimately was found not to be specifically related to NRs in terms of a global transcriptional effect. Furthermore, the biological role of these lncRNAs remains largely unknown. Currently, only a few of them have been tested

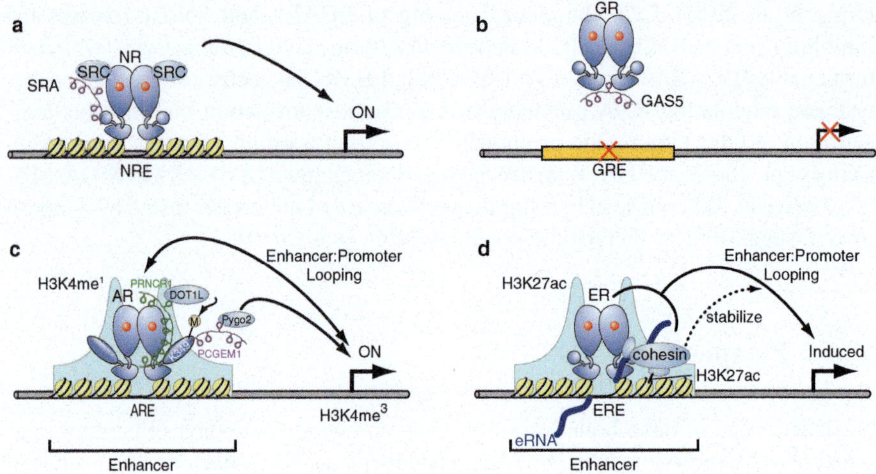

Fig. 13.1 Working model of NRs regulatory ncRNAs. (**a**) SRA interacts with NRs mediated by nuclear receptor coactivator (*SRC*), and activates target gene expression. (**b**) GAS5 interacts with the DBD of GR and competes with the GRE on GR-regulated genes, affecting GR transactivation. (**c**) PCGEM1 and PRNCR1 are DHT-dependent AR-interacting ncRNAs and mediate enhancer–promoter looping for AR target genes' activation. The interaction depends on appropriate AR post-translational modifications. (**d**) eRNA, in the estrogen signal system, stabilizes enhancer–promoter looping for ER target genes in cooperation with cohesin, which, in turn, activates gene expression

in a xenograft mouse model, based on shRNA-mediated stable knockdown. The physiological role of these lncRNAs and their correlation with specific diseases in patients remain fascinating issues that warrant further assessment.

Besides direct regulation on nuclear receptors, some lncRNAs are not related to the nuclear receptor itself but play an important role in breast or prostate cancer. These lncRNAs are likely to have a general role in cancer progress, regardless of nuclear receptor status. Indeed, their functional models are more related to general factors, such as PRC2 complex and DNA repair complex. In these cases, breast/prostate cancer cells that have been used for the study are more appropriate to be considered as cancer materials, instead of specified type of cancer.

In addition to regulatory lncRNAs in breast/prostate cancers, there is a bunch of lncRNAs that exhibit aberrant expression levels during pathogenesis but have not found a robust functional role and are likely to be a consequence of abnormal transcription/homeostasis of cancer cells. These lncRNAs could serve as cancer markers and broaden the way for clinical diagnosis—for example, PCA3. There will be more lncRNAs that are supposed to belong to this category. It will be beneficial to identify more of this kind of lncRNAs, together with proteins, which in turn form a signature for a specific cancer type or stage, and facilitate clinical diagnosis.

Why these ncRNAs are ectopically expressed in cancers remains a question. Some of them may, under a certain signal—possibly from estrogen or androgen itself—act in a feedback loop; some of them may be expressed because of genomic amplification or dysregulation of DNA methylation or transcription factors that are responsible for their transcription, or abnormal metabolism for RNA processing, maturation, and degradation. Further studies on the upstream events of how the hormone receptor regulatory ncRNAs are regulated will provide a deeper understanding of the intact signaling pathway, and also contribute to potential therapy.

Despite the encouraging results on the molecular mechanisms and in vivo studies of functional lncRNAs, eventually we will ask whether lncRNA-targeted diagnostic and therapeutic approaches are clinical applicable, which we hope, will be a new insight in cancer therapy.

References

Agoulnik IU, Weigel NL (2009) Coactivator selective regulation of androgen receptor activity. Steroids 74:669–674

Ahmadiyeh N, Pomerantz MM, Grisanzio C, Herman P, Jia L, Almendro V, He HH, Brown M, Liu XS, Davis M et al (2010) 8q24 prostate, breast, and colon cancer risk loci show tissue-specific long-range interaction with MYC. Proc Natl Acad Sci U S A 107:9742–9746

Al Olama AA, Kote-Jarai Z, Giles GG, Guy M, Morrison J, Severi G, Leongamornlert DA, Tymrakiewicz M, Jhavar S, Saunders E et al (2009) Multiple loci on 8q24 associated with prostate cancer susceptibility. Nat Genet 41:1058–1060

Angulo L, Perreau C, Lakhdari N, Uzbekov R, Papillier P, Freret S, Cadoret V, Guyader-Joly C, Royere D, Ponsart C et al (2013) Breast-cancer anti-estrogen resistance 4 (BCAR4) encodes a novel maternal-effect protein in bovine and is expressed in the oocyte of humans and other non-rodent mammals. Hum Reprod 28:430–441

Auchus RJ, Fuqua SA (1994) Hormone-nuclear receptor interactions in health and disease. The oestrogen receptor. Baillieres Clin Endocrinol Metab 8:433–449

Augustin H, Mayrhofer K, Pummer K, Mannweiler S (2013) Relationship between prostate cancer gene 3 (PCA3) and characteristics of tumor aggressiveness. Prostate 73:203–210

Bain DL, Heneghan AF, Connaghan-Jones KD, Miura MT (2007) Nuclear receptor structure: implications for function. Annu Rev Physiol 69:201–220

Beato M, Vicent GP (2013) A new role for an old player: steroid receptor RNA Activator (SRA) represses hormone inducible genes. Transcription 4:167–171

Beato M, Herrlich P, Schutz G (1995) Steroid hormone receptors: many actors in search of a plot. Cell 83:851–857

Beroukhim R, Mermel CH, Porter D, Wei G, Raychaudhuri S, Donovan J, Barretina J, Boehm JS, Dobson J, Urashima M et al (2010) The landscape of somatic copy-number alteration across human cancers. Nature 463:899–905

Bhan A, Hussain I, Ansari KI, Kasiri S, Bashyal A, Mandal SS (2013) Antisense transcript long noncoding RNA (lncRNA) HOTAIR is transcriptionally induced by estradiol. J Mol Biol 425:3707–3722

Bussemakers MJ, van Bokhoven A, Verhaegh GW, Smit FP, Karthaus HF, Schalken JA, Debruyne FM, Ru N, Isaacs WB (1999) DD3: a new prostate-specific gene, highly overexpressed in prostate cancer. Cancer Res 59:5975–5979

Caretti G, Schiltz RL, Dilworth FJ, Di Padova M, Zhao P, Ogryzko V, Fuller-Pace FV, Hoffman EP, Tapscott SJ, Sartorelli V (2006) The RNA helicases p68/p72 and the noncoding RNA SRA are coregulators of MyoD and skeletal muscle differentiation. Dev Cell 11:547–560

Chen FJ, Sun M, Li SQ, Wu QQ, Ji L, Liu ZL, Zhou GZ, Cao G, Jin L, Xie HW et al (2013) Upregulation of the long non-coding RNA HOTAIR promotes esophageal squamous cell carcinoma metastasis and poor prognosis. Mol Carcinog 52:908–915

Chooniedass-Kothari S, Emberley E, Hamedani MK, Troup S, Wang X, Czosnek A, Hube F, Mutawe M, Watson PH, Leygue E (2004) The steroid receptor RNA activator is the first functional RNA encoding a protein. FEBS Lett 566:43–47

Chooniedass-Kothari S, Hamedani MK, Troup S, Hube F, Leygue E (2006) The steroid receptor RNA activator protein is expressed in breast tumor tissues. Int J Cancer 118:1054–1059

Chooniedass-Kothari S, Hamedani MK, Auge C, Wang X, Carascossa S, Yan Y, Cooper C, Vincett D, Myal Y, Jalaguier S et al (2010a) The steroid receptor RNA activator protein is recruited to promoter regions and acts as a transcriptional repressor. FEBS Lett 584:2218–2224

Chooniedass-Kothari S, Vincett D, Yan Y, Cooper C, Hamedani MK, Myal Y, Leygue E (2010b) The protein encoded by the functional steroid receptor RNA activator is a new modulator of ER alpha transcriptional activity. FEBS Lett 584:1174–1180

Chung S, Nakagawa H, Uemura M, Piao L, Ashikawa K, Hosono N, Takata R, Akamatsu S, Kawaguchi T, Morizono T et al (2011) Association of a novel long non-coding RNA in 8q24 with prostate cancer susceptibility. Cancer Sci 102:245–252

Core LJ, Waterfall JJ, Lis JT (2008) Nascent RNA sequencing reveals widespread pausing and divergent initiation at human promoters. Science 322:1845–1848

Deblois G, Giguere V (2003) Ligand-independent coactivation of ERalpha AF-1 by steroid receptor RNA activator (SRA) via MAPK activation. J Steroid Biochem Mol Biol 85:123–131

Dennis AP, O'Malley BW (2005) Rush hour at the promoter: how the ubiquitin-proteasome pathway polices the traffic flow of nuclear receptor-dependent transcription. J Steroid Biochem Mol Biol 93:139–151

Emadi-Andani E, Nikpour P, Emadi-Baygi M, Bidmeshkipour A (2014) Association of HOTAIR expression in gastric carcinoma with invasion and distant metastasis. Adv Biomed Res 3:135

Emberley E, Huang GJ, Hamedani MK, Czosnek A, Ali D, Grolla A, Lu B, Watson PH, Murphy LC, Leygue E (2003) Identification of new human coding steroid receptor RNA activator isoforms. Biochem Biophys Res Commun 301:509–515

Feng Q, Wang H, Ng HH, Erdjument-Bromage H, Tempst P, Struhl K, Zhang Y (2002) Methylation of H3-lysine 79 is mediated by a new family of HMTases without a SET domain. Curr Biol 12:1052–1058

Ferreira LB, Palumbo A, de Mello KD, Sternberg C, Caetano MS, de Oliveira FL, Neves AF, Nasciutti LE, Goulart LR, Gimba ER (2012) PCA3 noncoding RNA is involved in the control of prostate-cancer cell survival and modulates androgen receptor signaling. BMC Cancer 12:507

Foulds CE, Tsimelzon A, Long W, Le A, Tsai SY, Tsai MJ, O'Malley BW (2010) Research resource: expression profiling reveals unexpected targets and functions of the human steroid receptor RNA activator (SRA) gene. Mol Endocrinol 24:1090–1105

Fu M, Wang C, Reutens AT, Wang J, Angeletti RH, Siconolfi-Baez L, Ogryzko V, Avantaggiati ML, Pestell RG (2000) p300 and p300/cAMP-response element-binding protein-associated factor acetylate the androgen receptor at sites governing hormone-dependent transactivation. J Biol Chem 275:20853–20860

Fu M, Rao M, Wang C, Sakamaki T, Wang J, Di Vizio D, Zhang X, Albanese C, Balk S, Chang C et al (2003) Acetylation of androgen receptor enhances coactivator binding and promotes prostate cancer cell growth. Mol Cell Biol 23:8563–8575

Fu X, Ravindranath L, Tran N, Petrovics G, Srivastava S (2006) Regulation of apoptosis by a prostate-specific and prostate cancer-associated noncoding gene, PCGEM1. DNA Cell Biol 25:135–141

Gadaleta RM, Magnani L (2014) Nuclear receptors and chromatin: an inducible couple. J Mol Endocrinol 52:R137–R149

Garcia-Bassets I, Kwon YS, Telese F, Prefontaine GG, Hutt KR, Cheng CS, Ju BG, Ohgi KA, Wang J, Escoubet-Lozach L et al (2007) Histone methylation-dependent mechanisms impose ligand dependency for gene activation by nuclear receptors. Cell 128:505–518

Ge X, Chen Y, Liao X, Liu D, Li F, Ruan H, Jia W (2013) Overexpression of long noncoding RNA PCAT-1 is a novel biomarker of poor prognosis in patients with colorectal cancer. Med Oncol 30:588

Glass CK, Rose DW, Rosenfeld MG (1997) Nuclear receptor coactivators. Curr Opin Cell Biol 9:222–232

Godinho MF, Sieuwerts AM, Look MP, Meijer D, Foekens JA, Dorssers LC, van Agthoven T (2010) Relevance of BCAR4 in tamoxifen resistance and tumour aggressiveness of human breast cancer. Br J Cancer 103:1284–1291

Godinho M, Meijer D, Setyono-Han B, Dorssers LC, van Agthoven T (2011) Characterization of BCAR4, a novel oncogene causing endocrine resistance in human breast cancer cells. J Cell Physiol 226:1741–1749

Godinho MF, Wulfkuhle JD, Look MP, Sieuwerts AM, Sleijfer S, Foekens JA, Petricoin EF 3rd, Dorssers LC, van Agthoven T (2012) BCAR4 induces antioestrogen resistance but sensitises breast cancer to lapatinib. Br J Cancer 107:947–955

Groskopf J, Aubin SM, Deras IL, Blase A, Bodrug S, Clark C, Brentano S, Mathis J, Pham J, Meyer T et al (2006) APTIMA PCA3 molecular urine test: development of a method to aid in the diagnosis of prostate cancer. Clin Chem 52:1089–1095

Gudmundsson J, Sulem P, Manolescu A, Amundadottir LT, Gudbjartsson D, Helgason A, Rafnar T, Bergthorsson JT, Agnarsson BA, Baker A et al (2007) Genome-wide association study identifies a second prostate cancer susceptibility variant at 8q24. Nat Genet 39:631–637

Gupta RA, Shah N, Wang KC, Kim J, Horlings HM, Wong DJ, Tsai MC, Hung T, Argani P, Rinn JL et al (2010) Long non-coding RNA HOTAIR reprograms chromatin state to promote cancer metastasis. Nature 464:1071–1076

Hah N, Danko CG, Core L, Waterfall JJ, Siepel A, Lis JT, Kraus WL (2011) A rapid, extensive, and transient transcriptional response to estrogen signaling in breast cancer cells. Cell 145:622–634

Hatchell EC, Colley SM, Beveridge DJ, Epis MR, Stuart LM, Giles KM, Redfern AD, Miles LE, Barker A, MacDonald LM et al (2006) SLIRP, a small SRA binding protein, is a nuclear receptor corepressor. Mol Cell 22:657–668

He JH, Zhang JZ, Han ZP, Wang L, Lv Y, Li YG (2014) Reciprocal regulation of PCGEM1 and miR-145 promote proliferation of LNCaP prostate cancer cells. J Exp Clin Cancer Res 33:72

Hsieh CL, Fei T, Chen Y, Li T, Gao Y, Wang X, Sun T, Sweeney CJ, Lee GS, Chen S et al (2014) Enhancer RNAs participate in androgen receptor-driven looping that selectively enhances gene activation. Proc Natl Acad Sci U S A 111:7319–7324

Hube F, Guo J, Chooniedass-Kothari S, Cooper C, Hamedani MK, Dibrov AA, Blanchard AA, Wang X, Deng G, Myal Y et al (2006) Alternative splicing of the first intron of the steroid receptor RNA activator (SRA) participates in the generation of coding and noncoding RNA isoforms in breast cancer cell lines. DNA Cell Biol 25:418–428

Hube F, Velasco G, Rollin J, Furling D, Francastel C (2011) Steroid receptor RNA activator protein binds to and counteracts SRA RNA-mediated activation of MyoD and muscle differentiation. Nucleic Acids Res 39:513–525

Huet T, Miannay FA, Patton JR, Thore S (2014) Steroid receptor RNA activator (SRA) modification by the human pseudouridine synthase 1 (hPus1p): RNA binding, activity, and atomic model. PLoS ONE 9:e94610

Hussein-Fikret S, Fuller PJ (2005) Expression of nuclear receptor coregulators in ovarian stromal and epithelial tumours. Mol Cell Endocrinol 229:149–160

Jin L, Li Y (2010) Structural and functional insights into nuclear receptor signaling. Adv Drug Deliv Rev 62:1218–1226

Kagey MH, Newman JJ, Bilodeau S, Zhan Y, Orlando DA, van Berkum NL, Ebmeier CC, Goossens J, Rahl PB, Levine SS et al (2010) Mediator and cohesin connect gene expression and chromatin architecture. Nature 467:430–435

Kawashima H, Takano H, Sugita S, Takahara Y, Sugimura K, Nakatani T (2003) A novel steroid receptor co-activator protein (SRAP) as an alternative form of steroid receptor RNA-activator gene: expression in prostate cancer cells and enhancement of androgen receptor activity. Biochem J 369:163–171

Kim TK, Hemberg M, Gray JM, Costa AM, Bear DM, Wu J, Harmin DA, Laptewicz M, Barbara-Haley K, Kuersten S et al (2010) Widespread transcription at neuronal activity-regulated enhancers. Nature 465:182–187

Kino T, Hurt DE, Ichijo T, Nader N, Chrousos GP (2010) Noncoding RNA gas5 is a growth arrest- and starvation-associated repressor of the glucocorticoid receptor. Sci Signal 3:ra8

Kirby R (2007) PCA3 improves diagnosis of prostate cancer. Practitioner 251:18, 21, 23

Korzus E, Torchia J, Rose DW, Xu L, Kurokawa R, McInerney EM, Mullen TM, Glass CK, Rosenfeld MG (1998) Transcription factor-specific requirements for coactivators and their acetyltransferase functions. Science 279:703–707

Kurisu T, Tanaka T, Ishii J, Matsumura K, Sugimura K, Nakatani T, Kawashima H (2006) Expression and function of human steroid receptor RNA activator in prostate cancer cells: role of endogenous hSRA protein in androgen receptor-mediated transcription. Prostate Cancer Prostatic Dis 9:173–178

Lanz RB, McKenna NJ, Onate SA, Albrecht U, Wong J, Tsai SY, Tsai MJ, O'Malley BW (1999) A steroid receptor coactivator, SRA, functions as an RNA and is present in an SRC-1 complex. Cell 97:17–27

Lanz RB, Razani B, Goldberg AD, O'Malley BW (2002) Distinct RNA motifs are important for coactivation of steroid hormone receptors by steroid receptor RNA activator (SRA). Proc Natl Acad Sci U S A 99:16081–16086

Lanz RB, Chua SS, Barron N, Soder BM, DeMayo F, O'Malley BW (2003) Steroid receptor RNA activator stimulates proliferation as well as apoptosis in vivo. Mol Cell Biol 23:7163–7176

Lee NK, Lee JH, Park CH, Yu D, Lee YC, Cheong JH, Noh SH, Lee SK (2014) Long non-coding RNA HOTAIR promotes carcinogenesis and invasion of gastric adenocarcinoma. Biochem Biophys Res Commun 451:171–178

Leygue E (2007) Steroid receptor RNA activator (SRA1): unusual bifaceted gene products with suspected relevance to breast cancer. Nucl Recept Signal 5:e006

Leygue E, Dotzlaw H, Watson PH, Murphy LC (1999) Expression of the steroid receptor RNA activator in human breast tumors. Cancer Res 59:4190–4193

Li L, Sun R, Liang Y, Pan X, Li Z, Bai P, Zeng X, Zhang D, Zhang L, Gao L (2013a) Association between polymorphisms in long non-coding RNA PRNCR1 in 8q24 and risk of colorectal cancer. J Exp Clin Cancer Res 32:104

Li W, Notani D, Ma Q, Tanasa B, Nunez E, Chen AY, Merkurjev D, Zhang J, Ohgi K, Song X et al (2013b) Functional roles of enhancer RNAs for oestrogen-dependent transcriptional activation. Nature 498:516–520

Liu Z, Wang W, Jiang J, Bao E, Xu D, Zeng Y, Tao L, Qiu J (2013) Downregulation of GAS5 promotes bladder cancer cell proliferation, partly by regulating CDK6. PLoS ONE 8:e73991

Liu Z, Merkurjev D, Yang F, Li W, Oh S, Friedman MJ, Song X, Zhang F, Ma Q, Ohgi KA et al (2014) Enhancer activation requires trans-recruitment of a mega transcription factor complex. Cell 159:358–373

Lonard DM, O'Malley BW (2012) Nuclear receptor coregulators: modulators of pathology and therapeutic targets. Nat Rev Endocrinol 8:598–604

Lu X, Fang Y, Wang Z, Xie J, Zhan Q, Deng X, Chen H, Jin J, Peng C, Li H et al (2013) Downregulation of gas5 increases pancreatic cancer cell proliferation by regulating CDK6. Cell Tissue Res 354:891–896

Malik S, Roeder RG (2005) Dynamic regulation of pol II transcription by the mammalian mediator complex. Trends Biochem Sci 30:256–263

Mangelsdorf DJ, Thummel C, Beato M, Herrlich P, Schutz G, Umesono K, Blumberg B, Kastner P, Mark M, Chambon P et al (1995) The nuclear receptor superfamily: the second decade. Cell 83:835–839

Marks LS, Fradet Y, Deras IL, Blase A, Mathis J, Aubin SM, Cancio AT, Desaulniers M, Ellis WJ, Rittenhouse H et al (2007) PCA3 molecular urine assay for prostate cancer in men undergoing repeat biopsy. Urology 69:532–535

Meijer D, van Agthoven T, Bosma PT, Nooter K, Dorssers LC (2006) Functional screen for genes responsible for tamoxifen resistance in human breast cancer cells. Mol Cancer Res 4:379–386

Mourtada-Maarabouni M, Hedge VL, Kirkham L, Farzaneh F, Williams GT (2008) Growth arrest in human T-cells is controlled by the non-coding RNA growth-arrest-specific transcript 5 (GAS5). J Cell Sci 121:939–946

Mourtada-Maarabouni M, Pickard MR, Hedge VL, Farzaneh F, Williams GT (2009) GAS5, a non-protein-coding RNA, controls apoptosis and is downregulated in breast cancer. Oncogene 28:195–208

Mourtada-Maarabouni M, Hasan AM, Farzaneh F, Williams GT (2010) Inhibition of human T-cell proliferation by mammalian target of rapamycin (mTOR) antagonists requires noncoding RNA growth-arrest-specific transcript 5 (GAS5). Mol Pharmacol 78:19–28

Murphy LC, Simon SL, Parkes A, Leygue E, Dotzlaw H, Snell L, Troup S, Adeyinka A, Watson PH (2000) Altered expression of estrogen receptor coregulators during human breast tumorigenesis. Cancer Res 60:6266–6271

Nakagawa T, Endo H, Yokoyama M, Abe J, Tamai K, Tanaka N, Sato I, Takahashi S, Kondo T, Satoh K (2013) Large noncoding RNA HOTAIR enhances aggressive biological behavior and is associated with short disease-free survival in human non-small cell lung cancer. Biochem Biophys Res Commun 436:319–324

Nie Y, Liu X, Qu S, Song E, Zou H, Gong C (2013) Long non-coding RNA HOTAIR is an independent prognostic marker for nasopharyngeal carcinoma progression and survival. Cancer Sci 104:458–464

Perissi V, Aggarwal A, Glass CK, Rose DW, Rosenfeld MG (2004) A corepressor/coactivator exchange complex required for transcriptional activation by nuclear receptors and other regulated transcription factors. Cell 116:511–526

Petrovics G, Zhang W, Makarem M, Street JP, Connelly R, Sun L, Sesterhenn IA, Srikantan V, Moul JW, Srivastava S (2004) Elevated expression of PCGEM1, a prostate-specific gene with cell growth-promoting function, is associated with high-risk prostate cancer patients. Oncogene 23:605–611

Pickard MR, Mourtada-Maarabouni M, Williams GT (2013) Long non-coding RNA GAS5 regulates apoptosis in prostate cancer cell lines. Biochim Biophys Acta 1832:1613–1623

Prensner JR, Iyer MK, Balbin OA, Dhanasekaran SM, Cao Q, Brenner JC, Laxman B, Asangani IA, Grasso CS, Kominsky HD et al (2011) Transcriptome sequencing across a prostate cancer cohort identifies PCAT-1, an unannotated lincRNA implicated in disease progression. Nat Biotechnol 29:742–749

Prensner JR, Iyer MK, Sahu A, Asangani IA, Cao Q, Patel L, Vergara IA, Davicioni E, Erho N, Ghadessi M et al (2013) The long noncoding RNA SChLAP1 promotes aggressive prostate cancer and antagonizes the SWI/SNF complex. Nat Genet 45:1392–1398

Prensner JR, Chen W, Iyer MK, Cao Q, Ma T, Han S, Sahu A, Malik R, Wilder-Romans K, Navone N et al (2014a) PCAT-1, a long noncoding RNA, regulates BRCA2 and controls homologous recombination in cancer. Cancer Res 74:1651–1660

Prensner JR, Sahu A, Iyer MK, Malik R, Chandler B, Asangani IA, Poliakov A, Vergara IA, Alshalalfa M, Jenkins RB et al (2014b) The IncRNAs PCGEM1 and PRNCR1 are not implicated in castration resistant prostate cancer. Oncotarget 5:1434–1438

Qiao HP, Gao WS, Huo JX, Yang ZS (2013) Long non-coding RNA GAS5 functions as a tumor suppressor in renal cell carcinoma. Asian Pac J Cancer Prev 14:1077–1082

Qiu JJ, Lin YY, Ye LC, Ding JX, Feng WW, Jin HY, Zhang Y, Li Q, Hua KQ (2014) Overexpression of long non-coding RNA HOTAIR predicts poor patient prognosis and promotes tumor metastasis in epithelial ovarian cancer. Gynecol Oncol 134:121–128

Remzi M, Haese A, Van Poppel H, De La Taille A, Stenzl A, Hennenlotter J, Marberger M (2010) Follow-up of men with an elevated PCA3 score and a negative biopsy: does an elevated PCA3 score indeed predict the presence of prostate cancer? BJU Int 106:1138–1142

Ribeiro RC, Kushner PJ, Baxter JD (1995) The nuclear hormone receptor gene superfamily. Annu Rev Med 46:443–453

Rinn JL, Kertesz M, Wang JK, Squazzo SL, Xu X, Brugmann SA, Goodnough LH, Helms JA, Farnham PJ, Segal E et al (2007) Functional demarcation of active and silent chromatin domains in human HOX loci by noncoding RNAs. Cell 129:1311–1323

Rosenfeld MG, Lunyak VV, Glass CK (2006) Sensors and signals: a coactivator/corepressor/epigenetic code for integrating signal-dependent programs of transcriptional response. Genes Dev 20:1405–1428

Rundlett SE, Miesfeld RL (1995) Quantitative differences in androgen and glucocorticoid receptor DNA binding properties contribute to receptor-selective transcriptional regulation. Mol Cell Endocrinol 109:1–10

Salagierski M, Verhaegh GW, Jannink SA, Smit FP, Hessels D, Schalken JA (2010) Differential expression of PCA3 and its overlapping PRUNE2 transcript in prostate cancer. Prostate 70:70–78

Sato K, Qian J, Slezak JM, Lieber MM, Bostwick DG, Bergstralh EJ, Jenkins RB (1999) Clinical significance of alterations of chromosome 8 in high-grade, advanced, nonmetastatic prostate carcinoma. J Natl Cancer Inst 91:1574–1580

Schneider C, King RM, Philipson L (1988) Genes specifically expressed at growth arrest of mammalian cells. Cell 54:787–793

Schumacher FR, Feigelson HS, Cox DG, Haiman CA, Albanes D, Buring J, Calle EE, Chanock SJ, Colditz GA, Diver WR et al (2007) A common 8q24 variant in prostate and breast cancer from a large nested case-control study. Cancer Res 67:2951–2956

Shi Y, Downes M, Xie W, Kao HY, Ordentlich P, Tsai CC, Hon M, Evans RM (2001) Sharp, an inducible cofactor that integrates nuclear receptor repression and activation. Genes Dev 15:1140–1151

Shi X, Sun M, Liu H, Yao Y, Kong R, Chen F, Song Y (2013) A critical role for the long non-coding RNA GAS5 in proliferation and apoptosis in non-small-cell lung cancer. Mol Carcinog DOI :10.1002/mc.22120

Smith CM, Steitz JA (1998) Classification of gas5 as a multi-small-nucleolar-RNA (snoRNA) host gene and a member of the 5′-terminal oligopyrimidine gene family reveals common features of snoRNA host genes. Mol Cell Biol 18:6897–6909

Sorensen KP, Thomassen M, Tan Q, Bak M, Cold S, Burton M, Larsen MJ, Kruse TA (2013) Long non-coding RNA HOTAIR is an independent prognostic marker of metastasis in estrogen receptor-positive primary breast cancer. Breast Cancer Res Treat 142:529–536

Spiegelman BM, Heinrich R (2004) Biological control through regulated transcriptional coactivators. Cell 119:157–167

Srikantan V, Zou Z, Petrovics G, Xu L, Augustus M, Davis L, Livezey JR, Connell T, Sesterhenn IA, Yoshino K et al (2000) PCGEM1, a prostate-specific gene, is overexpressed in prostate cancer. Proc Natl Acad Sci U S A 97:12216–12221

Sun M, Jin FY, Xia R, Kong R, Li JH, Xu TP, Liu YW, Zhang EB, Liu XH, De W (2014) Decreased expression of long noncoding RNA GAS5 indicates a poor prognosis and promotes cell proliferation in gastric cancer. BMC Cancer 14:319

Tani H, Torimura M, Akimitsu N (2013) The RNA degradation pathway regulates the function of GAS5 a non-coding RNA in mammalian cells. PLoS ONE 8:e55684

Tata JR (2005) One hundred years of hormones. EMBO Rep 6:490–496

Tepper CG, Boucher DL, Ryan PE, Ma AH, Xia L, Lee LF, Pretlow TG, Kung HJ (2002) Characterization of a novel androgen receptor mutation in a relapsed CWR22 prostate cancer xenograft and cell line. Cancer Res 62:6606–6614

van Gils MP, Hessels D, van Hooij O, Jannink SA, Peelen WP, Hanssen SL, Witjes JA, Cornel EB, Karthaus HF, Smits GA et al (2007) The time-resolved fluorescence-based PCA3 test on urinary sediments after digital rectal examination; a Dutch multicenter validation of the diagnostic performance. Clin Cancer Res 13:939–943

Wang D, Garcia-Bassets I, Benner C, Li W, Su X, Zhou Y, Qiu J, Liu W, Kaikkonen MU, Ohgi KA et al (2011) Reprogramming transcription by distinct classes of enhancers functionally defined by eRNA. Nature 474:390–394

Wu L, Murat P, Matak-Vinkovic D, Murrell A, Balasubramanian S (2013) Binding interactions between long noncoding RNA HOTAIR and PRC2 proteins. Biochemistry 52:9519–9527

Xing Z, Lin A, Li C, Liang K, Wang S, Liu Y, Park PK, Qin L, Wei Y, Hawke DH et al (2014) lncRNA directs cooperative epigenetic regulation downstream of chemokine signals. Cell 159:1110–1125

Xu B, Koenig RJ (2004) An RNA-binding domain in the thyroid hormone receptor enhances transcriptional activation. J Biol Chem 279:33051–33056

Xu L, Glass CK, Rosenfeld MG (1999) Coactivator and corepressor complexes in nuclear receptor function. Curr Opin Genet Dev 9:140–147

Xu B, Yang WH, Gerin I, Hu CD, Hammer GD, Koenig RJ (2009) Dax-1 and steroid receptor RNA activator (SRA) function as transcriptional coactivators for steroidogenic factor 1 in steroidogenesis. Mol Cell Biol 29:1719–1734

Xu B, Gerin I, Miao H, Vu-Phan D, Johnson CN, Xu R, Chen XW, Cawthorn WP, MacDougald OA, Koenig RJ (2010) Multiple roles for the non-coding RNA SRA in regulation of adipogenesis and insulin sensitivity. PLoS ONE 5:e14199

Yan Y, Skliris GP, Penner C, Chooniedass-Kothari S, Cooper C, Nugent Z, Blanchard A, Watson PH, Myal Y, Murphy LC et al (2009) Steroid Receptor RNA Activator Protein (SRAP): a potential new prognostic marker for estrogen receptor-positive/node-negative/younger breast cancer patients. Breast Cancer Res 11:R67

Yang L, Lin C, Jin C, Yang JC, Tanasa B, Li W, Merkurjev D, Ohgi KA, Meng D, Zhang J et al (2013) lncRNA-dependent mechanisms of androgen-receptor-regulated gene activation programs. Nature 500:598–602

Yao H, Brick K, Evrard Y, Xiao T, Camerini-Otero RD, Felsenfeld G (2010) Mediation of CTCF transcriptional insulation by DEAD-box RNA-binding protein p68 and steroid receptor RNA activator SRA. Genes Dev 24:2543–2555

Yin D, He X, Zhang E, Kong R, De W, Zhang Z (2014) Long noncoding RNA GAS5 affects cell proliferation and predicts a poor prognosis in patients with colorectal cancer. Med Oncol 31:253

Yoon JH, Abdelmohsen K, Kim J, Yang X, Martindale JL, Tominaga-Yamanaka K, White EJ, Orjalo AV, Rinn JL, Kreft SG et al (2013) Scaffold function of long non-coding RNA HOTAIR in protein ubiquitination. Nat Commun 4:2939

Zhao X, Patton JR, Davis SL, Florence B, Ames SJ, Spanjaard RA (2004) Regulation of nuclear receptor activity by a pseudouridine synthase through posttranscriptional modification of steroid receptor RNA activator. Mol Cell 15:549–558

Index

R. Kurokawa (ed.), *Long Noncoding RNAs*, DOI 10.1007/978-4-431-55576-6

T

U

W

X

Y

The manufacturer's authorised representative in the EU is Springer Nature Customer Service Centre GmbH, Europaplatz 3, 69115 Heidelberg, Germany. If you have any concerns regarding our products, please contact ProductSafety@springernature.com

Printed and bound by CPI Group (UK) Ltd, Croydon, CR0 4YY
15/07/2026
02167624-0001